The Physics of Sound and Music, Volume 1

A complete course text (Textbook)

Online at: https://doi.org/10.1088/978-0-7503-5212-3

The Physics of Sound and Music, Volume 1

A complete course text (Textbook)

Samya Bano Zain
Department of Physics, Susquehanna University, Selinsgrove, PA, USA

IOP Publishing, Bristol, UK

© IOP Publishing Ltd 2024

All rights reserved. No part of this publication may be reproduced, stored in a retrieval system or transmitted in any form or by any means, electronic, mechanical, photocopying, recording or otherwise, without the prior permission of the publisher, or as expressly permitted by law or under terms agreed with the appropriate rights organization. Multiple copying is permitted in accordance with the terms of licences issued by the Copyright Licensing Agency, the Copyright Clearance Centre and other reproduction rights organizations.

Certain images in this publication have been obtained by the author from the Wikipedia/Wikimedia website, where they were made available under a Creative Commons licence or stated to be in the public domain. Please see individual figure captions in this publication for details. To the extent that the law allows, IOP Publishing disclaim any liability that any person may suffer as a result of accessing, using or forwarding the images. Any reuse rights should be checked and permission should be sought if necessary from Wikipedia/Wikimedia and/or the copyright owner (as appropriate) before using or forwarding the images.

Permission to make use of IOP Publishing content other than as set out above may be sought at permissions@ioppublishing.org.

Samya Bano Zain has asserted her right to be identified as the author of this work in accordance with sections 77 and 78 of the Copyright, Designs and Patents Act 1988.

ISBN 978-0-7503-5212-3 (ebook)
ISBN 978-0-7503-5210-9 (print)
ISBN 978-0-7503-5213-0 (myPrint)
ISBN 978-0-7503-5211-6 (mobi)

DOI 10.1088/978-0-7503-5212-3

Version: 20240401

IOP ebooks

British Library Cataloguing-in-Publication Data: A catalogue record for this book is available from the British Library.

Published by IOP Publishing, wholly owned by The Institute of Physics, London

IOP Publishing, No.2 The Distillery, Glassfields, Avon Street, Bristol, BS2 0GR, UK

US Office: IOP Publishing, Inc., 190 North Independence Mall West, Suite 601, Philadelphia, PA 19106, USA

To my kids,

Hareem, Sami, Eman, and Fahmida.

I could not have asked for a better family.

Contents

Preface	xviii
Acknowledgements	xx
Author biography	xxi
Summary of chapters	xxii

Part I Introduction

1 Introduction — 1-1
1.1 The scientific method — 1-1
1.2 Units — 1-3
    1.2.1 Different systems of units — 1-3
    1.2.2 Converting units — 1-4
1.3 The International System of Units (SI units) — 1-4
    1.3.1 Derived units — 1-6
1.4 A few important concepts — 1-8
1.5 A quick review of vector and scalar quantities — 1-9
    1.5.1 Scalars — 1-9
    1.5.2 Vectors — 1-9
    1.5.3 Vector anatomy — 1-9
    1.5.4 The magnitude of a vector — 1-10
    1.5.5 Vector direction — 1-10
    1.5.6 The components of a vector — 1-10
    1.5.7 Finding a vector from its components — 1-11
    1.5.8 Adding vectors — 1-12
1.6 Distance versus displacement — 1-13
1.7 Speed versus velocity — 1-14
    1.7.1 The speed of an object — 1-14
    1.7.2 The velocity of an object — 1-16
1.8 Graphical representation of motion — 1-17
    Further reading — 1-22

2 Sound, music, and noise — 2-1
2.1 What makes a sound either music or noise? — 2-1
2.2 Some history of the science of sound — 2-2

		2.2.1 Branches of acoustics	2-2
2.3	How does it all work?		2-3
		2.3.1 What is a wave?	2-3
		2.3.2 The definition of a wave	2-4
		2.3.3 Types of waves	2-4
		2.3.4 The properties of waves	2-5
2.4	The properties of traveling waves		2-5
		2.4.1 Longitudinal waves	2-6
		2.4.2 Transverse waves	2-6
		2.4.3 Similarities and differences between various kinds of waves	2-8
		2.4.4 Waves that combine transverse and longitudinal motion	2-8
2.5	What is sound?		2-9
		2.5.1 About sound	2-9
		2.5.2 The propagation of sound waves	2-11
		2.5.3 The requirements for sound	2-12
		2.5.4 The makeup of a sound wave	2-12
		2.5.5 Waveforms	2-13
		2.5.6 Properties of sound waves	2-13
		2.5.7 Sources of sound	2-13
		2.5.8 Radiating patterns for sounds	2-13
		Further reading	2-15
3	**Music, history, and culture**		**3-1**
3.1	Music and life		3-1
3.2	Historic eras of music		3-1
		3.2.1 The prehistoric era	3-1
		3.2.2 The ancient musical era	3-2
		3.2.3 The early music of the world	3-2
		3.2.4 Greek music	3-2
		3.2.5 Roman music	3-3
		3.2.6 Arabic music	3-4
		3.2.7 Music of the Indian subcontinent	3-5
		3.2.8 Chinese music	3-5
		3.2.9 Renaissance music	3-6
		3.2.10 The modern musical era	3-7
3.3	Music and religion		3-9
		3.3.1 African music	3-9

	3.3.2 Native American music	3-9
	3.3.3 Jewish music	3-10
	3.3.4 Christian music	3-11
	3.3.5 Islamic music	3-11
3.4	Classes of musical instruments	3-12
	Further reading	3-14

Part II Sound production

4 Tension and deformations in a string 4-1

4.1	Energy and force	4-1
	4.1.1 Force	4-1
	4.1.2 Energy	4-2
	4.1.3 Forms of energy	4-2
	4.1.4 The conservation of energy	4-3
	4.1.5 Four basic forces	4-3
	4.1.6 The result of an applied force	4-4
4.2	Historic ideas about motion	4-5
	4.2.1 The 'no net force' condition	4-6
4.3	Newton's laws of dynamics	4-7
	4.3.1 Newton's first law (the law of inertia)	4-7
	4.3.2 Newton's second and third laws	4-8
	4.3.3 The unit of force	4-10
	4.3.4 Measuring forces	4-10
4.4	Categories of forces	4-11
	4.4.1 Calculating net force	4-13
	4.4.2 Another way to 'see' acceleration	4-15
4.5	Mass versus weight	4-16
	4.5.1 The difference between weight and mass	4-17
4.6	Tension	4-19
	4.6.1 The tension in a guitar string	4-19
	4.6.2 Factors that influence the frequency or pitch of a guitar string	4-19
	Further reading	4-21

5 Vibrating systems 5-1

5.1	Simple harmonic motion	5-1
	5.1.1 Restoring forces	5-1

	5.1.2	The simple harmonic oscillator	5-3
	5.1.3	Examples of simple harmonic motion	5-3
	5.1.4	'Seeing' sound waves	5-5
	5.1.5	An energy breakdown of an object undergoing simple harmonic motion	5-5
	5.1.6	Simple harmonic and circular motion	5-5
5.2	Standing waves		5-6
	5.2.1	Properties of standing waves	5-7
	5.2.2	The speed of standing waves	5-9
	5.2.3	The propagation of energy in standing waves	5-9
	5.2.4	Allowed and forbidden standing wave conditions	5-10
	5.2.5	Modes of vibrations	5-10
	5.2.6	Combining various modes of vibrations	5-11
5.3	The reflection of waves		5-12
	5.3.1	The laws of reflection	5-13
	5.3.2	Properties of a reflected wave	5-14
	5.3.3	Reflections of sound waves	5-16
5.4	Waves in stringed instruments		5-18
	5.4.1	Allowed frequencies of the standing waves in strings	5-18
5.5	Wave interaction: superposition and interference		5-20
	5.5.1	The phases of a wave	5-21
	5.5.2	The principle of wave superposition	5-21
	5.5.3	Types of wave interference	5-22
	5.5.4	Interference in music	5-24
	5.5.5	Interference between sound waves from two different sound sources	5-24
	5.5.6	Beats	5-27
	Further reading		5-30
6	**Damping and resonance in musical instruments**		**6-1**
6.1	Damping in oscillators		6-1
	6.1.1	Damped harmonic oscillators	6-2
	6.1.2	Energy and damping in vibrating systems	6-2
	6.1.3	Classes of damped oscillators	6-4
	6.1.4	Forces and frequencies in a vibrating motion	6-5
	6.1.5	Main types of damping	6-6
	6.1.6	Forced oscillation in a mass–spring system	6-7

6.2	Resonance	6-7
	6.2.1 The effects of resonance	6-8
	6.2.2 Seashells and resonance	6-9
6.3	Ways to drive a string at one of its resonances	6-10
6.4	Understanding resonance	6-11
	6.4.1 The Q-factor	6-12
	6.4.2 Damping and the Q-factor	6-12
6.5	Sympathetic vibrations	6-14
6.6	Resonances in musical instruments	6-14
	6.6.1 The Helmholtz resonator	6-15
	6.6.2 Chladni patterns	6-16
	6.6.3 Singing rods	6-17
	6.6.4 Singing wineglasses	6-19
	Further reading	6-21

Part III Sound propagation

7 Sound propagation — 7-1

7.1	Traveling waves	7-2
	7.1.1 Setting up a traveling wave	7-2
	7.1.2 The mathematical description of a wave in a string	7-3
	7.1.3 Properties of traveling waves	7-4
	7.1.4 The speed of a transverse wave in a string	7-5
	7.1.5 The energy and power of a wave traveling along a string	7-5
7.2	Periodic waves	7-7
7.3	The speed of sound waves	7-8
	7.3.1 The dependence of the speed of sound on temperature	7-10
	7.3.2 Other factors that influence the speed of sound	7-11
	7.3.3 The speed of sound in solids	7-11
	7.3.4 The speed of sound in liquids	7-13
	7.3.5 The speed of sound in gases	7-15
	7.3.6 The speed of sound in an ideal gas	7-16
	7.3.7 The speeds of sound in various media	7-16
7.4	Sound absorption	7-17
	Further reading	7-18

8 Factors that impact sound propagation — 8-1

8.1	Huygen's principle	8-1

8.2	The refraction of waves	8-1
	8.2.1 The properties of a refracted wave	8-3
	8.2.2 Refraction in surface water waves in the ocean	8-3
	8.2.3 The refraction of sound under various conditions	8-4
8.3	Diffraction	8-6
	8.3.1 Hearing around corners	8-7
	8.3.2 The diffraction equation	8-8
	8.3.3 Acoustic diffraction in amphitheaters	8-10
8.4	The Doppler effect	8-11
	8.4.1 The reason for the Doppler effect	8-11
	8.4.2 Some applications of the Doppler effect	8-15
	8.4.3 Shock waves	8-16
	Further reading	8-18

Part IV Sound reception

9 Sound power and sound intensity — 9-1

9.1	Power and pressure	9-1
	9.1.1 Pressure	9-1
	9.1.2 Work and energy	9-4
	9.1.3 Power	9-6
	9.1.4 Stress and strain	9-7
9.2	Sound waves	9-7
	9.2.1 The amplitude of sound waves	9-8
9.3	The intensity of sound waves (I)	9-8
	9.3.1 Intensity versus distance from a wave source	9-9
9.4	Decibels	9-11
	9.4.1 The sound intensity level (dB SIL)	9-12
	9.4.2 Sound intensity at the threshold of hearing I_0	9-13
	9.4.3 The sound intensity level as a function of the number of sources	9-14
9.5	The speed of sound versus particle velocity	9-15
9.6	Sound power (W)	9-16
9.7	Sound pressure level (dB SPL)	9-17
	9.7.1 Sound power (W) versus sound pressure (p)	9-18
	9.7.2 Measuring sound pressure levels	9-18
9.8	Summary	9-20
9.9	The sound power level (dB SWL)	9-21

9.9.1 The sound pressure level (SPL) versus the sound power level (SWL)	9-21
9.10 Loudness and loudness level	9-24
9.10.1 Loudness	9-24
9.10.2 Loudness as a function of distance	9-24
9.10.3 Other units for loudness and loudness level	9-25
Further reading	9-26

10 The human factor — 10-1

10.1 The ranges of human hearing and sight	10-1
10.1.1 The range of visible light	10-1
10.1.2 The range of hearing	10-2
10.1.3 Infrasound and ultrasound	10-2
10.2 Speech production in humans	10-4
10.3 Auditory systems	10-5
10.3.1 Signals processed in the ear (the peripheral auditory system)	10-5
10.3.2 Signals processed in the brain (the auditory nervous system)	10-8
10.4 Critical bands	10-10
10.4.1 Visual and auditory illusions	10-11
10.5 Bone conduction	10-11
Further reading	10-13

11 Psychoacoustics — 11-1

11.1 Hearing in humans	11-1
11.1.1 Spatial hearing or localization	11-2
11.1.2 Auditory masking	11-3
11.1.3 Auditory adaptation and auditory fatigue	11-4
11.1.4 Selectivity	11-4
11.1.5 Binaural hearing	11-5
11.1.6 Echolocation	11-6
11.2 The effect of noise on humans	11-6
11.2.1 Ear damage	11-6
11.2.2 Measuring hearing loss	11-7
11.3 Noise control	11-8
11.3.1 Sound absorption	11-9
11.3.2 Noise regulations	11-9
Further reading	11-10

12 The acoustics of rooms   12-1

- 12.1 Sound propagation   12-1
  - 12.1.1 Background noise   12-2
  - 12.1.2 Indoor sound propagation   12-2
- 12.2 The precedence effect   12-4
- 12.3 Room acoustics   12-5
  - 12.3.1 The acoustics of an ideal room   12-5
  - 12.3.2 The acoustics of a real room   12-5
  - 12.3.3 The absorption of sound by air   12-7
  - 12.3.4 The optimum RT60 for a room   12-8
- 12.4 Problems in acoustical design   12-9
- 12.5 The criteria for good acoustics   12-9
- 12.6 Designing spaces   12-10
- 12.7 Loudspeakers   12-14
  - 12.7.1 Loudspeaker placement   12-14
  - 12.7.2 Loudspeaker directivity   12-15
- 12.8 Outdoor sound systems   12-16
- Further reading   12-17

Part V   Of sound and music

13 Musical tones, pitch, timbre, and vibrato   13-1

- 13.1 Musical tones and pitch   13-1
  - 13.1.1 Pitch   13-2
  - 13.1.2 The history of pitch development   13-2
  - 13.1.3 Pitch versus frequency   13-3
  - 13.1.4 Pitch and loudness   13-3
  - 13.1.5 Pitch perception   13-4
  - 13.1.6 Concert pitch   13-4
  - 13.1.7 Relative pitch   13-5
  - 13.1.8 Absolute pitch   13-5
  - 13.1.9 Virtual pitch   13-5
- 13.2 Pitch perception theories   13-5
- 13.3 Vibrato   13-7
  - 13.3.1 Frequency modulation versus amplitude modulation   13-7
  - 13.3.2 AM versus FM radio signals   13-8
  - 13.3.3 TV signals   13-9

13.4	The just-noticeable difference (JND)	13-10
	13.4.1 Measuring the frequency JND	13-11
	13.4.2 Intensity and the JND	13-12
	13.4.3 Cents and mel scales	13-13
13.5	Timbre or tone quality	13-14
	13.5.1 The effect of timbre on loudness	13-15
	Further reading	13-16

14 A musician's graph paper and musical scales   14-1

14.1	Logarithms	14-1
14.2	The musical stave or staff	14-2
	14.2.1 The musical staff	14-2
	14.2.2 Important symbols on a musical staff	14-3
14.3	Musical scales	14-5
	14.3.1 Building a musical scale	14-6
	14.3.2 The scale step	14-6
	14.3.3 Harmonic series	14-8
14.4	Musical intervals	14-8
	14.4.1 Naming the intervals	14-8
	14.4.2 The octave	14-9
	14.4.3 Octave-repeating scales	14-10
	14.4.4 Scalar transposition	14-11
	14.4.5 Musical intervals	14-11
	14.4.6 Types of intervals	14-13
14.5	Various terms that are important to know	14-15
	14.5.1 False pattern recognition	14-17
	Further reading	14-18

Part VI Musical instruments

15 Stringed instruments   15-1

15.1	The history of stringed instruments	15-2
15.2	The introduction of energy into a string instrument	15-4
	15.2.1 Allowed frequencies of the standing waves on the string	15-4
	15.2.2 Changing the pitch in stringed instruments	15-5
15.3	Tuning	15-6
	15.3.1 Some methods of tuning	15-7

15.4	The guitar	15-7
	15.4.1 Guitar strings	15-8
	15.4.2 Acoustic guitars	15-10
	15.4.3 The solid-body electric guitar	15-12
15.5	The piano	15-13
	15.5.1 The development of the piano	15-13
	15.5.2 The piano as a vibrating system	15-14
	15.5.3 The production of sound in a piano	15-15
	15.5.4 The pitch of a vibrating string in a piano	15-16
	15.5.5 The decay of piano sounds	15-17
15.6	Bowed stringed instruments	15-18
	15.6.1 The violin	15-18
	15.6.2 Slip–stick action	15-22
	15.6.3 Regular stick–slip–stick–slip cycles	15-24
	Further reading	15-25

16 Percussion instruments — 16-1

16.1	Rhythms in everyday life	16-1
16.2	Various percussion instruments	16-2
16.3	Vibrations in a bar	16-3
16.4	Vibrations in plates and membranes	16-4
16.5	Membranophones	16-5
16.6	Bells	16-9
	16.6.1 The main parts of a bell	16-9
	16.6.2 Vibrations in bells	16-10
	16.6.3 Wavelengths in bells	16-11
	Further reading	16-12

17 Wind instruments — 17-1

17.1	Wind instruments	17-1
	17.1.1 Sound production in woodwind instruments	17-2
	17.1.2 Sound waves in pipes	17-3
17.2	The instruments of the woodwind family	17-6
	17.2.1 The flute	17-6
	17.2.2 The piccolo	17-7
	17.2.3 The oboe	17-7
	17.2.4 The clarinet	17-8

17.3 The pipe organ	17-9
17.4 The instruments of the brass family	17-11
17.4.1 The production of sound in brass instruments	17-14
17.5 The bagpipe	17-16
Further reading	17-18

Part VII Appendix

Appendix A: Review of mathematics   A-1

Appendix B: Unit conversions   B-1

Appendix C: Logarithms   C-1

Preface

The Physics of Sound and Music is, as the name implies, intended to be an introductory course in learning about the science of sound and music. This book is the outcome of a physics course taught at Susquehanna University, which is mainly provided for non-majors to fulfill their 'scientific explanations' portion of the central curriculum. This book has sufficient material for a one-semester, four-credit, hour-long course.

I recognize that mathematics is fundamental and indispensable to the study of physics. A little bit of familiarity with algebra is required to succeed in the course. However, I have avoided derivations of complex equations in the text. I have tried to establish, wherever possible, context and background. I feel that this manner of learning makes it easier for the reader to understand and grasp the validity of the particular physics concept/topic. For details of the mathematical proofs of the concepts involved, please check other sources.

Main ideas for understanding

As with my other books, when writing this book I pictured myself conversing with a student and explaining the fundamental physics of sound and its propagation. For this reason, I have tried to keep this book as conversational and down-to-earth as possible by (in part) avoiding too much unnecessary technical jargon. Readers should be able to pick it up and learn on their own. Worked examples are included as part of the text in nearly every section. Additional conceptual questions and exercises are included for further practice and understanding.

Mainly for instructors

In this text, I have put physics concepts before the historic development of the field of acoustics. Many concepts are repeated across multiple chapters. This repetition is intentional. I feel that reviews and re-reviews of important concepts never hurt. Some questions included at the end of sections may go just beyond the material in the chapter. Such content is intended to allow students to apply concepts they have learnt during the chapter and the book overall and hopefully increase their knowledge and understanding.

I have incorporated a lab manual with 'activities to do' as part of this textbook. I have tested most of these with students on our campus and I have found that they are useful in clarifying ideas that might be confusing for the non-physics majors to comprehend and/or appreciate. The majority of the activities included in this textbook are also intentionally kept simple and most can be done with materials that are easily acquired at most general stores and many do not require specialized physics or engineering equipment. The intention is to make these activities accessible to everyone around the world and make it easy for instructors to incorporate them in their own class and lab rooms as they see fit.

Mainly for students

Before we begin, I just have a few additional things to comment on. Remember, physics must *not* be memorized but rather understood. Know that understanding does not come quickly or easily; it takes patience and resilience. Do you remember how you learnt to play an instrument? Did you become an expert overnight, or did it take hours and hours of practice? Physics is just the same: you need to actively learn it; cramming overnight before the exam is not the ticket to understanding and hence success.

Key to success

As always, my advice throughout the years has not changed:
1. Manage your time and get organized.
2. Commit yourself. Be a good listener and take good notes.
3. Find a study partner.
4. Motivate your partner and yourself.
5. Actively seek help.
6. Set a dedicated time to study physics every day. Thirty minutes a day, every day will get you more than pulling an all-nighter the night before a quiz.

And most important of all remember, as Sam Ewing said: 'Hard work spotlights the character of people; some turn up their sleeves, some turn up their noses and some don't turn up at all.' Hence, just show up to class, be a good listener, ask questions, and take good notes. Read the book, do the examples and exercises, share your knowledge, communicate, and most important of all, work hard and enjoy the ride.

Samya Bano Zain
Professor, Susquehanna University
August 30, 2023

Acknowledgements

This work is based on over a decade of study, reflection, and critique. It hopefully represents a marked improvement in my understanding and teaching about acoustics. For that, I am grateful for the contributions of many, many people. They include, but are not limited to the numerous Susquehanna students I have had the pleasure to learn from, Dr Grosse, who first encouraged me to put words to paper, Dr Ken Brakke, Robert Everly, Dr Jeffrey Graham, for multiple informative and valuable physics and math conversations. I am also grateful to Dr Patrick Long, Department of Music, who graciously allowed me to attend his music theory class. A lot of technical musical improvements in this text are a direct result of his class and his wonderful teaching skills and I want to especially thank M. L. Klotz, Department of Psychology, for keeping me sane!

To my ever-patient husband, Zain, and my kids, Hareem, Sami, and Eman, for being (nearly always!) willing models and getting excited about whatever crazy idea I had for a picture that particular day! and to Fahmida, the newest addition to our family. A special thank-you to Sami's new phone and its picture quality, and last but not least, Hareem and Eman's contributions are truly appreciated; thank-you both for making artwork for me and letting me use your pictures in the book, it is deeply appreciated.

My heartfelt gratitude is also extended to my family: my dad Dr Iqbal and my mom Safia, who live in my heart, my brother Dr Aamir Iqbal, and my sister-in-law Sofia Aamir. I also want to thank my relatives-in-law, Rauf bhai, Hifza baji and Shazia, and Shoaib bhai, who welcomed me to their family almost 30 years ago and have become more than my brothers and sisters over the years. You have always believed in me, prayed for me and encouraged me to be my best self. Thank-you all for your unwavering support!

I am reminded of the quote by Sir Isaac Newton: *'If I have seen further than others, it is by standing upon the shoulders of giants.'* I guess I have met many giants in my life, and for that I am eternally grateful!

Author biography

Samya Bano Zain

Samya Bano Zain grew up in Lahore, Pakistan. In seventh grade she decided that she wanted to learn more physics, since physics explains how the universe works and additionally allowed her to avoid going into the medical field like the rest of her family. She completed her undergraduate work at the University of the Punjab, Lahore, Pakistan and was awarded the gold medal for the highest score in the annual final exam. With her husband and four-year-old son, she the went on to graduate work in physics at the university at Albany, State University of New York in 2001. She received her PhD in experimental particle physics in 2006 and continued as a postdoctoral fellow at the Albany High Energy Physics group. In the fall of 2008, she joined Susquehanna University as an assistant professor, where she was promoted to the rank of full professor last year in the department of physics.

When not working on physics, she can often be found with her nose in a book. She makes time to read every day (even when she must hide from her family in order to do so). She paints when the mood strikes her and tries to create Urdu and Arabic calligraphy (not as well as she would like!), embroiders to relax, and has recently taken up learning how to do mandala art. She loves visiting used bookstores and cannot pass up procuring more books and absolutely refuses to let any of her collection go!

Summary of chapters

When you study music you begin by learning about *sound*.

In this text, we describe the foundational concept of sounds and subdivide the study of the physics of sound and music into three main areas of study:
1. The creation of a sound.
2. Its transmission through a medium; on the Earth, this is most likely to be air.
3. Its detection by a listener.

Chapter 1—Introduction

Science is the observation and the precise description of natural phenomena, which is why we start this textbook with a discussion of the scientific method, a standard commonly agreed by the scientific community for the observation and exploration of natural phenomena. Math is the language of all sciences, so the rest of chapter 1 is intended to be a review of mathematical concepts. This will build the mathematical foundation for the rest of the book and is done by introducing the international system of units, followed by a review of vectors and scalars and the derivatives of positional vectors, namely velocity and acceleration.

Chapter 2—Sound, music, and noise

We start chapter 2 by defining the difference between music and noise. We introduce the science of sound (acoustics) and delve into the details of various kinds of waves. Longitudinal and transverse waves are introduced in detail along with examples of each kind of wave. We also discuss similarities and differences between waves. We then focus on the kind of wave that makes up sound and how it travels. This chapter lays the foundation for the rest of the book and introduces the reader to the fundamental concepts of sound.

Chapter 3—Music, history, and culture

Before we start the mathematical and scientific details of sound and music, this chapter allows us to develop the historical and cultural significance of music throughout the ages. Music is played and listened to all over the world and has been the medium for expressing emotions from happiness to pain and loss. In this chapter, we also briefly discuss some musical styles and instruments that were popular during certain historical eras and how they have developed into the musical genres and musical instruments of today.

Chapter 4—Tension and deformations in a string

This chapter discusses the vital importance of forces. There are two main results when a force is applied to an object: the object may accelerate in the direction of the applied force or deformation may be caused in the object itself. We discuss the details of each of these outcomes by introducing Newton's laws of motion (for acceleration) and elastic and inelastic deformations. In the course of this chapter, we also become familiar with concepts such as frictional forces, tensional forces, energy, types of energies, the conservation of energy, and the difference between mass and weight, and we end this chapter by examining the tension in a guitar string.

Chapter 5—Vibrating systems

Vibrating objects produce sounds as they disturb the medium around them; these vibrations travel through the medium as waves. Sound causes a disturbance of the air through which it moves. This means that if we are to understand the propagation of sound, we have to understand waves. We start our discussion of simple harmonic motion by introducing the harmonic oscillator and standing waves. Standing waves are set up in a medium when an incoming wave is reflected at a boundary, so this chapter details the laws of reflections and properties of reflected waves in sound. The chapter concludes with a detailed discussion of the following concepts: superposition, the interference of waves (including constructive and destructive interference), and the resulting beating phenomenon of sound waves.

Chapter 6—Damping and resonance in musical instruments

In this chapter, we discuss the phenomenon of damping and resonance present in physical systems, including musical instruments. Once the system is set into vibration, if no additional external force is provided, the amplitude of the vibrating system eventually reaches zero and the oscillating system comes to rest in its equilibrium state; this process is called '*damping.*' To keep the system vibrating, a continuously applied external force is required. If we provide the system with a small force, the overall effect is that the amplitude of the swing increases over time. This phenomenon is called '*resonance.*' This chapter discusses resonance in detail and ends with a detailed description of the effects of resonance in musical instruments, including the Helmholtz resonator, sympathetic vibrations, Chladni patterns, singing rods, and singing wineglasses.

Chapter 7—Sound propagation

Chapter 7 begins by describing traveling waves, which are waves that travel and transport energy from one place to another. The properties and mathematical details, such as mathematical equations of speed, energy, and power of a traveling wave, are also discussed. The second half of this chapter focusses on the propagation of sound waves. We know that vibrating objects produce sounds as they disturb the medium around them, which then travels from the source to the receiver. Here, we discuss the speed of these produced sound waves as they travel to the receiver along with their dependance on environmental conditions such as temperature, humidity, and the various forms of matter (solid, liquid, or gas) they travel through.

Chapter 8—Factors that impact sound propagation

In this chapter, we discuss some principles that govern wave behavior, starting with a phenomenon that might be considered fundamental to how waves propagate, called the Huygens principle. In addition, other common behaviors that all waves exhibit under standard conditions, such as refraction, diffraction, and the Doppler effect, are also discussed. Special emphasis is placed on the conceptual development and mathematical details of sound waves when each of these properties is addressed.

Chapter 9—Sound power and sound intensity

This chapter details commonly used but easily confused terms in acoustics: sound intensity level (SIL), sound pressure level (SPL), and sound power level (SWL). We start the discussion by restating the concepts of power, pressure, work, and energy from physics and then relating these to other sound concepts. The decibel scale is introduced and then the SIL, SPL, and SWL are defined in terms of the decibel scales. In addition, the differences between the terms sound intensity (I) and SIL, sound pressure (p) and SPL and sound power (W) and SWL are clarified. This chapter concludes by discussing the difference between the loudness and the loudness level of sound and introducing the loudness unit (called a sone) and the unit of loudness level (called a phon).

Chapter 10—The human factor

In this chapter, we attempt to discuss the human auditory system in some depth and explain how it is extraordinarily complex but amazingly remarkable. We start with the range of human hearing and discuss speech production in humans. We next examine the auditory system, which includes the human ear and the human brain. Details of the structure of a human ear are also given here along with an explanation of how a sound signal is processed in the ear (peripheral auditory system) and in the brain (auditory nervous system). The chapter concludes with a discussion of critical bands and bone conduction.

Chapter 11—Psychoacoustics

Psychoacoustics is the study of how sound is perceived and how it affects our bodies and minds. It is the branch of science that studies the psychological and physiological responses associated with sound, including noise, speech, and music. This chapter details concepts of sound and its perception, for example, sound localization, auditory masking, selectivity, binaural hearing, echolocation, auditory adaptation, and auditory fatigue. The chapter concludes with a discussion of the effect of noise on humans, including ear damage, hearing loss, noise control, and noise regulations in effect for most cities in the world.

Chapter 12—The acoustics of rooms

In this chapter, we address the acoustics of rooms, which is a subfield of acoustics that deals with sound behavior in closed spaces. We briefly discuss how good architectural design brings out the best acoustical environment. We start our discussion with sounds we encounter indoors, including reflections of sounds, echos, reverberant sounds, and the precedence effect. We then compare the acoustics of an ideal room to the acoustics of real rooms and examine the problems encountered in acoustical design for real rooms. The chapter ends with a discussion of the basic criteria for good acoustics and how to design the best acoustical environment depending on the primary use of each space.

Chapter 13—Musical tones, pitch, timbre, and vibrato

This chapter details the concepts of music, including pitch, musical tones, timbre, and vibrato. We start with the historic development of pitch and then the terms used to define pitches, such as relative, absolute, concert pitch, etc. The difference between pitch and frequency and the relationship between pitch and loudness are also discussed. The discussion of vibrato includes a study of frequency modulated (FM) and amplitude modulated (AM) radio waves. The chapter also discusses just-noticeable difference (JND) for sound and concludes by explaining timbre or tone quality.

Chapter 14—A musician's graph paper and musical scales

Chapter 14 starts with the introduction of the mathematical concept called the logarithm. Logarithms have very useful properties that help us when dealing mathematically with the frequency and amplitude responses of our ears. Graph paper that has one logarithmic axis and one linear axis is used to produce semi-logarithmic graphs. We use these (say a musical staff) to represent quantities that are functions of frequencies. A piano keyboard is based on an approximately logarithmic scale. This chapter continues with the details of the musical staff and how musical scales were developed and are currently used in the present era. The chapter ends with some important musical terms to know, such as consonance, dissonance, chord, temperament, etc.

Chapter 15—Stringed instruments

The history of music is as old as the history of man. Many early musical instruments were made from animal skins, bone, wood, and other materials. There is a considerable variation in the musical instruments found in nearly all cultures of the world; however, the basic principles are similar across the globe. In this chapter, we focus on stringed instruments such as the guitar and the piano and bowed stringed instruments such as the violin. Their construction along with the production of sound in each instrument are detailed, and special emphasis is given to the mathematical formulas used to determine the frequency (pitch) of the instrument.

Chapter 16—Percussion instruments

The history of percussive musical instruments is as old as the history of human culture. A percussion instrument is any object that produces a sound when it is hit, shaken, rubbed, scraped, or struck in any way which sets the object into vibration. Percussion instruments can be broadly divided into two categories: rhythmic percussion instruments, such as marimbas, snare drums, and timpani, and melodic percussion instruments, such as the steel drum, xylophones, etc. This chapter discusses the vibrations in various percussion instruments from bars to plates and membranes. Standing waves in Chaldini plates are also discussed with examples.

Chapter 17—Wind instruments

Chapter 17 focusses on the wind instrument family. Historically, it is most likely that ancient wind instruments were made out of animal bones or dried wood and were mostly one-note instruments. These days, wind instruments are broadly grouped into three main classes: (i) brass or lip reed instruments, for example, horns or trumpets; (ii) air reed instruments such as the flute; and (iii) wind instruments with a cane reed, for example, the clarinet. In this chapter, we discuss each in some detail, and in order to do that, we explain sound waves in pipes for both the open–open and open–closed pipe cases. Also discussed in this chapter are the details of the construction and the mechanisms of sound production in instruments that belong to the wind instrument family.

Part I

Introduction

Chapter 1

Introduction

Physics is the observation and description of natural phenomena that occur around us. It serves as a foundation for other sciences and is the study of matter, energy, and waves, from the smallest of particles, such as quarks and electrons, to the largest objects, such as galaxies. The basic goal of physics is to find laws that govern the universe and to formulate said laws in the most precise way possible. Physicists look for patterns in natural phenomena and use the language of mathematics to construct physical theories to describe the universe around us. This description might not be perfect, but it is remarkable how well we still do with our descriptions.

1.1 The scientific method

The scientific method has been developed over the years and has been in effect since the 17th century. The scientific method is a general procedure for experimentation used to explore, observe, and answer questions. The main purpose of the scientific method is to discover cause-and effect-relationships. However, it is worth mentioning that the exact hierarchy of the scientific method may change depending on the subject under consideration. For example, a botanist may be able to to obtain direct observational data with which to test their hypotheses. On the other hand, paleontologists studying dinosaurs cannot obtain direct evidence by observation with which to test their hypotheses.

In this text, I will give you a general description of the scientific method with which most of the scientific community agrees. A generalized scientific method must consist of the following minimum number of steps, as shown in figure 1.1:

1. **Develop a question.** The first step is to develop a question about something you are interested in. It could be a direct observation or an indirect observation, but it has to be posed as an inquiry: what, when, how, where, etc.

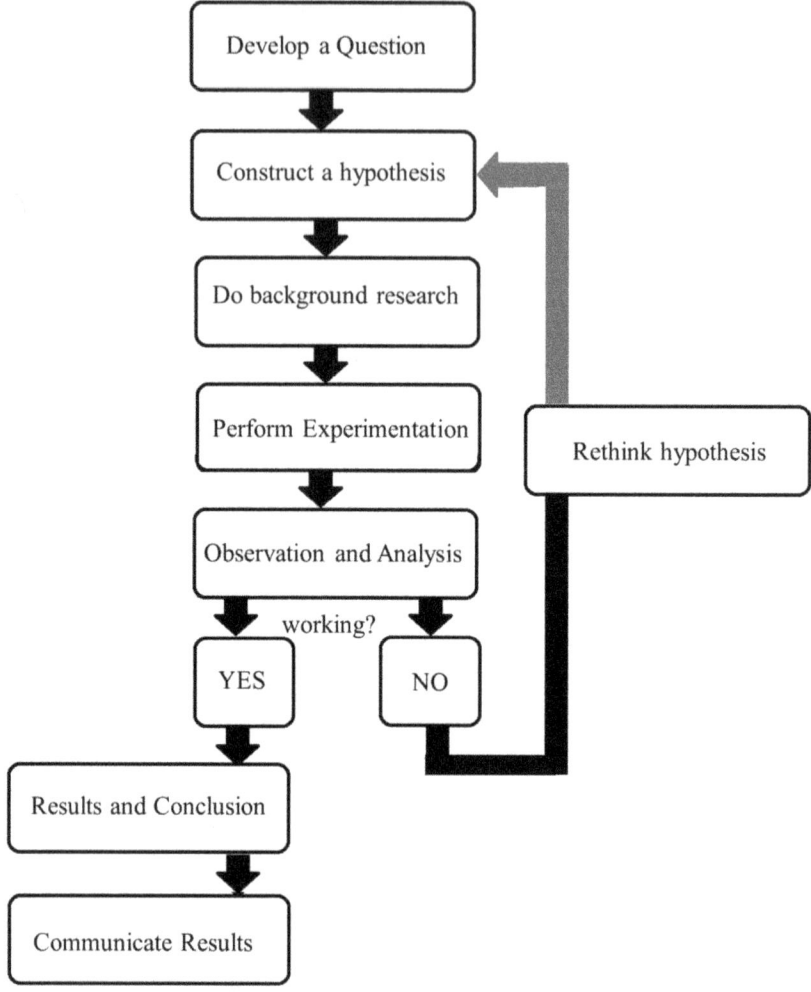

Figure 1.1. The scientific method in a box.

2. **Hypothesis.** The next task is to develop a hypothesis, basically an educated guess. A hypothesis then allows you to make some predictions about your posed question.
3. **Background research.** Another way to ensure that your hypothesis is an educated guess is to conduct background research about your question. This includes searching for other studies undertaken about your research question.
4. **Experimentation.** Next comes the task of experimentation. In order to test your hypothesis, you must come up with an experiment that allows you to answer your question. It is vital at this point to recognize that your experiment must be repeatable and it must also be performed while varying as few variables as possible.

5. **Observation and analysis.** Your experiment must be followed by data collection and data analysis to test whether the experiment supports your hypothesis. This must be done very precisely and accurately. Sometimes, this step will lead to good results and conclusions that validate your hypothesis. Other times, it may lead to the development of a completely new question and a new hypothesis. Both of these are desirable outcomes for a scientist.
6. **Results and conclusions.** Whether your hypothesis was supported by data or not, the results are generally communicated to the scientific community by publishing them as a final report in a scientific journal or as a presentation at a scientific meeting.

1.2 Units

At the heart of it all, physics is an experimental science and experiments involve measurements. Measurements in physics involve many varied and sometimes complex quantities, such as force, energy, momentum, magnetic field strength, capacitance, etc.

Surprisingly, these complex measurements can be expressed using only a few fundamental quantities, for example, the measurement of time (t), length (L), and mass (m). Historically, the scientific community decided that these units, e.g. seconds, meters, and kilograms, are convenient to use in physical experiments and decided that these will be called *fundamental units*, or SI units which is abbreviated from the French, le Système international d'unités (translated as *the International System of Units*). However, not everyone around the globe agrees on these standard units, and the history of these debates is quite interesting.

1.2.1 Different systems of units

There are three systems of units currently in use around the world:
1. **The MKS system:** also called the metric system, uses meters (m), kilograms (kg), and seconds (s) as the base units for measurements. This system is the most used worldwide.
2. **The CGS system:** also called the Gaussian system, uses centimeters (cm), grams (gm), and seconds (s) as base units for measurements.
3. **The US-engineering system:** also called the British system, is the system of measure most commonly used in the US. In this system, the unit of length is a foot and the unit of time is a second. This system does not have a base unit for mass; instead, it uses the unit of the pound-mass, symbolized as (lb-mass).

We may have different systems of units that are used in various forms; however, one thing the whole scientific community agrees upon is that we should always be able to convert from one system of units to another given a certain *dictionary* of units. I will not go into detail about each of the units, but it is worth knowing a few common conversions, as shown in table 1.1.

Table 1.1. Some common conversions.

Units of	MKS system	CGS system	US-engineering
Length (L)	Meter (m)	Centimeter (cm)	Foot (ft)
Mass (m)	Kilogram (kg)	Gram (gm)	Pound-mass (lb-masss)
Time (t)	Second (s)	Second (s)	Second (s)
Temperature	Kelvin (K)	Kelvin (K)	Fahrenheit (F)

Table 1.2. Some common conversions.

Unit	Conversion
Length	1 inch (1") = 2.54 cm
	1 mile = 1.609 km.
	1 foot = 12 inches
Mass	1 kg = 1000 gms
	1 u = 1.66×10^{-27} kg
Time	1 year = 365.24 days
	1 day = 24 hours
	1 hour = 3600 s
Angle	1° = 0.017 45 rad
	360° = 2 π rad
	1 rad = 57.3°

1.2.2 Converting units

When dealing with units other than those of the SI systems, it is always a good idea to convert to a single consistent set of units. A good rule of thumb to remember for general problem-solving is to recognize that all quantities to be added, subtracted, multiplied, or divided must be expressed in the same units. For this text, we will be concentrating on SI units which are the current standard around most of the world. For reference, some common conversions are given in table 1.2.

1.3 The International System of Units (SI units)

The International System of Units (SI Units) specifies a set of base units and derived units. The symbol of the base SI units are written in lowercase; for example, meter (m), kilogram (kg), and second (s). Derived units are associated with derived quantities such as the newton (N), which is a unit of weight or force (kg m s^{-2}) (chapter 4), and the watt (W), which is a unit of power (kg m^2 s^{-3})(chapter 9).

1.3.0.1 Base SI units: time—the second
What is time and what is the arrow of time? is a deep philosophical question, one which great philosophers have spent eons pondering about. However, all we must

remember is that time is measured in seconds in SI units. A brief summary of the conversion factors for units of time is given in table 1.3.

1.3.0.2 Base SI units: mass—the kilogram
Mass is something we are familiar with in terms of how heavy something is to lift or to move. Big stones are heavier to lift, hence they have more mass; feathers are easier to lift, hence they have less mass. However, the concept of mass has a very deep relationship to both length and time.

Historically, mass has been defined by an artifact, a platinum-iridium cylinder stored at the International Bureau of Weights and Measures, in France. The mass of this cylinder was defined to be exactly one *kilogram*. However, on November 16, 2018, at a session of the 26th General Conference on Weights and Measures in Versailles, France, representatives of countries from around the world voted to redefine several units of measurements, including the kilogram. Hence on May 20, 2019, the mass of the kilogram was redefined and was replaced by a fundamental constant of nature known as *Planck's constant*[1], whose numerical value is exactly $6.626\ 070\ 15 \times 10^{-34}$ kg m^2 s^{-1}.

However, it is important to remember that this change did not affect the actual weight of the kilogram; a quarter pounder burger still has a quarter-pound of beef. However, this redefinition of mass makes it easier to measure very, very small masses, which is vital for industries that deal with tiny quantities, for example, pharmaceuticals. Some conversions of the kilogram are given in table 1.4.

Table 1.3. Some time conversions.

1 hour	= 60 minutes	= 3660 s
1 millisecond	= 1 ms	= 10^{-3} s
1 microsecond	= 1 μs	= 10^{-6} s
1 nanosecond	= 1 ns	= 10^{-9} s

Table 1.4. Some mass conversions.

1 kilogram	= 1 kg	= 10^3 g
1 gram	= 1 gm	= 10^{-3} kg
1 milligram	= 1 mg	= 10^{-6} kg
1 microgram	= 1 μg	= 10^{-9} kg
1 nanogram	= 1 ng	= 10^{-12} kg

---

[1] Planck's constant (*h*), named after Max Planck, a German theoretical physicist, relates the energy carried by a photon (*E*) to the frequency (*f*) of the photon ($E = hf$). Since energy and mass are equivalent, Planck's constant relates mass to frequency. Planck's constant is of fundamental importance in quantum mechanics.

Table 1.5. Some length conversions.

1 kilometer	= 1 km	= 10^3 m
1 centimeter	= 1 cm	= 10^{-2} m
1 millimeter	= 1 mm	= 10^{-3} m
1 micrometer	= 1 μm	= 10^{-6} m
1 nanometer	= 1 nm	= 10^{-9} m

1.3.0.3 Base SI units: length—the meter

Despite the fact that length and time appear to be independent, there is a deep connection between them. The details of the connection needs a special course in special relativity. But for now, we will only concern ourselves with a few conversions of units of length as given in table 1.5.

Example 1.1. A wooden log has a mass of 5 kg and is 0.5 m long. Find its mass in grams and length in centimeters.
Solution:
The mass in grams for the wooden log:

$$5 \text{ kg} = 5 \text{ kg} \times 1000 \frac{\text{gms}}{\text{kg}} = 5000 \text{ gm.}$$

The length in centimeters for the wooden log:

$$0.5 \text{ m} = 0.5 \text{ m} \times \frac{100 \text{ cm}}{1 \text{ m}} = 50 \text{ cm.}$$

1.3.1 Derived units

The SI specifies a set of seven base units from which the other SI units of measurement are derived and hence called *derived units*. Each derived unit can be expressed as the product of one or more base units. For example, the SI derived unit of area is the square meter (m^2), and that of density is the kilogram per meter cubed ($kg \ m^{-3}$).

1.3.1.1 Important facts about derived quantities

Some important facts about derived quantities include:
1. The dimension of any derived physical quantity is some product of other base quantities, for example, density is kilograms per meter cubed.
2. When the values of the base quantities change, the derived values also change by the same factor, for example, when mass is doubled, the density of the object also doubles.
3. When two or more derived quantities of the same units are added, their sum also has the same units. For example, the sum of two speeds given in ($m \ s^{-1}$) also has units of ($m \ s^{-1}$).

4. The products and ratios of derived quantities usually have units that are not the same as those of the original quantities. For example, force (in units called newtons) per unit area (expressed in m^2) is N m^{-2}, which is called a pascal (Pa), which is a unit of pressure.

Example 1.2. Density (ρ), or mass per unit volume, is the concentration of matter in an object and is a derived unit given in kg m^{-3} in the MKS System. Mathematically, the density of a uniform sample is expressed as:

$$\rho = \frac{m}{V} \tag{1.1}$$

where m is the mass of the substance and V is the volume of the sample. The density of water is 997 kg m^{-3}, whereas fresh snow is 10% less dense and platinum has density about 21 times that of water. Convert the density of water to:
1. CGS units
2. US-engineering or British units

Solution:
1. In the CGS units, the base units are grams and centimeters, and 1 kg = 1000 gms and 1 m = 100 cm. Hence:

$$997 \frac{\text{kg}}{\text{m}^3} = \frac{997 \text{ kg}}{1 \text{ m}^3} \times \frac{1000 \text{ gms}}{1 \text{ kg}}^1 \times \frac{1}{1000^3} \frac{\text{m}^3}{\text{cm}^3}$$
$$= 0.997 \frac{\text{gms}}{\text{cm}^3}.$$

2. Similarly, 0.454 kg = 1 lb-mass and 1 foot = 0.305 m. Hence:

$$997 \frac{\text{kg}}{\text{m}^3} = \frac{997 \text{ kg}}{1 \text{ m}^3} \times \frac{1}{0.454} \frac{\text{lb} - \text{mass}}{\text{kg}} \times \frac{0.305^3}{1} \frac{\text{m}^3}{\text{ft}^3}$$
$$= 62 \frac{\text{lb} - \text{mass}}{\text{ft}^3}.$$

Review Questions 1.1.
1. What, if any, are the advantages of using SI units?
2. Could you think of setting up your own system of fundamental units? Which units do you think you can use if you are not allowed to use the standard length, time, and mass?
3. What do we mean by coordinate systems, and why are they important?
4. A student wrote in their report that '*the speed of light is numerically equal to 3×10^6.*' Why did they not get full credit for their answer?

Exercises 1.1.
1. Calculate your height in each of the three systems of units, i.e. the MKS, CGS, and British units.
2. One kilometer is approximately
   (a) two miles
   (b) half a mile
   (c) a quarter mile
3. Given that the volume of a cube is numerically given by (length × width ×height), by what factor does the volume change if we double the length, width, and height of each side?
   (a) 16
   (b) 8
   (c) 4
   (d) 2
4. If the time between human heartbeats is 0.8 s, how many times does a human heart beat in one hour? in one day?

1.4 A few important concepts

Before we begin a formal discussion of the topic at hand, namely the physics of music, let us first define a few important physics concepts we will use throughout the text.

1. **Coordinate system:** a coordinate system is an artificially imposed grid that fills all space under consideration, and it allows us to solve problems by making quantitative measurements. For example, latitudes and longitudes are not actual lines drawn on planet Earth but are artificial grid lines placed on models of the Earth or used on maps that help with navigation around the globe. In addition, a coordinate system is used to identify the location of a particular particle or an event.
2. **Origin:** an origin is an agreed-upon reference point in a particular coordinate system. All measurements are made with respect to the origin in our chosen coordinate system.
3. **Three dimensions of space:** since we live in a three-dimensional space, it is obvious that we require three coordinates to locate an object relative to a user-defined origin. In general, Cartesian coordinates (x, y, z) are used to solve problems, but spherical and cylindrical coordinates are also used in physics. However, in this text, we will limit ourselves to mostly solving all problems in the Cartesian coordinate system.
4. **Complete description:** in classical physics, we need to know the position of a particle. Hence, for a *complete description* of the particle state, we must know the mass of the object and its position in space at any given time. This means

that measurements of position, time, and mass are fundamentally important to our understanding of physical systems.
5. **An event:** if an action occurs at a particular point in space at a particular time, then in physics we call it an *event*. For example, the instantaneous contact between colliding cars involved in an accident is called *an event*.
6. **Measuring an event:** in order to measure an event, problem solvers, you have to do the following:
    (a) Set an origin, call it a 'user-defined origin' (O).
    (b) Define a coordinate system with respect to the user-defined origin.
    (c) Set the orientation of the grid axis.
    (d) Measure the distance between the grid lines.
    (e) Measure the coordinates of the event with respect to the user-defined origin and the coordinate system.

1.5 A quick review of vector and scalar quantities

1.5.1 Scalars

In physics, quantities that can be completely described by just a numerical value are called scalar quantities. A scalar quantity has a magnitude (size) but no direction. For example, the money in your pocket, your age, your mass, and the temperature of a room are all scalar quantities.

1.5.2 Vectors

Quantities that require both a magnitude and a direction to completely describe them are called vector quantities. A good example of a vector quantity is your weight measured when you get on a weighing scale, since it always points to the center of the Earth and depends on the value of acceleration due to gravity (g). In other words, vector quantities are scalars with a direction. (Recall Vector, the evil genius from the movie 'Despicable Me.' Vector is so called because he commits crimes with both magnitude and direction!). A vector can be formally defined as:

A quantity that has a magnitude and a direction relative to a chosen reference frame. Vector quantities are represented pictorially by an arrow whose head represents the direction of the quantity and whose length represents its magnitude.

Familiar examples of a vector include velocity, acceleration, and force.

1.5.3 Vector anatomy

Any given vector, say $\vec{A}$, is specified by defining its magnitude and its direction relative to a chosen reference frame. Vectors are represented in textbooks by boldface letters **a** or **b** or by an arrow above a symbol, such as $\vec{a}$ or $\vec{b}$. In this text, I will use the latter symbolism (an arrow above a symbol, $\vec{A}$) to represent vector quantities.

A typical example of a vector in three dimensions in unit vector notation[2] is written as:

$$\vec{A} = (a_x, a_y, a_z) = a_x\hat{\imath} + a_y\hat{\jmath} + a_z\hat{k}, \tag{1.2}$$

where $\hat{\imath}$, $\hat{\jmath}$, and $\hat{k}$ are the unit vectors in the x-, y- and z-directions, respectively.

If $\vec{v}$ is a velocity vector, then it is mathematically written as

$$\vec{v} = (v_x, v_y, v_z) = v_x\hat{\imath} + v_y\hat{\jmath} + v_z\hat{k}, \tag{1.3}$$

where $\vec{v}_x$ is the velocity of the particle in the x-direction, $\vec{v}_y$ is the velocity of the particle in the y-direction, and $\vec{v}_z$ is the velocity of the particle in the z-direction.

1.5.4 The magnitude of a vector

The magnitude of a vector ($\vec{a}$) is represented as $|\vec{a}|$ or a. Please note that the magnitude of a vector is a number and gives no directional information about the vector $\vec{a}$.

1.5.5 Vector direction

In order to describe vector quantities, it is important for everyone to agree on a way to describe the direction of a vector. Conventionally, *up* on a map refers to the northward direction and *right* on a map refers to the eastward direction. The counterclockwise (CCW) convention describes the direction of any vector, and it results in a positive angle.

1.5.6 The components of a vector

A vector is a quantity described by length and direction. It is also useful to describe vectors in terms of their *components*, especially in situations that deal with vectors mathematically.

For the vector $\vec{a}$ shown in figure 1.2, the vector is described by a magnitude or length $|\vec{a}|$ from a specific user-defined origin and its direction is specified by an angle θ, measured with respect to (w.r.t.) a user-defined reference line, generally the positive x-axis. Note that we will use a or $|a|$ to indicate the length of vector $\vec{a}$. The x- and y-components of the vector $\vec{a}$ are then mathematically written as

$$\begin{cases} a_x = a\cos\theta \\ a_y = a\sin\theta. \end{cases} \tag{1.4}$$

In this form, the vector $\vec{a}$ is the vector sum of the two components

---

[2] A unit vector in vector space is a vector of length one. In general, any vector can be represented as a sum of multiples of unit vectors. A unit vector is often denoted by an addition of a *hat* on the vector: example, $\hat{\imath}$ usually represents the x-direction in Cartesian coordinates, while $\hat{\jmath}$ represents the y-direction and $\hat{k}$ represents the z-direction.

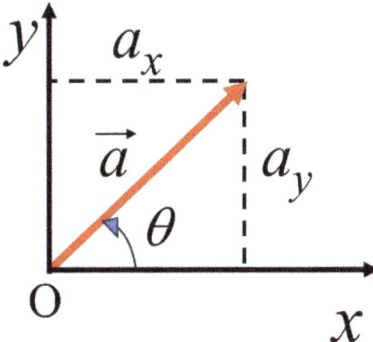

Figure 1.2. The components of a vector.

$$\vec{a} = a_x \hat{i} + a_y \hat{j}, \quad (1.5)$$

where $\hat{i}$ and $\hat{j}$ are unit vectors along the x-axis and the y-axis, respectively.

What we have really done is broken up our vector into two perpendicular vectors that are each parallel to one of the coordinate axes of the xy coordinate system. This process is called the *decomposition of a vector* into its components.

Example 1.3. What are the magnitudes and directions of the vectors given in figure 1.3?
Solution:
Refer to figure 1.3.
1. For vector ($\vec{a}$) represented by a red arrow:
    $\vec{a}$: magnitude = (8, 6) ; direction = north-east
2. For vector ($\vec{b}$) represented by a blue arrow:
    $\vec{b}$: magnitude = (−5, 9) ; direction = north-west
3. For vector ($\vec{c}$) represented by a green arrow:
    $\vec{c}$: magnitude = (6, −6) ; direction = south-east

1.5.7 Finding a vector from its components

As mentioned in section 1.5.6, if we know the magnitude $|\vec{a}|$ and the angle θ of any vector ($\vec{a}$), we can find its components a_x and a_y. The inverse is also true. Given components a_x and a_y, we can find the magnitude and the angle of the vector by using the equations

$$\begin{cases} a = \sqrt{(a_x)^2 + (a_y)^2} \\ \theta = \tan^{-1}\left(\dfrac{a_y}{a_x}\right) \end{cases}. \quad (1.6)$$

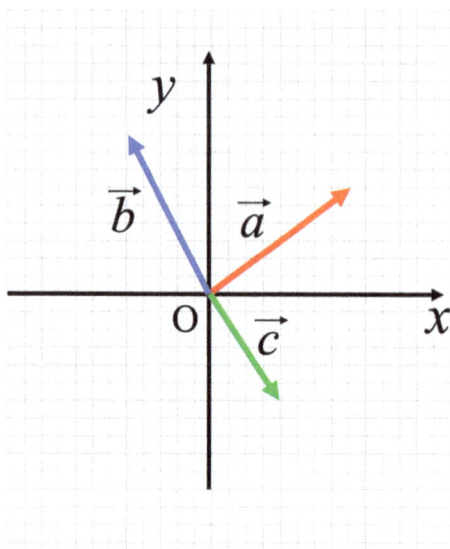

Figure 1.3. Example vectors.

The signs of the components are very important in these calculations. They determine the quadrant in which the vector is placed, which in turn tells us its direction.

1.5.8 Adding vectors

Vector addition is different from scalar addition since vectors have a direction associated with them. If you have a garage sale over the weekend and make $40 on Saturday and $50 on Sunday, your 'net' income for the weekend is $90. This is an example of the addition of scalar quantities. The word 'net' implies addition.

The same word 'net' is used for vector addition; however, it has a very different implication when it comes to the addition of vectors. For example, suppose you drive 4 miles east and then go 3 miles north, your net displacement is not 7 miles; rather, it is 5 miles, as shown in figure 1.4. This is because vectors add differently than scalars. This can be explained using Pythagoras' theorem.

Pythagoras, a Greek mathematician (570–495 BCE) is credited with the discovery and proof of the Pythagorean theorem, which allows us to add vectors. The Pythagorean theorem states that the 'net' displacement $\vec{C}$ equals the sum of the first displacement $\vec{A}$ and the second displacement $\vec{B}$ and is mathematically written as

$$\vec{C} = \vec{A} + \vec{B}. \tag{1.7}$$

We use the Pythagorean theorem to calculate the magnitude of the net displacement as follows:

$$|\vec{C}| = \sqrt{A^2 + B^2} = \sqrt{4^2 + 3^2} = 5 \text{ miles}, \tag{1.8}$$

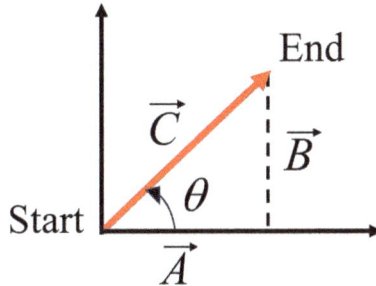

Figure 1.4. A displacement vector.

and we can find the direction of the net displacement as follows:

$$\theta = \tan^{-1}\left(\frac{B}{A}\right) = \tan^{-1}\left(\frac{3}{4}\right) = 37°. \tag{1.9}$$

Hence, the car's 'net' displacement is,

$$\vec{C} = \vec{A} + \vec{B} = 5 \text{ miles, } 37° \text{ north of east.} \tag{1.10}$$

The net displacement is also called the *resultant vector,* which means that any two vectors can be added together in any order as long as you follow the rules of vector addition.

Example 1.4. A minivan travels due east on a level road for 32 km, then turns due north at an intersection and travels 47 km before stopping. Find the vector that indicates the resulting location of the car.
Solution:
The magnitude of the resulting vector is:

$$|r| = \sqrt{(47)^2 + (32)^2} = 56.8 \text{ km.}$$

The magnitude of the resulting vector is 56.8 km, and it is directed north-east.

1.6 Distance versus displacement

The branch of physics that deals with motion is called kinematics. In order to talk about motion, we must first define a few important terms, such as displacement and distance.

Distance: distance is a scalar quantity. In very simple terms, it is a numerical description of how far apart two objects are. In other words, the actual physical path traversed from an initial position to a final position is called the distance traveled.

Displacement: displacement is the shortest distance from an initial point to a final point. Displacement is a vector quantity and hence it has magnitude as well as direction. The displacement vector is determined by the starting and ending points of the interval traversed and not by the physical path traveled between the two points.

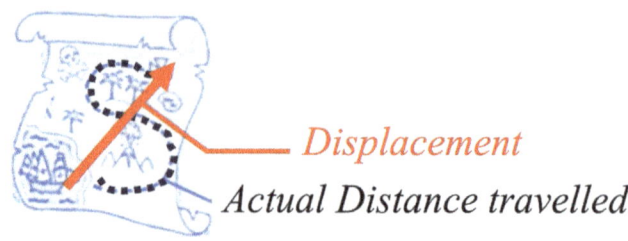

Figure 1.5. Distance (which is the actual distance traversed by the pirates) and displacement vector (which is the shortest distance between the ship and the treasure).

In other words, a displacement vector represents the length and direction of a straight path from some initial position to some final position, as seen in figure 1.5.

Mathematically, suppose the initial position is ($\vec{r}_1$) and the final position is ($\vec{r}_2$); then the displacement vector is given by:

$$\vec{S} = \vec{r}_2 - \vec{r}_1 = \Delta \vec{r}. \qquad (1.11)$$

The symbol Δ represents the change in a quantity that takes place between some initial state and some final state. The magnitude of the displacement may or may not be equal to the distance traveled; however, the initial and the final points of the journey are the same, as we observe in figure 1.5.

Example 1.5. A cat is trying to get from the south side to the north side of a 10 meter wide road during rush hour heavy traffic. She first moves 3 meters eastwards, then moves 4 meters north, and then 3 meters east again. She then stops to catch her breath, then continues onwards and goes 2 meters north and then 6 meters west and then goes 4 more meters north. Does she make it safely across the road?
Solution:
Draw the figure and then evaluate the step-by-step movement of the cat.

1.7 Speed versus velocity

1.7.1 The speed of an object

1. **The speed of an object:**
   Speed describes how fast an object is moving. The speed of an object is the numerical value of the distance traveled by an object divided by the time taken by the object to cover this distance. Speed is also called the rate of change of distance and it is always a positive number (a scalar). Mathematically, speed is given by:

$$\text{speed} = \frac{\text{distance traveled}}{\text{time taken}}. \qquad (1.12)$$

2. **The average speed of an object:**
    The average speed of an object is the distance traveled by the object divided by the length of the interval. Mathematically, it is given by

$$\text{average speed} = \frac{\text{total distance traveled}}{\text{elapsed time}} = \frac{\Delta x}{\Delta t}, \quad (1.13)$$

   where Δx is the distance traveled in the x-direction and Δt is the time interval.

3. **Instantaneous speed of an object:**
    The instantaneous speed is the value of the average speed in the limit in which the length of the time interval approaches zero. The speedometer reads the instantaneous speed of your car and is directionally independent. It only tells you the magnitude of your speed, but it does not tell you the direction. However, it helps you predict where the car will be at a later time, as long as the car keeps traveling at the same speed. For example, a car moving at 55 m.p.h travels 55 miles from its original location in 1 hour, but we cannot say in which direction it will be at that time. Mathematically, the instantaneous speed is given by

$$\text{instantaneous speed} = \lim_{\Delta t \to 0} \frac{\text{total distance traveled}}{\text{elapsed time}} = \frac{\mathrm{d}x}{\mathrm{d}t}. \quad (1.14)$$

Example 1.6. Suppose the given speed on a highway is 55 miles per hour. Convert this speed to m s^{-1}. (Hint: 1 mile = 1.6 km)
Solution:
To convert the speed into meters per second, we have to do the following:

$$55 \frac{\text{miles}}{\text{h}} = \frac{55 \text{ miles}}{1 \text{ h}} \times \frac{1609 \text{ m}}{1 \text{ miles}} \times \frac{1 \text{ h}}{3600 \text{ s}}$$
$$= 24.58 \frac{\text{m}}{\text{s}}.$$

Example 1.7. How far does a car travel in 15 minutes if it is traveling at 20 m s^{-1}?
Solution:
We can find the distance traveled by a car traveling at 20 m s^{-1} using equation (1.13):

$$\text{total distance traveled} = \text{average speed} \times \text{elapsed time}.$$

But first we have to convert minutes into hours as follows:

$$20 \frac{\text{m}}{\text{s}} \times \frac{60 \text{ s}}{1 \text{ min}} = 1200 \frac{\text{m}}{\text{min}}$$

Total distance traveled in 15 min = $1200 \frac{\text{m}}{\text{min}} \times 15 \text{ min} = 18\,000 \text{ m}$.

Hence, the total distance traveled in 15 min is 18 000 meters or 18 km.

Example 1.8. Light travels at 3.0×10^8 m s^{-1} and a light-year is a measure of length (not a measure of time) equal to the distance that light travels in one year.

1. Compute the conversion factor between light-years and meters.
   Solution:
   We have to convert light-years to meters.
   $$\text{speed} = \frac{\text{distance}}{\text{time taken}}$$
   $$3 \times 10^8 \frac{\text{m}}{\text{s}} = \frac{\text{distance}}{1 \text{ year}}.$$

   We have to find the distance light travels in a year; to do so, we have to convert years to seconds:
   $$1 \text{ year} = 1 \text{ year} \times \frac{365 \text{ days}}{1 \text{ year}} \times \frac{24 \text{ hour}}{1 \text{ day}} \times \frac{3600 \text{ s}}{1 \text{ hour}} = 3.1536 \times 10^7 \text{ s}.$$

   Putting this value into above equation gives
   $$3 \times 10^8 \frac{\text{m}}{\text{s}} = \frac{\text{distance}}{3.1536 \times 10^7 \text{ s}}$$
   $$3 \times 10^8 \text{ m s}^{-1} \times 3.1536 \times 10^7 \text{ s} = \text{distance}.$$

   We find that the distance of 1 light-year converted to meters is 9.45×10^{15} m.

2. Find the distance in meters to Gliese 581g, a potential Earth-like planet in the habitable zone of its star, which lies at an expected (predicted) distance of 20.2 light-years. Gliese 581g is potentially capable of supporting Earth-like life because it has a temperature in the range at which liquid water can exist on its surface. (Hint 1: use your answer from part 1 to answer part 2; 1 light year = $9.460\,5284 \times 10^{15}$ m = $9.460\,5284 \times 10^{12}$ km = $5.87849981 \times 10^{12}$ miles.)
   Solution:
   The distance to Gliese 581g is calculated as 20.2 light-years $\times (9.45 \times 10^{15})$ $\frac{\text{meter}}{\text{light} - \text{year}} = 1.91 \times 10^{17}$ m.

1.7.2 The velocity of an object

1. **The velocity of an object**
   The velocity of an object is the 'vector speed' of an object. To define velocity, we need to define the speed of the object and a direction associated with the movement. Mathematically, velocity is given by
   $$\text{velocity} = \frac{\text{displacement}}{\text{time taken}}. \tag{1.15}$$

   The velocity of an object is also defined as the rate of change of displacement.

2. **The average velocity of an object**
   The average velocity of an object in any interval is the change in displacement ($\Delta \vec{r}$) divided by the time duration (Δt) over which the displacement occurs, given by

$$\text{average velocity} = \frac{\Delta \vec{r}}{\Delta t}. \quad (1.16)$$

   The average velocity does not tell us the specifics of the motion since it only tells us the displacement (shortest distance) value; however, the average speed gives us a detailed description of the motion as it tells us the speed at every point in the path provided that the object is traveling. Also, the average speed is generally unrelated to the magnitude of the average velocity. An Indy 500 race car has an average velocity equal to zero around a lap!!!!

3. **The instantaneous velocity of an object**
   The Instantaneous velocity is the value of the average velocity in the limit when the duration of time interval approaches zero. The magnitude of the instantaneous velocity is the speed and the direction of the instantaneous velocity is the direction in which the motion is occurring:

$$\text{instantaneous velocity} = \lim_{\Delta t \to 0} \frac{\Delta \vec{r}}{\Delta t} = \frac{d\vec{r}}{dt}. \quad (1.17)$$

   Please note that since velocity is a vector quantity, we have to use the vector laws in order to add, subtract, or multiply the velocities.

Example 1.9. A car moves 20 km east and 60 km west in 2 hours.
1. What is its average speed?
2. What is its average velocity?

Solution:
   Using equation (1.13), we can find the average speed as follows.
   1. Average speed:

$$\text{average speed} = \frac{\text{total distance traveled}}{\text{elapsed time}} = \frac{(20 + 60) \text{ km}}{2 \text{ h}} = 40 \text{ km h}^{-1}.$$

   2. We can use equation (1.16) to find the average velocity.

$$\text{velocity} = \frac{\text{displacement}}{\text{time taken}} = \frac{(+20 - 60) \text{ km}}{2 \text{ h}} = -20 \text{ km h}^{-1},$$

   where the negative sign indicates that the final position of the car is west of the origin.

1.8 Graphical representation of motion

To describe the motion of a particle, we generally employ two methods, mathematical equations and/or a graphical method.

1. **Mathematical equations:** mathematical equations are better for solving problems related to motion. Mathematical equations permit precision in our data collection and results.
2. **Graphical method:** graphs or histograms provide more physical insight into the behavior of an object and allow us to extrapolate values by following trend lines.

In order to describe the motion of an object, we first define a reference point that is usually called the origin (O). In most instances, the origin is the point at which we define the positional vector of the object to be zero. For motion along the x-axis, $+3$ m means that our object moved 3 m east of 'O' and -5 m implies that the object moved 5 m towards the west.

Various common situations we encounter in life are discussed below using graphs.
1. **An object at rest:** when a particle is at rest, its position does not change with respect to time. The particle remains at coordinate 'A' at all times. Mathematically, this can be expressed as

$$x(t) = A,$$

as shown in figure 1.6.
2. **An object moving at a constant speed:** an object traveling at a constant speed is shown in figure 1.7. Its distance increases linearly over time, while its velocity remains constant. This is like a car driving down a highway with cruise control turned on at 55 m.p.h. The car maintains its speed at 55 m h^{-1} and is 55 miles away one hour later.

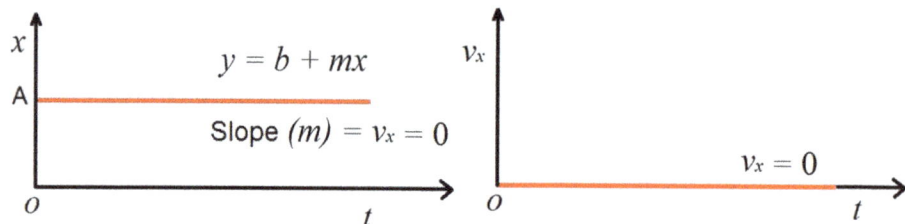

Figure 1.6. An object at rest.

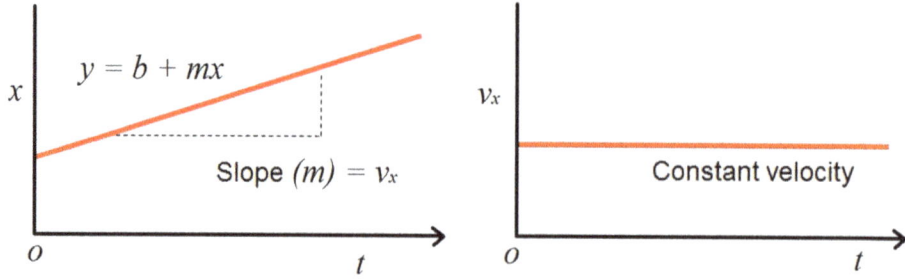

Figure 1.7. An object moving at a constant speed.

3. **An object with an increasing velocity:** an object increasing in velocity is said to be undergoing accelerated motion. The acceleration can also be considered to be the rate of change of velocity. The SI unit for acceleration is meters per second squared (m s^{-2}):

$$\text{acceleration} = \frac{\text{velocity}}{\text{time interval}}.$$

An object decreasing in velocity is said to be undergoing decelerated motion. Deceleration is expressed mathematically by adding a negative sign to the value. This is why acceleration is a vector quantity. An example of accelerated motion is discussed in example 1.8.3.

Example 1.10. The motion of a car: suppose a car is stopped at a red light. The signal then turns green and the car speeds up. It accelerates and moves for time (t) and attains the posted speed of 30 m.p.h. The car maintains this speed for a while until the next traffic light turns red and the driver applies the brake. The car slows down and eventually comes to rest before the signal. Its motion is represented in figure 1.8.

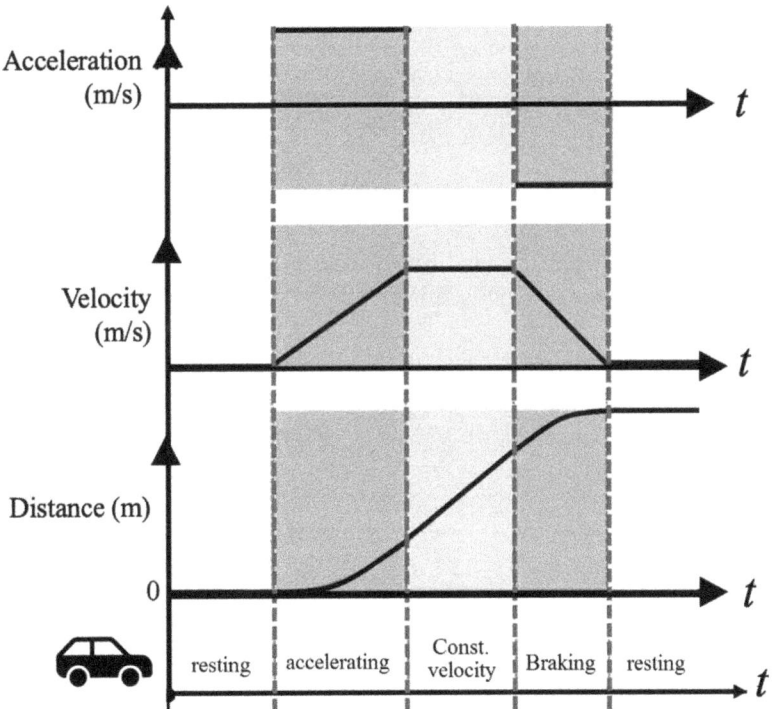

Figure 1.8. The accelerated motion of a car starting from rest.

Example 1.11. Suppose a car travels according to the given data:
1. Plot the values on graph paper.

Time (s)	0	1	2	3	4	5	6	7
Position (m)	0	25	50	75	100	125	150	175

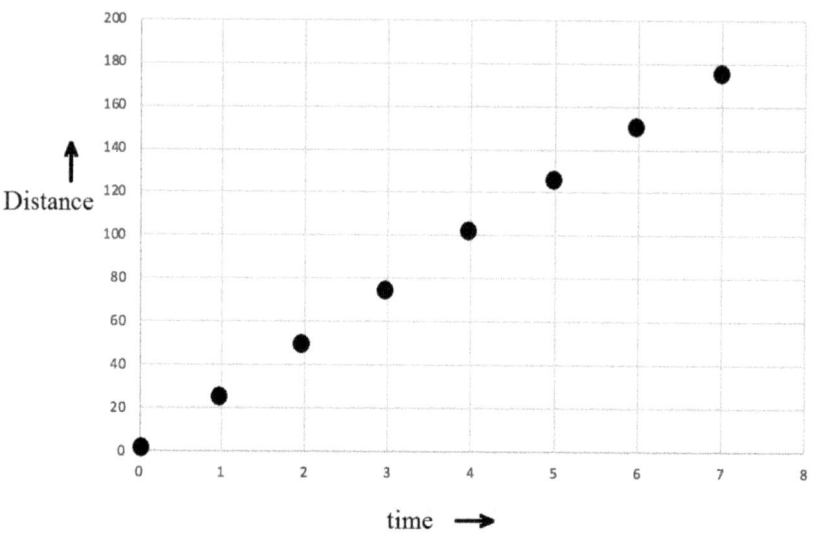

2. What is the total distance traveled by the car?
   Solution:
   The total distance traveled by the car = 175 m.
3. What is the average velocity during the first two seconds?
   Solution:
   The average velocity during the first two seconds
   $$v = \frac{50 \text{ m}}{2 \text{ s}} = 25 \text{ m s}^{-1}.$$
4. What is the average velocity of the whole trip? The average velocity of the whole trip
   $$v = \frac{175 \text{ m}}{7 \text{ s}} = 25 \text{ m s}^{-1}.$$
5. Is this a constant velocity or an accelerated motion?
   Solution:
   From the graph produced in part 1, we can see that it is constant velocity motion.

Example 1.12. Suppose a car travels according to the given data:
1. Plot the values on graph paper.

Time (s)	0	1	2	3	4	5	6	7
Position (m)	0	10	25	75	125	200	300	450

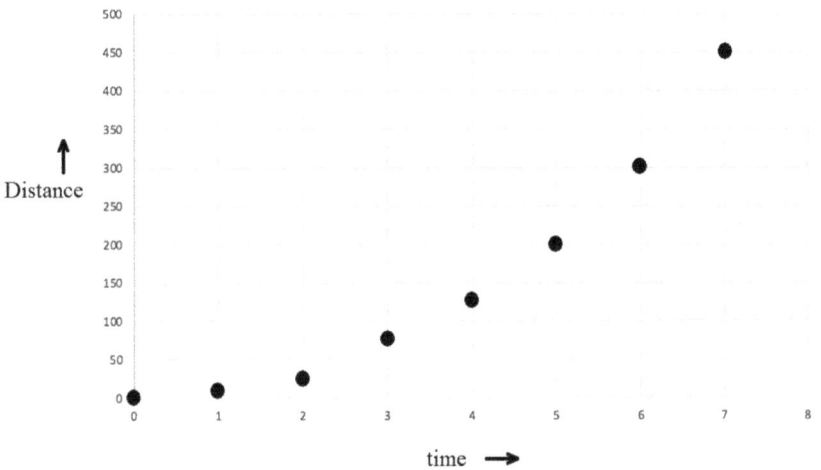

2. What is the total distance traveled by the car?
   Solution:
   The total distance traveled by the car = 450 m.
3. What is the average velocity during the first two seconds?
   Solution:
   The average velocity during the first two seconds
   $$v = \frac{25 \text{ m}}{2 \text{ s}} = 12.5 \text{ m s}^{-1}.$$

4. What is the average velocity of the whole trip? The average velocity of the whole trip
   $$v = \frac{450 \text{ m}}{7 \text{ s}} = 64.3 \text{ m s}^{-1}.$$

5. Is this a constant velocity or an accelerated motion?
   Solution:
   From the graph produced in part 1, we see that it is an accelerated motion.

Review Questions 1.2.
1. What is the difference between scalars and vectors? Give examples.
2. What do scientists mean by the components of a vector and why are they useful?
3. Can a zero magnitude vector have a direction?
4. Can the average speed ever be greater than the magnitude of the average velocity? Explain in your own words with an example.

Exercises 1.2.
1. Two vectors are given: $\vec{a} = 5\hat{\imath} + 3\hat{\jmath}$ and $\vec{b} = -3\hat{\imath} + 2\hat{\jmath}$.
   (a) Calculate $(\vec{a} + \vec{b})$ mathematically and graphically.
   (b) Calculate the magnitude of the final vector and the angle θ that this vector makes with the positive x-axis.
   (c) Find a vector $\vec{c}$ such that $\vec{a} + \vec{b} + \vec{c} = 0$.
2. Given the following vectors: $\vec{a} = 3\hat{\imath} - 3\hat{\jmath} + 1\hat{k}$ and $\vec{b} = 4\hat{\imath} + 9\hat{\jmath} + 2\hat{k}$, find:
   (a) $\vec{a} + \vec{b}$,
   (b) $\vec{a} - \vec{b}$.
3. An eagle observes a mouse moving in the following pattern, 3 m north, 2 m west and finally 5 m south.
   (a) Construct a vector diagram that represents this motion.
   (b) How far and in which direction would the eagle have to fly in a straight line to arrive at the same final point, given that they both start from the same initial point?
4. A bicyclist starts from rest and accelerates to a final velocity of $20\,\text{m s}^{-1}$ in 20 seconds. He then applies the brake and comes to rest in five seconds. What is his average acceleration in each case?
5. A car increases its speed from $50\,\text{mi h}^{-1}$ to $55\,\text{mi h}^{-1}$. In the same time, a person on a bicycle increases their speed from $5\,\text{mi h}^{-1}$ to $10\,\text{mi h}^{-1}$. Compare their accelerations.

Further reading

- Ohanian H 1985 *Physics* (New York: W.W. Norton & Company)
- Resnick R, Halliday D and Krane K 2002 *Physics* vol 1 5th edn (New York: Wiley)
- French A P 1971 *Vibrations and Waves* (Boca Raton, FL: Chapman and Hall)

IOP Publishing

The Physics of Sound and Music, Volume 1
A complete course text (Textbook)
Samya Bano Zain

Chapter 2

Sound, music, and noise

In this chapter, we introduce sound as either being music or noise. We will discuss this dichotomy but will not delve into the details; for a more detailed discussion, please consult an aesthetics book (or an aesthetician) of your choice.

2.1 What makes a sound either music or noise?

In order to understand the difference between music and noise, let us first start by defining what we mean by the word *sound*. The word *sound* means:
1. An auditory sensation in the ear.
2. A disturbance in a medium (such as a solid, liquid, or gas) that causes this sensation.

With this definition, we can now differentiate between music and noise. Let me start by asking you some simple questions:
1. Do you like music?
2. What kind of music do you generally like?
3. What kind of music are you listening to right now?
4. But maybe the most important question of them all: *would your parents consider your choice to be music as well?*

It is obviously not easy to objectively define the difference between noise and music. To avoid a discussion of some sillier examples of *music*, let us define some basic criteria for music. I will claim that we need at least these basic rules for a sound to be called music:
1. There must be some rhythm in the sound we claim to be music.
2. There must be a presence of tones that combine with other tones, preferably an orderly and harmonic manner.

Thus noise can be defined as mostly non-harmonic sound that is generally *unpleasant* to listen to. It is often associated with loud, chaotic sounds that do not blend well with each other.

However, the distinction between noise and music is subtler. If all sounds that consist of rhythmic tones are called music, then a jackhammer tearing up a road could be said to produce music. Should we then call that jackhammer a musical instrument? Most people agree that the sound produced by a jackhammer is not music. Similarly, if our definition of noise consists of loud, unpleasant sounds, then the sound produced by cymbals (in a marching band) is noise. However, this is obviously false as well.

Hence, the moral of the story is that we need to be careful when talking about *pleasing* sounds. Music that is pleasing to one person may not be pleasing to all.

Music Defined...

We need to define music somehow, so here is an attempt; I do not claim that it is absolute or complete:

1. *Music is any set of sound/sounds organized in patterns (tones/rhythm) in context that a human ear can recognize and appreciate.*
2. *Music is a simple rhythmic beating that makes a listener feel a particular emotion, such as happiness, sadness, etc.*
3. *Music is a source of stress relief and/or a form of cultural expression.*

2.2 Some history of the science of sound

Since the very beginning, mathematicians and scientists have been interested in the science of the production and the transmittance of sound. The science of sound is called **'acoustics.'** Merriam-Webster's dictionary defines *acoustics* as:

'*the science that deals with the production, control, transmission, reception, and effects of sound.*'

It is a common misconception that acoustics deals strictly with music and musical instruments. Acoustics is much more than just the study of musical instruments; it involves the study of architectural spaces, physics, physiology, ultrasounds (for medical imaging), electroacoustic communication, and neuroscience. A very interesting way to interpret and appreciate the vast scope of acoustics is to study of *Lindsay's Wheel of Acoustics*, created by R Bruce Lindsey in *J. Acoust. Soc. Am. V* **36** 2242 (1964). This wheel subdivides the field of acoustics starting from the four broad fields of Earth Sciences, Engineering, Life Sciences, and the Arts. The outer circle lists the disciplines studied to prepare for a career in acoustics. The inner circle lists the fields within acoustics that the various disciplines lead to.

2.2.1 Branches of acoustics

The principal branches of acoustics are music, musical engineering, environmental acoustics, and architectural engineering. Musical acoustics deals with the design and use of musical instruments, including the study of human anatomy and physiology (for sound production) and the study of how musical sounds affect listeners. Environmental acoustics deals primarily with noise control, and it focuses on

controlling noise produced by factories, machinery, traffic, etc. Engineering acoustics deals with development of sound production, recording, and effective reproduction equipment. Based on this discussion, we may subdivide the field of acoustics into the following:
1. Architectural acoustics
2. Physical acoustics
3. Musical acoustics
4. Psychoacoustics
5. Electro-acoustics
6. Noise control
7. Shock and vibration
8. Underwater acoustics
9. Speech
10. Physiological acoustics

Example 2.1. Explain in your own words why you think each of the branches of acoustics mentioned above is necessary and important.

2.3 How does it all work?

Sound travels in the form of waves, so to understand the physics of sound, we must first understand the physics of waves. The way that waves travel is called *wave propagation* or *wave motion*.

2.3.1 What is a wave?

If I were to ask you to close your eyes and think of a wave and describe it, what would you say? Most people get a mental picture of an ocean wave, such as the ones seen in figure 2.1. Most people can also tell you about their experience of riding a

Figure 2.1. Ocean waves.

wave or being too close and being swept away by an ocean wave. But describing a wave in physics or mathematics is a whole different story. Let us attempt to do so in this section.

The space all around us is occupied by waves. There is not an inch of space on Earth that can be called a *wave-free zone*. All communication, whether speaking, listening, or seeing, depends on waves. The type of wave may be different, but they are all waves nonetheless. General examples of waves are sound waves, light waves, water waves, radio waves, x-rays, seismic waves caused by earthquakes, etc.

Two of our five basic senses are wave detectors, namely the ears and the eyes. The ear is an air pressure wave detector (for sound waves) and the eyes are electromagnetic wave detectors (for light waves). Touch can also be classified as a wave detector, since we can feel through vibrations.

2.3.2 The definition of a wave

In general, we can say that all waves carry information from one point to another and may be defined as *propagation of a wave away from its source in the middle*, as shown in figure 2.2. Waves carry energy from one point to another without requiring matter to travel between the points. Therefore, a wave can be simply defined as follows:

> '*A wave is a transfer of energy without a transfer of matter.*'

2.3.3 Types of waves

There are three main types of waves: mechanical, electromagnetic, and matter waves.

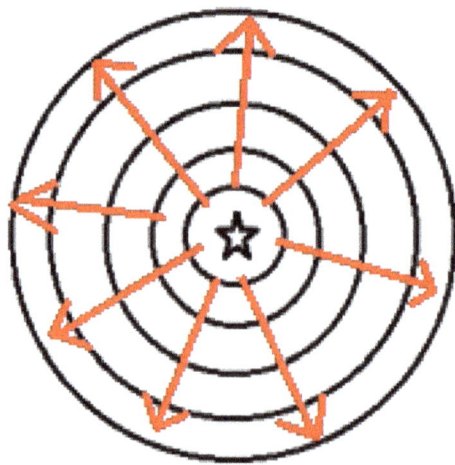

Figure 2.2. Propagation of a wave away from its source in the middle.

1. **Mechanical waves:** mechanical waves are the most familiar in our everyday experience. Common examples include water waves, sound waves, and seismic waves. All these waves have two key features:
   (a) They are governed by Newton's laws, discussed in chapter 4, and
   (b) They can only exist within a material medium, such as metal, wood, air, or water.
2. **Electromagnetic waves:** electromagnetic waves are less familiar to us, since most are not visible to humans. However, we use them all the time; common examples include microwaves, x-rays, visible light, etc. These waves do not need a material medium to exist or to travel. All electromagnetic waves travel through vacuum at the same speed, called the speed of light (c), which is numerically equal to 2.99×10^8 m s^{-1}.
3. **Matter waves:** although commonly used in modern technology, matter waves are the most unfamiliar to us. Matter waves are associated with fundamental particles, such as electrons, protons, etc.

2.3.4 The properties of waves

Based on the most fundamental definition of a wave, we may define some general properties of a wave as follows:
1. Waves transport energy from one place to another.
2. Waves transport information from one place to another.
3. Some waves propagate through a medium, for example, sound and ocean waves. The medium itself is not transported and reverts back to its undisturbed state after the wave has passed.
4. Some waves do not need a medium to propagate, for example, electromagnetic waves.

A good example of energy that is carried by waves is seismic waves that are produced by earthquakes. In an earthquake, seismic waves travel away from the epicenter of the earthquake. They travel through the Earth (in the form of body waves) and along the Earth's crust[1] (in the form of surface waves). Even though fault movement is the cause of the seismic wave, most earthquake damage is not caused by the movement of the fault but is rather caused by the motion of the Earth's crust resulting from the seismic wave. Please note that the material through which the seismic waves travel is not transported, and unless the amplitude of the wave is very large (which causes deformation in the Earth's surface), both the body and the crust of the Earth are essentially left unchanged after the waves have passed through.

2.4 The properties of traveling waves

There are two main ways in which waves travel: longitudinally and transversely. Based on these, we can define longitudinal and transverse waves.

---

[1] Note: the accuracy of pure geological terminology is sacrificed for the purpose of brevity.

2.4.1 Longitudinal waves

Longitudinal waves are the waves that travel away from the source and cause the medium to move back and forth in the same direction as the motion of the wave itself. If the wave moves through air, the motion of the air molecules is parallel to the direction of the wave. The motion is said to be *longitudinal*, and the wave is called a *longitudinal wave*.

A way to visualize the movement of a longitudinal wave is by using a Slinky toy. Set up the Slinky such that it is stretched between its two ends. If you now push the Slinky along its length such that the coils of the Slinky are compressed, you will observe compression traveling down the length of the Slinky, as depicted in figure 2.3.

2.4.2 Transverse waves

Transverse waves are waves in which the medium moves perpendicularly to the direction of motion of the wave, as seen in figure 2.4. *Transverse waves* are what we most easily recognize as *waves*.

To set up a transverse wave, position a Slinky toy again stretched between its two ends. If the Slinky is jiggled sideways or to and fro perpendicular to its length, you will observe a motion that travels down the length of the Slinky. This motion is called *transverse*, and hence the wave is called a *transverse wave*.

Other examples of transverse waves include ripples in water and the vibration of a piano string (figure 2.6). In a water wave, the actual water molecules move up and down while the wave moves horizontally. In figure 2.5, a duck is floating on the surface of a pond. If you throw a rock into the pond, water waves move outwards from that spot and create an expanding ripple. When the waves pass the duck, the duck bobs up and down. It does not move with the wave; once the wave passes, the

Figure 2.3. Longitudinal waves, i.e. regions of compression, are depicted in black, whereas rarefactions are shown in gray.

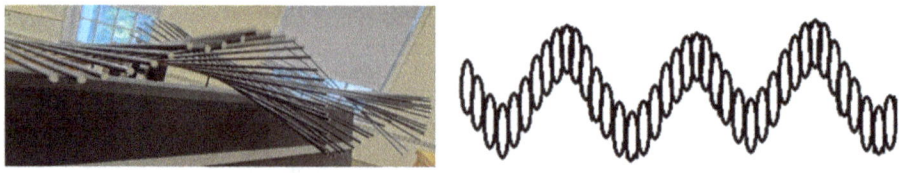

Figure 2.4. In transverse waves, the wave propagates to the right, while the medium moves perpendicularly to the motion of the wave.

Figure 2.5. Transverse waves in water.

Figure 2.6. A wave in a guitar string.

duck still floats where it was before. The wave set up by the rock hitting the water is an example of a transverse wave.

Both transverse and longitudinal waves are called '**traveling waves**' because they both travel from one point to another, for example, from one end of a string to the other end. Note that it is the wave that moves from end to end, not the material (string) through which the wave moves.

Example 2.2. Explain in your own words why this statement is not correct: 'Vibration in a guitar string is a sound wave.'
  Solution:
When strings in musical instruments such as a guitar are plucked or bowed, the waves set up are transverse waves. These transverse waves are not the longitudinal sound waves we hear. It is important to note that the transverse motion of the strings

causes longitudinal sound waves to be produced in the surrounding air by setting up compressions and rarefactions in the surrounding air. These longitudinal sound waves are what reach the listener and are therefore heard as the sound of the instrument.

2.4.3 Similarities and differences between various kinds of waves

The similarities between various kinds of waves include the following properties:
1. All waves can be reflected, refracted, or diffracted.
2. All waves have energy that they transport from one place to another.

The differences between various kinds of waves include the following:
1. Light/radio waves do not need a medium to travel through.
2. Sound waves require a medium such as a solid, liquid, or gas for propagation.
3. Waves of different types propagate with widely varying speeds:
    (a) Light/radio waves travel at 3×10^8 m s^{-1} (186 000 miles s^{-1}).
    (b) Sound waves travel at about 344 m s^{-1} in air at room temperature.
    (c) Water waves travel a few feet s^{-1}.

2.4.4 Waves that combine transverse and longitudinal motion

1. **Seismic waves**

    A sudden release of energy in the Earth's crust due to the plate tectonics (movement) in the Earth's lithosphere (the rocky layer of the Earth) results in an earthquake. As a result of this motion, seismic waves are initiated, and they travel through the Earth. The speed at which these waves travel depends on the depth as well as the density and elasticity of the Earth's crust. In general, the speed increases with depth from about 5 km s^{-1} (about 3 miles s^{-1}) in the Earth's crust to up to 13 km s^{-1} (about 9 miles s^{-1}) deep in the mantle. Two types of seismic waves are described below.

    (a) **Body waves:** there are two types of body waves:
    i. *P-waves (Primary waves)*: P-waves are the initial pressure waves produced by an earthquake. They are the fastest seismic waves and travel at about 4–8 km s^{-1}, through the Earth as longitudinal waves. Humans generally do not feel P-waves, however there are studies that show that animals, especially frogs and snakes, can feel these waves sometimes well in advance of the S-waves. Some theories attribute animals' ability to feel P-waves to their four legs. One theory says that since P-waves are pressure waves, humans do not feel them, but animals that have four legs can feel them passing from their hind legs to their front legs.

ii. *S-waves (Secondary waves)*: these waves typically follow P-waves during an earthquake and are thus called secondary or S-waves. S-waves average about 2–5 km s^{-1} and travel along the Earth's crust as transverse waves. In an S-wave, particles in the interior of the crust vibrate perpendicularly to the direction of the wave propagation. These are the waves responsible for the most damage and are felt by humans.
(b) **Surface waves:** surface waves are analogous to water waves and travel along the Earth's surface. In a surface wave, the motion of the ground is a combination of the longitudinal and transverse components of the seismic wave.

2.5 What is sound?

The term 'sound' has many meanings. In acoustics, it can mean a disturbance in an elastic medium that propagates through the medium or it can mean a stimulus that causes an auditory sensation. These definitions cause some ambiguity in interpreting the meaning of the word 'sound.'

A physicist would define sound using the first definition, whereas a psychologist may define sound based on the second definition. In physics, the terms 'sound' and 'acoustic' are used interchangeably. Both denote the propagation of longitudinal waves, regardless of whether they are heard or not. In this book, I will also use 'sound' and 'acoustic' interchangeably.

Sound is a longitudinal wave; hence, it has all the general characteristics of a wave. For a sound to occur as a mechanical event, four things are necessary: (i) an energy source, (ii) a vibrating object, (iii) an elastic medium and (iv) a receiver. For example, for a guitar to make music, we first need someone to pluck the string, which is the energy source; second, we need the strings to vibrate; third, we need air as an elastic medium through which the sound travels; and fourth, we need an audience to hear the music.

Simply put, the term 'sound' can be thought of in three basic ways:
1. How it is created (sound production), discussed in chapters 4, 5, and 6.
2. How it travels from one place to another (sound propagation), discussed in chapters 7, 8, 9, and 12.
3. How it affects the senses and emotions of the listener (sound perception), discussed in chapters 10 and 11.

2.5.1 About sound

When any sound is traced back to its source, we find that it is almost always caused by a vibrating body. A body is said to be vibrating when it has a back-and-forth pendulum-like motion, for example, the prong of a tuning fork, or a pendulum in a clock. If you rest your fingertips lightly against the lump (Adam's apple) in the front of your neck and sing, hum, or talk, you will feel vibration associated with sound you are producing. This back-and-forth motion causes variations of pressure in the surrounding medium, for example, air or water. The surrounding medium vibrates

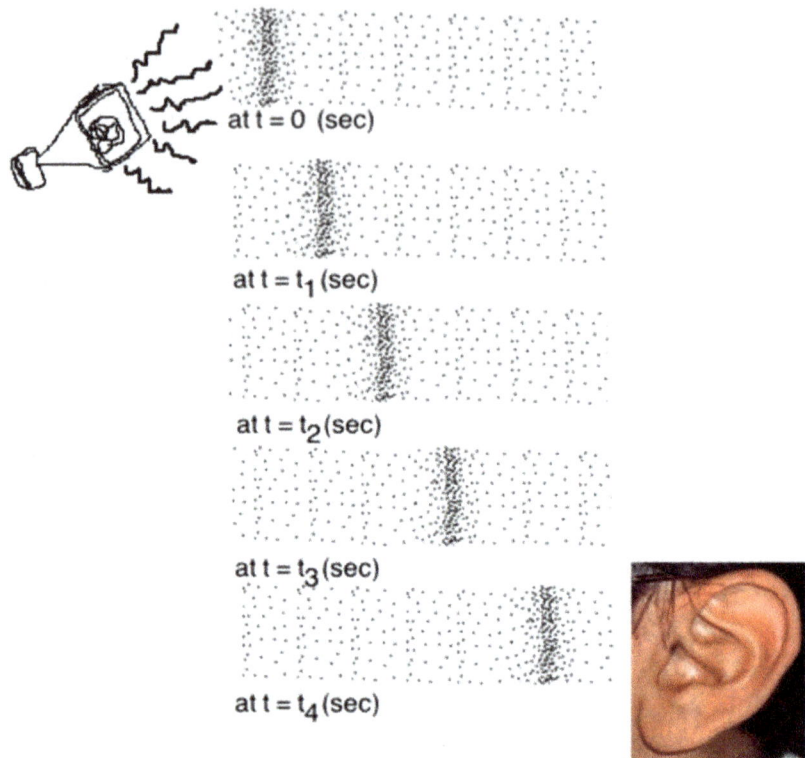

Figure 2.7. As the source vibrates it sets up vibrations in the surrounding medium. These vibrations travel to the listener's (in this case Sami's) ear and are heard by the listener. In this figure, the black dots represent air molecules. As the sound wave moves through the air, these air molecules are compressed and then return to their original state once the sound wave has passed.

with the same frequency as the source. The vibrating air then causes the human eardrum to vibrate with the same frequency. This frequency signal (also called the pitch) is transferred to the brain and is interpreted as sound. Figure 2.7 shows the motion of a sound wave as it leaves the source at time ($t = 0$ s) and approaches the listener at later times. The black dots in the figure represent the distribution of air molecules within a sound wave. It is clear from figure 2.7 that as a sound wave passes, the air molecules compress and then return to their original state once the sound wave has passed.

Please note that even though sound vibrations travel through the air, we cannot see or feel them moving. So, for now, we will try to *see* sound with our imagination; later, we will use an oscilloscope to 'see' the sound. In addition, when looking at figure 2.7, please keep in mind that sound waves are three-dimensional, and real sound waves travel outwards in ever-expanding spheres from the source, like those shown in figure 2.8.

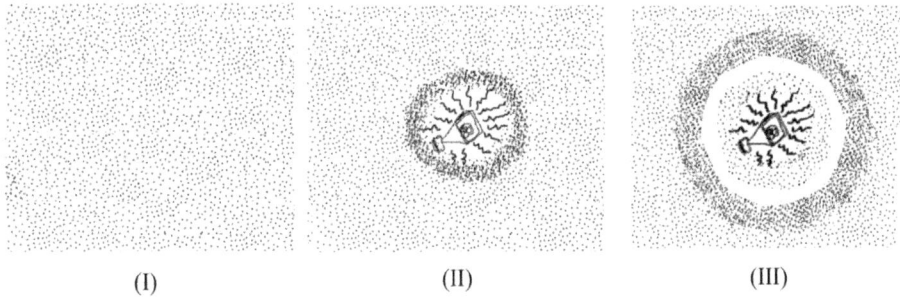

Figure 2.8. The propagation of sound in 3D space. (I). Normal atmospheric conditions. (II). The introduction of a sound source and the resulting change in atmospheric pressure. (III). The sound wave propagates outwards.

2.5.2 The propagation of sound waves

A sound wave propagates when particles of a medium collide with each other. A collision is an event in which momentum (defined as mass times velocity) or kinetic energy (defined as $1/2\ mv^2$) is transferred from one object to another. There are two general types of collisions in physics: elastic and inelastic collisions.

1. **Elastic collision:** an elastic collision occurs when two objects 'bounce' apart when they collide. In an elastic collision, both momentum and kinetic energy are conserved and almost no energy is lost.
2. **Inelastic collision:** an inelastic collision occurs when two objects collide and do not necessarily bounce away from each other. In an elastic collision, momentum is conserved, because the total momentum of both objects is the same before and after the collision. However, kinetic energy is not conserved.

During the propagation of sound, the molecules of the air do not actually travel from one place to another. If the molecules of air traveled, we would call it wind, not sound! Each individual molecule only moves a small distance before it bumps into an adjacent molecule and causes the adjacent molecule to vibrate. The energy lost per collision is small enough that we can approximate this as an elastic collision. This bouncing creates a ripple effect that carries the sound energy in all directions.

It may surprise you to learn that propagation, which seems to be the most abstract part of acoustics, is the simplest for physicists. Sound waves traveling through the air obey a very simple *linear* mathematical equation, given in equation (7.2). The physical meaning of linearity is that many different sounds can travel through the same space at the same time, and each is unaffected by the presence of other sound waves.

Physicists find it much easier to calculate and explain linear phenomena than they do any others. Sound production and perception are nonlinear (actually logarithmic) in nature, and the nonlinear nature of these concepts makes it a little harder to understand these areas of sound.

2.5.3 The requirements for sound

In order for sound to, for a lack of a better word, 'be there,' we need four things:
1. An energy source.
2. A vibrating material to set up or initiate the sound waves.
3. A medium to carry the waves.
4. A receiver to detect them.

If one of the four things mentioned above is not there, we cannot say that sound was ever there. This is one of the classic problems in the world of philosophy: *'if a tree falls in the forest, does it make a sound?'*

Maybe, from a sound perspective, there is no sound because there is no one there to acknowledge it, but from an acoustic standpoint there is still a sound in the woods because the falling tree causes a disturbance in the surrounding air.

2.5.4 The makeup of a sound wave

A sound wave is a longitudinal wave motion caused by a mechanical disturbance in an elastic medium. In other words, we can say that sound is an organized movement of molecules caused by a disturbance in the medium. The propagation of sound waves comprises of a series of compressions and rarefactions.

1. **Compressions:** when a vibrating body moves forward towards a listener, it strikes the air directly in front of it, crowding the particles of air together and causing the air molecules to compress; these are called *compressions*. Some books refer to compression as 'condensation,' but in this text, I will always call regions with a greater concentration of particles compressions. These particles move forward and strike other particles in their path, at which point they impart their energy to the next set of molecules in their path, resulting in a net energy transfer.
2. **Rarefactions:** when a vibrating body moves away from a listener, it leaves a partial vacuum behind it called *rarefaction*. Such regions also have smaller concentrations of particles. The surrounding particles quickly fill this partial vacuum due to the greater pressure outside the vacuum.

An undisturbed medium, say air, consists of particles that are in constant random motion called Brownian motion. If you introduce a sound source into the medium, the vibrating sound source alternately pushes and pulls the surrounding particles of the medium, creating regions of greater concentration called *compressions* and regions of lesser concentrations of particles behind the compressions, called *rarefactions*. This means the particles of the medium move in an organized manner in response to an external stimulus that creates changes in the density in the medium. Hence, the sound energy is transferred from the vibrating object through the medium to the listener in the form of density changes. We called this the *sound energy*.

2.5.5 Waveforms

Physicists have devised many ways to visualize sound vibrations (sound energy). The initial attempts included phonographs and ordinary microscopes that could be used to inspect their grooves. Nowadays, we study sound waves by monitoring the *waveforms* or shapes of sound waves as they move using an oscilloscope to visualize sound. An oscilloscope is a device that converts incoming signals (in this case acoustical signals) to electrical signals and then displays the waveform of the wave on the screen, as we see in chapter 5, figure 5.3. It is the simplest way to see the details of simple and complex vibrations.

Each musical instrument produces a different waveform corresponding to the notes played. This is closely related to the *color* of the tone. The color of the tone is our perception of the quality of the incoming tone of sound. We will discuss the color of the tone further in chapter 13.

2.5.6 Properties of sound waves

1. Sound waves, like all other waves, carry information and transport energy from one point to another.
2. Sound waves only travel through matter and do not travel in vacuum. When you see movies or TV shows about battles in outer space, please consider that you can see the explosions but should not be able to hear the explosion from a distance; sound effects are only added for dramatic effect in the movie!

2.5.7 Sources of sound

Anything that imparts a disturbance to an elastic medium is a source of sound. The main sources of sound may be categorized as follows.

1. **Vibrating bodies:** the most understood and familiar source of sound is the vibrating body. The vibration of a human's vocal cords results in speech and hence in the production of sound in humans. Other sources include:
    (a) vibrating strings, as in a piano or a violin,
    (b) vibrating columns of air, as in a trumpet or flute,
    (c) vibrating solids, as a struck drumhead.
2. **Heat sources:** an exploding firecracker produces a bang due to rapid heating. Similarly, a boom of thunder is due to the rapid heating of air by a bolt of lightning.
3. **Supersonic speed:** when a plane or any other sound source traveling faster than the speed of sound (at a supersonic speed) passes overhead, the compressions and rarefactions are heard at the same time, which results in a sonic boom.

2.5.8 Radiating patterns for sounds

Sound sources can be broadly divided into two main types: point sources and line sources.

Figure 2.9. Radiating patterns corresponding to: (a) a point source, (b) a line source.

1. **Point sources:** most sources in the world, including humans, can be approximated as point sources. In point sources, sound expands away from the source in an enlarging three-dimensional sphere, as shown in figure 2.9(a). Think of sound as the spreading ripples created on a water surface when you throw a rock into water, just in 3D.
2. **Line sources:** line sources are tall and narrow. Sound radiates outwards from the source in an expanding cylinder, as seen in figure 2.9(b). The advantage of a line source is that if it is placed vertically in a room, the sound it produces spreads outwards through the room parallel to the source. It does not spread in all directions like a point source and therefore has fewer reflections in a room or a theater as compared to a point source.

Example 2.3. Can you think of any other sound sources? Do they fit into the categories above, or can you think of another possibility?

Review Questions 2.1.
1. What do we call the science of sound?
2. What are the two kinds of waves?
3. What are the differences between a longitudinal wave and a transverse wave? Give examples.
4. Can you describe the differences and similarities between various kinds of waves (give examples)?
5. Define some general properties of a wave in its most basic form.

Exercises 2.1.
1. Compare and contrast longitudinal and transverse waves (give examples). Can you name a type of wave that does not require a medium for propagation?
2. Can you hear sounds in outer space?
3. Can you think of another geometry that might be used for the radiating patterns of sound? Do you think that it will be usable in the 3D space we live in?

Further reading

- Resnick R, Halliday D and Krane K 2002 *Physics* vol 1 5th edn (New York: Wiley)
- Rossing T D, Moore R F and Wheeler P A 2002 *The Science of Sound* 3rd edn (Upper Saddle River, NJ: Pearson Education)
- Hall D E 1990 *Musical Acoustics* 2nd edn (Belmont, CA: Wadsworth Publishing Co.)

IOP Publishing

The Physics of Sound and Music, Volume 1
A complete course text (Textbook)
Samya Bano Zain

Chapter 3

Music, history, and culture

3.1 Music and life

All people of the world, including the most isolated tribal people, have music and musical instruments. It is thought that the history of music may be as old as the history of humanity. Humanity has always found a way to express itself using the art of sound. The first music may have been conceived in Africa and then evolved to become a fundamental constituent of human life as people migrated to different parts of the world. The music of a people is influenced by all aspects of their culture and their values, including the societal and economic organization of the people, the climate of the region, and the materials readily available in the living spaces. Throughout history, music has been the medium for expressing emotions from happiness to pain and loss.

Music is played and listened to all over the world. However, the attitudes toward music, musical performers, and composers differ between regions and also time periods. In some cultures, musicians are revered as spiritual leaders and in others they are thought of as mere entertainers.

3.2 Historic eras of music

In this section we will talk briefly about the history of humanity and music.

3.2.1 The prehistoric era

Prehistoric music is music produced by prehistoric cultures. The origin of music is unknown, as it occurred prior to recorded history. Some suggest that the origin of music likely stems from naturally occurring sounds and rhythms. The earliest form of music was most likely the human voice, which can make a vast array of sounds from singing, humming, and whistling through to clicking, coughing, and yawning.

The earliest instruments were possibly percussion instruments, likely drum-like instruments, in which sticks and animal skins were used extensively. These instruments are thought to have been used in various spiritual and/or religious ceremonies. There was no notation or writing for this kind of *music*, and its sounds can only be extrapolated from the music of South American indigenous peoples and African natives who still adhere to some of their ancient religious practices.

Prehistoric music is also referred to as folk music, indigenous, or traditional music. The history of musical development in present-day Iran (Persian music) dates back to the prehistoric era. The great legendary king, Jamshid, is credited with the invention of music. Music in Iran can be traced back to the days of the Elamite empire (2500–644 BCE). Fragmentary documents from various periods of the country's history establish that the ancient Persians possessed an elaborate musical culture.

3.2.2 The ancient musical era

The prehistoric era is considered to have ended with the development of writing, and with it, by definition, prehistoric music. *Ancient music* is the name given to the music that followed prehistoric music. The *oldest known song* was written in a cuneiform script from Ugarit[1] dated to 4000 years ago. The name cuneiform means 'wedge shaped,' and cuneiform is one of the earliest known systems of writing, known by its wedge-shaped marks on clay tablets made by using a blunt reed as a stylus. The oldest song was deciphered by Prof. Anne Draffkorn Kilmer from the University of California at Berkeley, and was found to be composed in harmonies of thirds and written using a Pythagorean tuning of the diatonic scale.

By studying Egyptian hieroglyphs, archaeologists have theorized that the Egyptians had created harps and flutes by 4000 BCE and that double-reeded clarinets had been developed by 3500 BCE.

3.2.3 The early music of the world

The early music era may include contemporary but traditional or folk music, including Persian music, Indian music, Greek music, Roman music, the music of Mesopotamia and Egypt, and Muslim music.

3.2.4 Greek music

The music of Greece is as diverse and as celebrated as Greek history. Greek compositions originated during the Byzantine period, and all throughout Greek antiquity there was a continuous development of music, which clearly shows up in the language of the people, the rhythms, the structure, and the melody of Greek music.

In ancient Greece, men usually performed choruses for entertainment and celebrations. Music was an important part of the education process, and music education started as early as six for young boys. This practice resulted in musical

---

[1] Ugarit, which was first settled in the Neolithic period (about 6500 BCE), lies about six miles north of the Syrian port of Latakia on the Mediterranean coast. It was accidentally discovered in 1928 CE due to the discovery of an ancient tomb at the small Arab village of Ras Shamra.

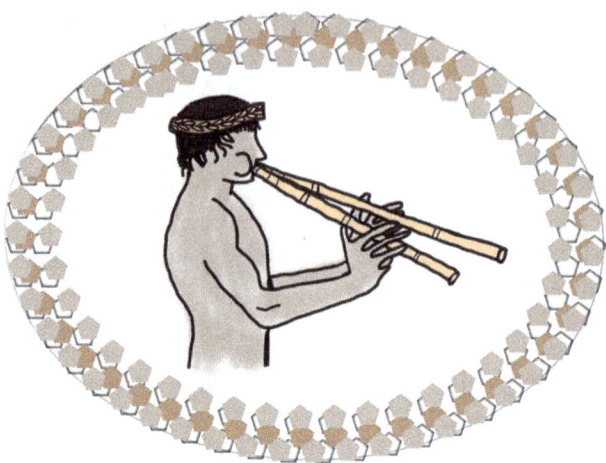

Figure 3.1. Paying the aulos.

literacy and hence the development of musical theory. Greek music theory included Greek musical modes that eventually became the basis for western religious music and classical music.

Greek instruments included the double-reed aulos and the plucked string instruments called the lyre and kithara. The Greek aulos had two double-reed pipes, as in a modern oboe, but these were not joined and were generally played with a mouthband, as shown in figure 3.1, to hold both pipes steadily between the player's lips.

3.2.5 Roman music

The music of ancient Rome was a part of Roman culture from earliest times. Music was an integral part of funeral services, and the tibia (Greek aulos), a woodwind instrument, was played at sacrifices to ward off bad influences. Songs were an integral part of almost every social occasion.

Roman art depicts various woodwinds, brass, percussion, and stringed instruments popular in Roman culture. Roman-style instruments are found in all parts of the empire, which indicates that music was among the aspects of Roman culture that spread throughout the Roman provinces.

Some Roman wind instruments included the Roman tuba, which was a long, straight bronze trumpet with a detachable, conical mouthpiece. The cornu, which is Latin for 'horn,' was a long tubular metal wind instrument that curved around the musician's body, shaped rather like an uppercase G, and possibly produced a low, clarinet-like sound.

Roman stringed instruments included the lyre, which was borrowed from the Greeks and was basically an early harp, which had a frame of wood or tortoise shell and a numbers of strings stretched from a cross bar to the sounding body, and the lute, the ancestor of the guitar, which had three strings. The lute was not as popular as the lyre but was easier to play.

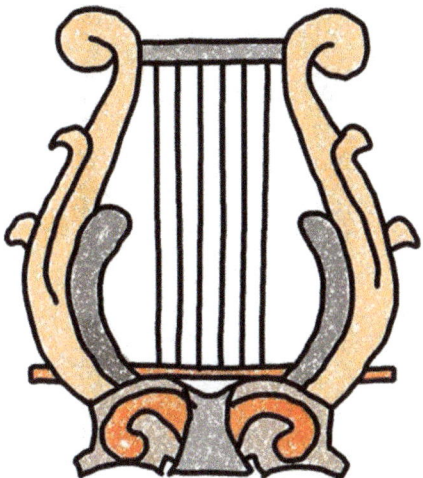

Figure 3.2. A Roman cithara.

The cithara, shown in figure 3.2, was the main musical instrument of ancient Rome and was played in nearly all forms of musical displays from popular music to professional settings. Larger and heavier than a lyre, the cithara was a loud, sweet, and piercing instrument that could be accurately tuned.

Roman percussion instruments included various instruments, such as the *scabellu*, which means the clapper of multiple rattles, bells, and tambourines. A *scabellu* constructed from wood or metal was used to keep time during performances and songs. Other percussion instruments such as timpani, castanets, the Egyptian sistrum, and brazen pans also served assorted purposes in Roman music, including as backgrounds for rhythmic dance, celebratory rites, and military and hunting music.

3.2.6 Arabic music

In history of civilizations, Islam stands as a cultural link between the Dark Ages and the Renaissance period. *Islamic Civilization* is purely a sociological association. Islam spread over a quarter of the then known geographical world; it gave rise to a specific mode of lifestyle and became a central idea to live by for all observers of the religion. Islam included many cultures from far-East Asia to Central Asia, Africa, and Europe.

The religion of Islam began in Arabia, and the visible signs of music and religion in ancient Arabia confirm that the Arabs of the Arabian Peninsula had indeed inherited and were conservators of the Mesopotamian cultural heritage. Music was largely delegated to women, especially in the upper elite class. Women of certain upper-class households were employed as *qainat* (singing women) for the purpose of entertainment. The male *mughani* (singing men), *mitrib* (musicians), and *alati* (instrumentalists) were written about by Ibn Musa al-Nasibi (860 CE) in his Kitab al-Aghani (book of songs).

Figure 3.3. Sain Zahoor playing the ektara.

Since Islam is a multiethnic religion, the musical expression of its adherents is as diverse as the Muslim population. The indigenous musical styles of these areas have shaped the devotional music enjoyed by contemporary Muslims. Sain Zahoor Ahmed, a Pakistani folk musician performing with a decorated ektara, is shown in figure 3.3. The ektara is a lute with a single string. Sain Zahoor also plays a three-stringed lute. In addition, he has a unique dress, the long kurta (shirt) and anklet bells, called *ghungaroo*, which add a dance component to the musical performance.

3.2.7 Music of the Indian subcontinent

Some musical instruments used in the Indian subcontinent are shown in figure 3.4. The basic concept of this music includes *shruti* (microtones), *swara* (notes), *raag* (melodies improvised from basic grammars), and *taal* (rhythmic patterns used in percussion). The seven basic swaras of the scale are named shadja, rishabh, gandhar, madhyam, pancham, dhaivat, and nishad, and are shortened to Sa, Re, Ga, Ma, Pa, Dha, and Ni. Collectively, these notes are known as the *sargam*. The tone Sa is not associated with any particular pitch but rather it refers to the tonic (beginning) of a piece or scale. The tonal system in Indian music divides the octave into 22 segments called *shrutis*, which are not all equal; however, each is roughly equal to one quarter of a whole tone in Western music.

3.2.8 Chinese music

Music in ancient China was used in religious ceremonies, announcements, dance, and entertainment. The ancient Chinese were the first to divide musical instruments into eight categories based on the material used to construct the instrument. The categories were silk, bamboo, hide, clay, gourd, metal, stone, and wood. Chinese

Figure 3.4. Various musical instruments of the Indian subcontinent: (left) the tabla; (middle) the sitar; and (right) the Indian harmonium.

Figure 3.5. A mural from the tomb of Xu Xianxiu in Taiyuan, Shanxi province, 571 CE during the Northern Qi Dynasty, showing male court musicians playing stringed instruments, either the liuqin or pipa, and a woman playing a konghou (harp). This image, 'paintings on north wall of Xu Xianxiu Tomb,' has been obtained by the author(s) from the Wikimedia website, where it is stated to have been released into the public domain. It is included within this book on that basis.

stringed instruments used wood and strings made of silk. Woodwind instruments were primarily made from bamboo, whereas metals were used to make bells, gongs, and chimes. Drums, including hand-held rattle drums, were made from hide. Stone and clay were also used to make chimes. It is interesting to note that these categories are still in use in China. The mural depicted in figure 3.5 is from the tomb of Xu Xianxiu in Taiyuan, Shanxi province, 571 CE, and it shows some of the instruments of the time. Chinese instruments produce a unique sound that is very different from those of any other culture (figure 3.5).

3.2.9 Renaissance music

The music written in Europe during the Renaissance period (from about 1400 to 1600 CE) is called Renaissance music. The music of the period was significantly influenced by the changing ideas about the value of a human life and the recovery of the artistic heritage of ancient Greece and Rome. Relative political stability in Europe, along with a flourishing system of music education in many churches, allowed for the training of hundreds of singers and composers. These musicians were highly sought-after throughout Europe, particularly in Italy, where churches and aristocratic courts hired them as composers and teachers. The development of printing also aided in the distribution of music on a wide scale during this era.

By the end of the 16th century, opera had begun in Florence, Italy as a deliberate attempt to resurrect the music of ancient Greece. Music, freed from medieval constraints, effectively became an instrument of both personal and cultural expression for renaissance composers. Many familiar modern instruments, including the violin, the guitar, and keyboard instruments, were born and developed during the Renaissance era.

3.2.10 The modern musical era

The evolution of more advanced instruments was slow and steady. Some important eras in the evolution of modern musical era are mentioned here; please refer to other texts for a more in-depth description of each era. The Renaissance period was followed by the Baroque era from about 1600–1760 CE, the classical era from 1730–1820 CE, and the romantic era from 1815–1910 CE.

1. **The baroque era:** the word *baroque* derives from the Portuguese word *barroco*, meaning misshapen pearl, a negative description of the ornate and heavily ornamented music of this period. *Baroque* is generally used to describe a broad range of styles from a wide geographic region, mostly in Europe, from about 1600 to 1760 CE. During this period, composers and performers used more elaborate music than ever before. They were able to make changes in musical notation and develop new playing techniques. Baroque music expanded the size, range, and complexity of instrumental performance and also established opera, cantata, oratorio, concerto, and sonata as musical genres. Many musical terms and concepts from this era are still in use today. Baroque instruments include the harpsichord, the bass viol, lute, violin, guitar, etc.
2. **The classical era:** the classical era is generally considered to be about 1730–1820 CE. The early classical period was developed by the Mannheim School. The father of the school is considered to be the Czech composer Jan Václav Antonín Stamic, generally known as Johann Stamitz. The composers of the Mannheim school introduced a number of new ideas into the orchestral music, such as sudden crescendos and diminuendos.

    During the classical period, composers began producing public operas in their own native languages, whereas prior to this, operas were generally only performed in Italian. The symphony orchestra, the most widely known medium for classical music, includes members of the stringed, woodwind, brass, and percussion families of instruments, whereas the concert band consists of members of the woodwind, brass, and percussion families.

3. **The romantic era:** the late 19th century saw a dramatic expansion in the size of the orchestra and the role of concerts in the urban society. Famous composers from the second half of the century include Johann Strauss II, Brahms, Liszt, Tchaikovsky, Verdi, and Wagner. In the Romantic period, approximately between 1815–1910 CE, music became more expressive and emotional, expanding to encompass literature, art, and philosophy.

   During the Romantic period, music often took on a nationalistic purpose. For example, Jean Sibelius Finlandia has been credited with the rise of the nation of Finland, which gained independence from Russia. In part, the romantic period was a revolt against social and political norms.

4. **The modern era:** the truly modern era began around 1890–1930 CE. The 20th century saw a revolution in music as new technologies were developed to record, reproduce, and distribute music. Early in the century, radio gained worldwide popularity and music became accessible to the masses for the first time. This brought with it a new freedom and worldwide accessibility, which led to experimentation with new musical styles that challenged the conventional rules of music. Music composers started exploring and experimenting by blending sounds and instruments from diverse backgrounds, which led to the development of many genres of music, such as the development of jazz music. Jazz evolved from a melding of African American and European musical roots.

   The invention of musical amplification and electronic instruments in the mid-20th century revolutionized popular music and helped to accelerate the development of new forms of music. With the advent of television, musical performances also became progressively visual (figure 3.6). A typical high school concert band from 2023 is shown in figure 3.7.

Figure 3.6. The modern era: (left) a phonograph, (top) a recordable compact disc (CD), and (right) an audio cassette tape and a Walkman.

Figure 3.7. A typical high school concert band.

3.3 Music and religion

Religious music, also called sacred music, is music performed or composed for religious use. Religious music is defined by the intent of the author and the objective of their work. Religious music seeks to arise pious and religious feelings in the listener. In principle, only music free of commercial interests and that which does not produce the draw of worldly things can be recognized as religious music.

3.3.1 African music

In Africa, songs and music are associated with life events such as births, deaths, weddings, prayers, work, wars, etc. Music is present at all key moments in a person's social life, not only as a way of participating in the group but also as a source of information on the nature of the gathering (figure 3.8).

African traditional music has a particularly sacred aspect. Sacred Christian and Muslim music arose from the meeting between African traditions and specific religious morals. The evangelization of sub-Saharan Africa dates back many centuries. The official language of the Catholic Church was Latin, which was not spoken in Africa, so the missionaries translated religious songs into local languages. However, holy Christian music that is specifically African only dates back a few decades. It started in the early 1970s after a visit by Pope Paul VI to Kampala, during which, he encouraged Africans to '*be authentically Christian and authentically African.*'

Regarding Islam, there is no specific African Muslim music. In most cases, the use of music is limited to songs in praise of Allah and to those that honor the glory of Prophet Muhammad (p.b.u.h.).

3.3.2 Native American music

Native American music created and performed by Native Americans in the United States and the First Nations people of Canada plays a vital role in the history of the people of the lands. Even though the style and the purpose of Native American

Figure 3.8. African music.

Figure 3.9. Assorted Native American musical instruments.

music varies greatly between tribes, the legacy of orally passing on ancestral customs to new generations depends on the music of the tribe and plays a crucial role in the education of new generations.

Native American ceremonial music traditionally originates from either spirits, deities, or from individuals of the tribes. Singing and percussion are the most important aspects of traditional Native American music; examples are shown in figure 3.9. Percussion, especially drums and rattles, is commonly used to keep the rhythm for singers.

3.3.3 Jewish music

Jewish music is the music and melodies of the Jewish people. The history of religious Jewish music spans the evolution of cantorial, synagogal, and Temple melodies from

Biblical times. The earliest synagogal music was based on the system used in the Temple in Jerusalem. The Temple orchestra consisted of twelve basic instruments: the kinnor (lyre), nevel (harp), shofar (ram's horn), trumpet (chatzoteroh), three varieties of pipes, a tziltzal (cymbal) made of copper, and a choir of twelve male singers. Even though much is known about Temple instruments, we do not know a lot about the tunes that were played. Traditions exist for both religious music, such as that sung at the synagogue and domestic prayers, and secular music, such as klezmer.

In the nineteenth century, religious reform led to the composition of synagogue music in the styles of classical music, and modern Jewish composers have often been influenced by the different traditions of Jewish music.

3.3.4 Christian music

Historians have inferred that the early Church was most likely a singing church, in which songs were not only used to accompany worship, but that they were the main form of worship itself. However, no early manuscripts from the first century actually exist, so this fact has not been validated. However, we can theorize that synagogue tunes were adjusted for Christian use, and the general rules were continued.

There are two main categories of Christian sacred songs: the Psalms and Hymns. The psalms were initially used as the prayerbook of the Church; they are sung in unison, and their music is based on simple melodic patterns called 'tetrachords.' However, hymns that were modeled on the Psalms are generally syllabic (one note per syllable) as opposed to being melismatic (many notes per syllable). Christian hymns are used to praise God and many of them, either directly or indirectly, refer to Jesus Christ.

3.3.5 Islamic music

Islamic music is Muslim religious music. The heartland of Islam is the Middle East, North Africa, the Horn of Africa, Iran, Central Asia, and South and East Asia. Since Islam is a multiethnic religion, the musical expression of its adherents is vastly diverse. The indigenous musical styles of these areas have shaped the devotional music of Muslims throughout the world.

Types of devotional recitation and music include:
1. **The nasheed:** nasheeds are religious songs sung in various melodies by Muslims and are generally not accompanied by any musical instrument. Nasheeds include *hamd*, a song in praise of God, and *naat*, sung in the praise of Prophet Muhammad (p.b.u.h[2]).
2. **The zikr:** Sufi worship services are often called *zikr*. The music of Sufiism best known in the West includes the chanting and rhythmic dancing of the whirling dervishes in Turkey, as depicted in figure 3.10, and the Qawwali, shown in figure 3.11, primarily performed in Pakistan and India.

---

[2] Peace be upon him.

Figure 3.10. A whirling dervish during a multicultural performance at Susquehanna University.

Figure 3.11. Pakistani Qawwali music. Credit: Humayun Rauf (2023).

3.4 Classes of musical instruments

For people that do not play a musical instrument, it is humbling to pick one up and try to create something that resembles a melody or a tune. This gives a novice musician a true appreciation for talented musicians. However, it is not necessary to be a musician to understand the physics of music and musical instruments. All musicians create music by making standing waves in their instruments.

The guitar player makes standing waves in the strings of the guitar that reverberate throughout the body of the guitar, and the drummer does the same in the skin of the drumhead but creates standing waves in two dimensions. The trumpet blower and flute player both create standing waves in the column of air within their instruments in three dimensions. Most kids have done the same thing, producing a

tone as they blow over the top of a bottle. If you understand standing waves, you can understand the physics of musical instruments. Musical instruments may be broadly divided into three main classes:
- **chordophones** (strings), discussed in detail in chapter 15.

  A stringed instrument is a musical instrument that produces sound by means of vibrating strings. The most common stringed instruments are guitars (both acoustic and electric), violins, violas, cellos, double basses, banjos, mandolins, ukuleles, and harps.
- **aerophones** (open and closed pipes), discussed in detail in chapter 17.

  Wind instruments may be grouped into three main classes:
  - brass or lip reed instruments,
  - air reed instruments,
  - cane reed instruments.

In all wind instruments, a coupled system is formed between the pipe or column of air. In the case of air reed instruments (like the flute), the mouth forms one part of the coupled system, the lips and the mouthpiece form another part, and the air column forms the third.
- **idiophones** (vibrating rigid bars and pipes), discussed in detail in chapter 16.

  Percussion instruments include tuning forks, chimes, xylophones, marimbas, etc. A percussion instrument is any object which produces a sound by being hit, shaken, rubbed, scraped, or by any other action which sets it into vibration.

Review Questions 3.1.
1. What is the oldest song known to humanity? Where was it discovered?
2. How did the musical evolution mimic human evolution? Explain in your own words.
3. When did the truly modern era of music and musical instruments begin?
4. Name some ancient musical instruments that are still used in the modern era.

Exercises 3.1.
1. Write down the details of a particular historic era of music of your choice not mentioned in this chapter.
2. Write some details about some religious music.
3. Name the various classes of instruments. Explain each class in a few words.
4. Explain in your own words the effect (if any) that nature has on the musical instruments of people around the world.

Further reading

- Berg R E and Stork D G 1994 *The Physics of Sound* 2nd edn (Englewood Cliffs, NJ: Prentice-Hall)
- Rossing T D, Moore R F, Wheeler P A 2002 *The Science of Sound* 3rd edn (Upper Saddle River, NJ: Pearson Education)
- Emanuel D C and Letowski T 2009 *Hearing Science* (Philadelphia, PA: Wolters Kluwer)
- Hall D E 1991 *Musical Acoustics* 2nd edn (Pacific Grove, CA: Brooks/Cole)

Part II

Sound production

IOP Publishing

The Physics of Sound and Music, Volume 1
A complete course text (Textbook)
Samya Bano Zain

Chapter 4

Tension and deformations in a string

May the force be with you!

Classical physics is a causal discipline, which means that it attempts to find the cause of a system's behavior. Kinematics, as discussed in chapter 1, is the branch of physics that describes the motion of objects in space without considering the cause of the motion. The focus of kinematics is the relationship between displacement, velocity, acceleration, and time.

In this chapter, we will consider the causes behind a change in the state of an object and the energy that has to be applied to the object to do so. When energy is applied to an object, the result is the movement of the object. The branch of physics that deals with energy and force is called *dynamics*. In other words, dynamics is the study of the effects of forces that result in a change in the state of an object or a system of objects.

4.1 Energy and force

In this text, I will introduce the concepts of force and energy at the same time, since force and energy are inextricably linked to each other.

4.1.1 Force

In physics, force ($\vec{F}$), expressed in units of newtons (N), is described as either a 'push' or a 'pull.' All human activity involves forces, including but not limited to running, lifting, eating, writing, and even understanding. In other words, when two objects interact with each other, they exert forces on each other. Forces have both a magnitude and a direction, thus they must be described using vectors, previously defined in chapter 1. This means that for every force, we must specify the magnitude of the force and the direction in which it acts.

4.1.2 Energy

Energy (E) is defined as the ability of an object to do work. In physics, work is said to be done on an object when an applied force ($\vec{F}$) moves it by a certain displacement ($\vec{x}$). Work is mathematically given by

$$\text{work } (W) = \vec{F} \cdot \vec{x}. \tag{4.1}$$

This means that when we want an object to do more work, we need to provide it with more energy. Another way to look at this is to see work as *used energy*. To a physicist, this means that both energy and work must have the same units, called **'joules'** (J).

4.1.3 Forms of energy

Energy has many forms, for example, kinetic energy, potential energy, thermal energy, electrical energy, nuclear energy, etc. Since we are talking about the physics of music in this text, I will limit our discussion to kinetic and potential energies only.

- **Kinetic energy:** kinetic energy (KE) is the energy an object possesses due to its speed. It is mathematically expressed as

$$\text{KE} = \frac{1}{2}mv^2, \tag{4.2}$$

  where m is the mass of the object and v is its speed. Since the mass of an object generally remains the same under normal conditions, the kinetic energy is directly proportional to the velocity of the object. This means that when the speed of the object is higher, the kinetic energy of the object is also higher.

- **Potential energy:** potential energy (PE) is the energy stored in an object due to its position within a physical system. Potential energy could be due to an object's position in a gravitational field, or it could be elastic potential energy due to a compressed spring in a mass–spring system.
    - The potential energy of an object of mass (m) exerted on the Earth due to the force of gravity is mathematically given by

$$\text{PE} = mgh, \tag{4.3}$$

      where g is the acceleration due to gravity and h is the physical height of the object. The higher an object is above the surface of the Earth, the greater its potential energy.
    - The potential energy of a compressed spring is

$$\text{PE} = \frac{1}{2}kx^2, \tag{4.4}$$

      where k is called the spring constant and x is the amount by which the spring is compressed.

4.1.4 The conservation of energy

One of the most important laws in physics is the law of conservation of energy. The law of conservation of energy states that for closed systems,[1] energy cannot be created or destroyed; however, it can be converted from one form to another. Mathematically, conservation of energy is written as

$$\Delta E = E_f - E_i = 0, \qquad (4.5)$$

where E_f represents the total final energy of the system and E_i represents the total initial energy of the system under consideration.

Conservation of energy means that the total energy of a closed system cannot disappear. But one form of energy can be converted into another, that is, kinetic energy can be converted into potential energy and potential energy may be converted into kinetic energy. For example, when you rub your hands together on a cold day, you are actually converting kinetic energy (the motion of your hands) into heat energy that warms you. In fact, what may seem like a 'loss of energy' in a system is actually energy being converted into heat or sound or both.

Example 4.1.
1. Jill pushes a box of mass 3 kg with a force of 10 N and it moves a distance of 3 m. How much work has she done on the box?
   Solution:
   Use equation (4.1) to find the work that Jill does to the box:

   $$W = \vec{F} \cdot \vec{x} = (10 \text{ N})(3 \text{ }m) = 30 \text{ J}.$$

2. If the final speed of the box is 2 m s^{-1}, find the kinetic energy of the box.
   Solution:
   Use equation (4.2):

   $$\text{KE} = \frac{1}{2}mv^2 = \frac{1}{2}(3 \text{ kg})(2 \text{ m s}^{-1})^2 = 6 \text{ J}.$$

4.1.5 Four basic forces

There are four basic forces that include the gravitational force, the electromagnetic force, the weak nuclear force, and the strong nuclear force. The weak and strong nuclear forces are essential in the structure of matter and are responsible for holding matter together. However, we do not experience them in everyday life because they act over distances smaller than the size of the nucleus (10^{-15} m).

---

[1] A closed system is a physical system that does not allow the transfer of mass or the transfer of energy across its boundaries.

As humans, in everyday life we mainly experience the gravitational and electromagnetic forces. Electromagnetic forces act between charged particles and are a combination of electrical and magnetic forces. Electrical forces are forces that cause static electricity, and magnetic forces are those that affect a compass needle. Electromagnetic forces are long-range forces and can be either attractive or repulsive. Gravitational force, on the other hand, is a solely attractive (pull) force that all masses in our universe experience. In astronomical systems, the gravitational force (numerically equal to Gm_1m_2/R^2) determines the motions of moons, planets, stars, and galaxies. In this formula, $G\,(= 6.67 \times 10^{-11}\,\text{N m}^2\,\text{kg}^{-2})$ is the gravitational constant, m_1 is the mass of object 1, m_2 is the mass of object 2, and R is the distance between the objects. Our weight (mg) is the gravitational pull due to the entire Earth acting on us, where m is the mass and $g\,(= 9.8\,\text{ms}^{-2})$ is the acceleration due to the force of gravity pulling us toward the Earth.

In this textbook, I will limit our discussion to gravitational forces, and we will mostly concern ourselves with forces between objects on Earth.

4.1.6 The result of an applied force

There are two main results of applying a force to an object: the object may accelerate in the same direction as the applied force, or the force may cause deformation in the object itself.

4.1.6.1 Force and acceleration

Let us start by considering motion in one dimension, which means we must specify a force component (positive or negative) relative to one direction. The force exerted on an object is nearly always due to another object in its surroundings. The other objects apart from the object under consideration are called the *environment* of our object. Hence, when analyzing forces in a problem, we have to:
 1. Describe each force by the body on which it acts, and
 2. Describe the environment responsible for the applied force.

For example, suppose a child is pushing a toy car on the floor; the following forces are involved:
- a pushing force exerted *on* the car by the child;
- a frictional force exerted *on* the car by the floor;
- a gravitational force exerted *on* the car by the Earth, pulling it downwards;
- a normal force exerted *on* the car by the floor, acting upwards.

4.1.6.2 Force and deformation

Another way that an applied force affects an object on which it acts is to physically change the shape of the object. A change in the shape of an object due to the application of a force is called a *deformation*. We can categorize deformations by dividing them into two broad categories:
- **Elastic deformation**: under elastic deformation, the object returns to its original shape once the force (called stress) is removed. Elastic deformation

occurs when the deformation is small, and it is observed that in elastic deformation, the size of the deformation remains proportional to the force. Elastic deformation obeys Hooke's law. In equation form, Hooke's law is given by,

$$\vec{F} = k\Delta L, \tag{4.6}$$

where ΔL is the change in length and k is called the spring constant.

Elasticity is a measure of how difficult it is to stretch an object. In other words, elasticity is a measure of k. When k is small, the object is considered to be very elastic, for example, rubber. Rubber stretches a lot when subjected to only a small force but returns to its original shape once the external force is removed. When k is large, the object is rigid and considered not to be elastic.

- **Inelastic deformation**: under inelastic deformation, the object does not return to its original shape once the force is removed. Inelastic deformation can either be:
  1. *Ductile deformation*, in which the strain is irreversible, or
  2. *Fracture* or *brittle deformation*, in which the strain is irreversible and the material also breaks apart.

4.2 Historic ideas about motion

Since ancient times, common wisdom and observation was that an external *force* was needed to keep objects in motion. For an object to move in a straight line, scientists believed that an external agent had to constantly push it, otherwise it would *naturally* come to rest, and a body at rest was thought to be in its *natural state*. According to this idea, if we were to remove all external influences, all moving objects would naturally come to rest. In order to test this idea experimentally, we would have to find a way to free a body from all the influences of its environment, or, in other words, free the body from all the forces that are acting on it. This is very hard to do on Earth, but we can try to make the forces as small as possible.

Let us try an experiment: say we push a block of mass (m), across a surface, as shown in the figure 4.1(top). We find that as we push the block, it starts to move in the direction of the applied force. However, when we stop pushing it, the block almost immediately comes to rest. This observation agrees with the statement of the ancient philosophers. However, if we lubricate the surface with a bit of water as shown in figure 4.1(middle), we note that the block slides farther under the influence of a similar push force than it did without the water. We can then lubricate the surface by pouring some oil between the block and the surface, as we see in figure 4.1(bottom). We observe that in this case the block slides even further than when we had lubricated it with water and much further than when there was no lubrication at all.

Conclusion: we can extrapolate from the above observations that if all resistive (especially friction) could be eliminated, the body would continue to move indefinitely in a straight line at a constant velocity. Hence, we can say that an external

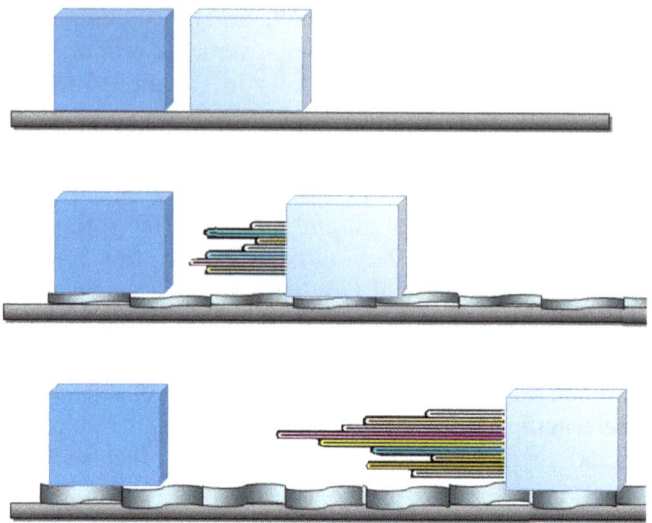

Figure 4.1. The effect of decreasing external forces on an object.

force is needed to set a body in motion, but no external force is needed to keep a body moving at a constant velocity. This observation leads directly to Newton's first law of motion.

4.2.1 The 'no net force' condition

On planet Earth, it is difficult to imagine a situation in which no external force acts on a body. For example:
1. The force of gravity always acts on an object on Earth, and it is impossible to ignore as long as you are on planet Earth.
2. Resistive forces such as friction and air resistance act on bodies in motion on both the ground and in the air. These resistive forces were unrecognized by the ancient philosophers, which led them to the obvious conclusion that the *natural state* of all objects was indeed the state of rest.

So how do we satisfy the 'no net force' condition and still remain on Earth? Fortunately, we have a solution that does not require us to go into outer space. To study a motion that is free of external forces, as far as the overall translational motion of a body is concerned, according to physics there is no distinction between a body on which no external force acts and a body on which the sum or resultant of all the external forces is zero, as shown in figure 4.2.

In figure 4.2, the symbol F_N represents the normal force, F_{appl} is the applied force, F_g is the force of gravity pulling the object towards the center of the Earth and F_s is the frictional force that acts oppositely to the direction of motion. The sum or resultant of all the external forces is called the *net force* acting on the body. The net

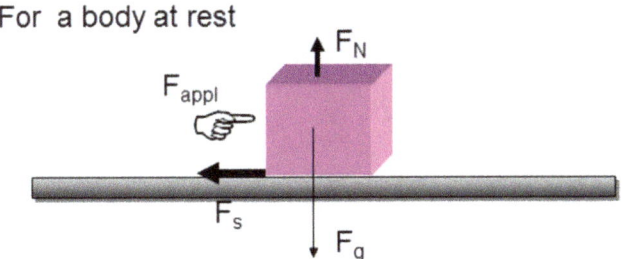

Figure 4.2. The sum of forces on an object at rest.

force is determined by the vector sum of all the forces that act on the body. It is important to note that in figure 4.2, even though four forces act on the block, the *net force* is zero and the object remains at rest. The 'no net force' condition is achieved by arranging forces such that they counteract other forces acting on the body. For example, the upwards normal force F_N is balanced by the downwards force $F_g(=mg)$ and the applied force F_{appl} is counteracted by static friction; thus, the object remains at rest.

4.3 Newton's laws of dynamics

Dynamics is the branch of physics that studies of the causes of motion. It is built on the Newton's three laws of motion, named in recognition of Sir Isaac Newton, one of the greatest scientists and mathematicians that has ever lived.[2] In this section, we will discuss Newton's three laws of motion.

4.3.1 Newton's first law (the law of inertia)

All objects resist any change to their state of motion. An object at rest remains at rest unless acted on by an unbalanced force, whereas an object in motion continues along its trajectory at the same speed and in the same direction unless acted upon by an unbalanced force (figure 4.3).

This means that there is a natural tendency for objects to keep on doing what they are doing and that in the absence of an unbalanced force, an object maintains its state of motion or its state of rest. This is called Newton's first law of motion. Newton's first law is also called *the law of inertia*, in which inertia is a measure of an object's resistance to changes in its motion (figure 4.4).

An object is in *translational equilibrium* if the net force acting on it is zero. If the acceleration $\vec{a}$ is zero, then the sum of the forces (mathematically represented as $\Sigma \vec{F}$)

---

[2] Sir Isaac Newton was born in England on December 25, 1643 CE and lived to be 85 years old, which, during that time period, was quite a long time. Isaac Newton was raised by his grandmother and he attended Trinity College in Cambridge, where he became interested in math, physics, and astronomy. While at college, he wrote about his ideas on motion, gravity, the diffraction of light, and forces. He spend his life developing these ideas and his lifelong work earned him a knighthood at the age of 62.

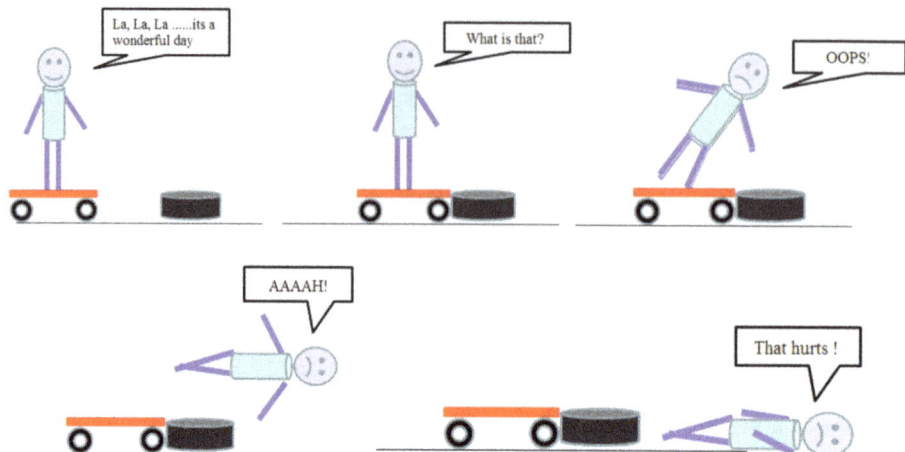

Figure 4.3. Inertial example 1: a cartoonist's rendition of motion involving inertia. The boy falls forward when the skateboard stops because an object wants to maintain its movement.

Figure 4.4. Inertial example 2: (a) A toy car moves towards the wall at speed (v) with pennies on its roof; (b) the toy car hits the wall; (c) upon impact, the car rebounds backwards; however, the pennies fall forward.

acting on the object is also zero. This is possible for the body under the following conditions:

1. When the speed of the object is zero, the equilibrium is called *static equilibrium*.
2. When the speed of the object is nonzero but its speed is constant, in which case, it is called *dynamic equilibrium*.

4.3.2 Newton's second and third laws

According to Newton's first law, the absence of a net force leads to the absence of acceleration. But what about the case in which forces are present? Common experience is that applying a force on a body accelerates it. This force can either be a push force or a pull force.

Newton's second law: When the vector sum of the applied force is greater than the sum of all other forces acting on the object, say friction and the force of gravity, we observe that the object accelerates in the direction of the applied force. As the magnitude of the applied force increases, the object accelerates faster and attains a faster velocity. We may generalize this phenomenon and say:

Figure 4.5. An action–reaction pair.

the acceleration produced in an object is directly proportional to the force applied.

$$\text{the vector sum of forces} = \text{mass} \times \text{acceleration}$$
$$\Sigma \vec{F} = m\,\vec{a}, \quad (4.7)$$

where m is the mass of the object and $\vec{a}$ is the acceleration produced in it. This is a statement of Newton's second law of motion. The mass of the object is the constant of proportionality; it is a scalar quantity and is measured in kilograms (kg). The greater the mass of the object being accelerated, the greater the amount of net force needed to accelerate the object.

Newton's third law:

Newton's third law states that *for every action there is an equal and opposite reaction.*

In other words, when a body, say object 1, exerts a force ($\vec{F}$) on another body, say object 2, object 2 also exerts an equal and opposite force on object 1 ($-\vec{F}$). These two forces are equal in magnitude but opposite in direction. For example, figure 4.5 shows Eman holding a book. The weight ($= mg$) of the book is due to the force of gravity pulling on the book. In order to hold it up, Eman has to apply an upwards force that balances the weight of the book. Hence the action–reaction pair in this case consists of the force of gravity acting downward and Eman's hand force acting upwards. Since the book is not moving, all the forces are balanced, and the net force is zero.

Example 4.2. In figure 4.6, Eman pushes against a wall; the wall in turn applies an equal and opposite (in direction) force on Eman. Since both Eman and the wall do not move, the two forces are considered to be equal and opposite. Mathematically, if

Figure 4.6. Example 4.3.1: Newton's third law.

we take $\vec{F}_{EW}$ as the force with which Eman pushes on the wall and $\vec{F}_{WE}$ as the reaction force of the wall pushing back on Eman, then Newton's third law can be written as

$$\vec{F}_{EW} = -\vec{F}_{WE}. \qquad (4.8)$$

4.3.3 The unit of force

In SI units, the unit of force is called a '**newton**' (N). One newton, by definition, is a force that causes one unit of acceleration ($= 1 \text{ m s}^{-2}$) when applied to a unit mass ($=1$ kg). Mathematically, a newton is

$$1 \text{ newton (N)} = 1 \text{ kg}\frac{\text{m}}{\text{s}^2}. \qquad (4.9)$$

4.3.4 Measuring forces

A force is measured by two basic methods:
1. How much acceleration the force produces in an object.
2. How much the force changes the size or the shape of an object held in place, for example, a spring, or the impact of the racket on the tennis ball. This is also called deformation.

Example 4.3.
1. If a mass of 10 kg is accelerated at 10 m s^{-2}, what force is applied?
   Solution:
   $$\Sigma\vec{F} = m\vec{a} = 10 \text{ kg} \times 10 \text{ m s}^{-2} = 100 \text{ N}.$$

2. If a force of 5 N is applied to a mass of 10 kg, what is the resulting acceleration?
   Solution:
   $$\vec{a} = \frac{\Sigma \vec{F}}{m} = \frac{5 \text{ N}}{10 \text{ kg}} = 0.5 \text{ m s}^{-2}.$$

4.4 Categories of forces

Forces can be divided into two broad categories: long-range forces and contact forces.

1. **Long-range forces**: long-range forces are forces between two objects that need not be in direct contact with each other. Examples include the gravitational force that keeps us grounded on Earth and the attractive force between the Earth and the Sun that keeps the Earth in a stable orbit around the Sun. We call the magnitude of the force that the Earth exerts on us our weight. Another common example of a long-range force is the force between electrical charges.

2. **Contact forces**: contact forces are forces that arise due to the direct contact between an object and its environment. These include but are not limited to, pull forces (say those applied by a rope), push forces (applied through direct contact or through rods), surface forces (resulting from direct contact between objects), etc. Contact forces last only as long as the two objects are in direct contact with each other. A baseball bat hitting an incoming ball applies a rather large force for a very short interval of time, called 'impulse.' The time interval of this impact lasts only as long as the bat and ball are in direct contact. Before and after the direct contact, the ball is a projectile (a free-falling object), and the only force acting upon it is gravity.

   I will limit our discussion to some common contact forces that are directly related to our text only. These include the following:

   (a) **The normal force (N)**: the normal force (symbol N) is an example of a contact force. The normal force acts perpendicularly to two surfaces in contact and continues to act as long as the two surfaces are in direct contact with each other. The normal force prevents objects from passing through one another, for example, a book resting on a table has a normal force equal to the weight of the book.

   (b) **Tension (T)**: tension (symbol T) is defined as the force that is transmitted through a rope, string, or a wire. The tensional force is directed over the length of the wire and pulls equally on bodies at both ends of the wire. For example, if we take a mass (m) and hang it from a ceiling with a string, as seen in figure 4.7(left), the hanging mass experiences a pull towards the Earth due to the force of gravity (F_g). But it does not fall because there is a string attached to it which pulls it upwards. The string provides an equal and opposite (in direction) force that counters

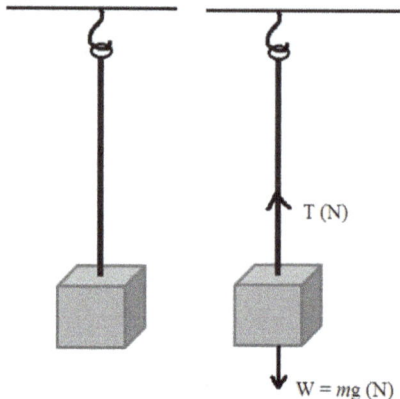

Figure 4.7. Tensional force: (left) a mass hanging from a string; (right) forces acting on the system.

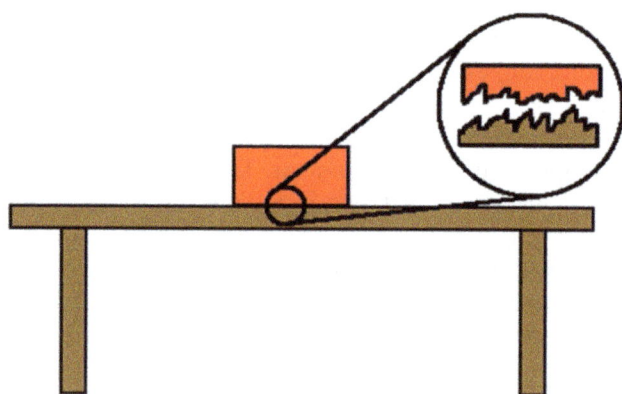

Figure 4.8. A microscopic view of frictional force.

the force of gravity. Since F_g always points downwards, tension should always point upwards, as shown in figure 4.7(right).

Example 4.4. In figure 4.7, predict and comment in the space provided what happens if we increase the mass to $2m$, then to $3m$, then to $4m$ and so on?
Solution:

(c) **Frictional forces** (f_s and f_k): frictional force is another type of contact force. Frictional forces act parallel to the contact surfaces and always oppose the direction of motion of the object itself. Frictional forces arise due to microscopic imperfections in the two surfaces in contact, as shown in figure 4.8. It is the force that prevents you from slipping off the chair you may currently be sitting in. Frictional forces may broadly be divided into two categories:

   i. **Static friction** (f_s): static friction acts to prevent objects from sliding. Static friction is due to the formation of partial chemical bonds between the surfaces of the objects in contact. The force of static friction is:

   $$f_s \leqslant \mu_s N, \qquad (4.10)$$

   where μ_s is the coefficient of static friction and N is the normal force.

   ii. **Kinetic friction** (f_f): kinetic friction makes sliding objects slow down. In general, kinetic friction is always less than static friction. The mathematical formula for the force of kinetic friction is:

   $$f_k = \mu_k N, \qquad (4.11)$$

   where μ_k is the coefficient of kinetic friction and N is the normal force.

In the static case, the object does not move, so all forces on the object must cancel. In the kinetic case, a net force is present, but the acceleration of the object always decreases because kinetic friction resists the applied force. Remember the beginning of this chapter (section 4.2); can you see why the ancient philosophers thought that a constant force was needed to keep an object moving? What happens when you stop pushing on a sliding object?

4.4.1 Calculating net force

Force is either a pull or a push, which is mathematically expressed by assigning a direction to the magnitude of a force. Hence, force is a vector quantity, so we have to include not only the magnitude of the force but also the direction of the force in order to completely describe the effect of the applied force on an object. This means that we cannot simply add forces together but rather we need to employ the method for vector addition, as discussed in chapter 1.

When more than one force acts on an object, the resulting motion of the object is determined by the net force acting on the object. This net force is the vector sum of all the forces acting on the object. Suppose there are three forces $\vec{F}_1$, $\vec{F}_2$, and $\vec{F}_3$ acting on an object. The net force acting on the object is then given by

$$\vec{F}_{net} = \vec{F}_1 + \vec{F}_2 + \vec{F}_3. \qquad (4.12)$$

Example 4.5. Consider a box resting on a table, as seen in figure 4.9. Write the force equations for the x-axis and the y-axis by applying Newton's second law.

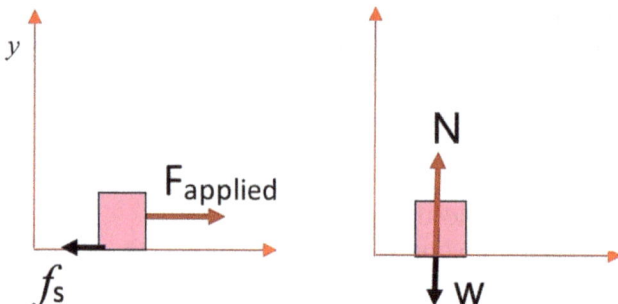

Figure 4.9. Forces in: (left) the *x*-direction and (right) the *y*-direction.

Solution:
Resting implies that the acceleration of the box is zero, hence Newton's second law gives us $\Sigma F = 0$. We can then write force equations for the x and y axes by applying Newton's second law, as follows:

For the *y*-component of force:

$$\Sigma F_y = N - mg$$
$$0 = N - mg \qquad (4.13)$$
$$\Rightarrow N = mg.$$

For the *x*-component of force:

$$\Sigma F_x = F_{applied} - f_s$$
$$0 = F_{applied} - f_s \qquad (4.14)$$
$$\Rightarrow F_{applied} = f_s.$$

Example 4.6. A small wooden box of mass 5 kg is resting on top of a table, as shown in figure 4.8. A force is applied to the box in the positive *x*-direction, but it is resisted by a static frictional force of $\mu_s = 0.1$. Find the net force acting on the box.

Solution:
To solve this question, we must first write the force equations for the x and y axes by applying Newton's second law:

$$\begin{cases} \Sigma \vec{F}_x = \vec{F}_{applied} - \vec{f}_s \\ \Sigma \vec{F}_y = \vec{N} - mg. \end{cases}$$

Since the box is not moving, both ΣF_x and ΣF_y must be zero and the above equations reduce to

$$\begin{cases} 0 = \vec{F}_{\text{applied}} - \vec{f}_s \Rightarrow \vec{F}_{\text{applied}} = \vec{f}_s \\ 0 = \vec{N} - mg \Rightarrow \vec{N} = mg. \end{cases}$$

From equation (4.10), $f_s \approx \mu_s N$ and we can write $f_s \approx \mu_s mg$,

$$\vec{F}_{\text{applied}} = \vec{f}_s = \mu_s mg = (0.1)(5\text{ kg})(9.8\text{ m s}^{-2}) = 4.9\text{ N}.$$

Example 4.7. A car starts from rest and accelerates to 55 m.p.h. on a level highway. The forces that act on it include thrust ($\vec{T}_t$), air drag ($\vec{D}$), friction between the tires and the road ($\vec{f}_k$), the normal force ($\vec{N}$), and its weight ($\vec{W} = mg$).
1. Calculate the magnitude and the direction of the net force on the car.
   Solution: To solve this question, we must first write the force equations for the x and y axes by applying Newton's second law. The force equations for this motion are

$$\begin{cases} \Sigma F_x = \vec{T}_t - \vec{f}_k - \vec{D} \\ \Sigma F_y = N - mg. \end{cases}$$

During acceleration, according to Newton's second law, $\Sigma F_x = m\vec{a}$, hence the above equation becomes

$$\begin{cases} m\vec{a} = \vec{T}_t - \vec{f}_k - \vec{D} \\ 0 = N - mg. \end{cases}$$

2. Once the car reaches 55 m.p.h., the cruise control is engaged. Rewrite the above equations for this condition.
   Solution: Once cruise control is engaged, the acceleration goes to zero, hence $\vec{a} = 0$, so the above equation reduces to

$$\begin{cases} 0 = \vec{T}_t - \vec{f}_k - \vec{D} \\ 0 = N - mg. \end{cases}$$

4.4.2 Another way to 'see' acceleration

Acceleration is produced when a net (unbalanced) force acts on a mass. The greater the mass, the greater the net force required to accelerate the object. So, in terms of force, acceleration on an object may be written as follows:

'The acceleration of an object is directly proportional to the net force applied on it divided by its mass.'

$$\vec{a} = \frac{\Sigma \vec{F}}{m} \qquad (4.15)$$

where $\Sigma \vec{F}$ is the vector sum of all forces acting on the body, and $\vec{a}$ is the vector acceleration. A couple of things to keep in mind are:
1. When the magnitude of the force is large, it produces a larger acceleration in a mass.
2. On the other hand, the more massive an object is, the more force is required to produce the same acceleration.

Force applied in the direction of motion increases the speed of the object. This is called *acceleration* or a positive acceleration. For example, pressing the accelerator pedal accelerates (or speeds up) a car. Force applied oppositely to the direction of motion decreases the speed of an object. This is called *deceleration* or negative acceleration; it occurs, for example, when the brakes are applied in a car.

Review Questions 4.1.
1. State Newton's laws of motion.
2. What three quantities are related by Newton's second law?
3. What two types of motion occur when the net force is acting on an object is zero?
4. What is another name for the first law of motion? How does it tell us why we need to wear seatbelts?

4.5 Mass versus weight

So far, we have only considered cases in which forces with different magnitudes act on an object. In this section, we will extend our discussion to include the effect of applying the same force to different objects. Everyday experience tells us that it is easier to move a bicycle than a car by applying the same force. This implies that the same force produces different accelerations when applied to different bodies. The property of an object that affects our ability to accelerate it is called the 'mass' of the object. Hence, mass is:

'*the property of a body that resists change in its motion.*'

From subsection 4.4.2, we know that the acceleration of an object is proportional to the net force acting on the object and that the acceleration has the same direction as the force applied. That means that if we apply more force, we get a larger acceleration of an object. However, the same force applied to two different objects may not produce the same effect. Objects with different masses respond differently to the application of the same magnitude of force. An object with greater mass undergoes a smaller acceleration and an object with smaller mass undergoes a larger acceleration. This is why it is easier to accelerate a bicycle than a car.

Exercises 4.1.
1. You go fishing with your friends and manage to catch the largest trout. If you hold it up for everyone to appreciate, identify all the forces acting on the rod.
2. A hummingbird is able to hover almost motionless beside a flower by rapidly beating its wings up and down. If we calculate that the force with which the air pushes the bird upwards is 0.3 N, calculate the weight of the hummingbird in grams?
3. A cubic box slides across a rough table. If the coefficient of kinetic friction between the box and table is 0.3, what is the acceleration of the box?
4. A box rests on a carpeted floor. The normal force the floor exerts on the box is 250 N.
   (a) You push horizontally on the box with a force of 120 N, but it refuses to budge. What can you say about the coefficient of friction between the box and the floor?
   (b) If you must push horizontally on the box with a force of 150 N to just start it sliding, what is the coefficient of static friction?
   (c) Once the box is sliding, you only have to push with a force of 120 N to keep it sliding. What is the coefficient of kinetic friction?
   (d) Compare the numerical values of the coefficients of static and kinetic friction. What do they tell you?

4.5.1 The difference between weight and mass

In normal conversations, we use the terms mass and weight interchangeably; however, weight and mass in physics are not the same thing. Mass is a scalar quantity (basically a number) and its SI units are kilograms. Mass is independent of the physical location of the object and is defined as

'the quantity of matter inside the object.'

On the other hand, weight is a force that is due to the force of gravity attracting the object. The SI unit of weight is the same as the unit of force, newtons (N). The weight of an object is directly proportional to the mass (m) of the object. In addition, the weight of an object depends on its physical location on Earth, and it tells us how hard the Earth pulls the object towards itself. Since weight depends on mass, we say:

'weight is the magnitude of the gravitational attractive force acting on an object.'

$$\vec{F} = W = mg, \quad (4.16)$$

where g is the acceleration due to gravity and is numerically equal to 9.8 m s^{-2} on the surface of the Earth. Please note that the value of g changes as you move around the Earth; some of these values are given in table 4.1.

The value of g on the Moon is one-sixth the value of g on Earth, so a 1 kg mass that weighs 9.8 N on the Earth only weighs 1.66 N on the Moon. On the other hand, the

Table 4.1. Variations in the value of acceleration due to gravity (g) on the surface of the Earth.

Location on the Earth's surface	Latitude (°)	g (m s^{-2})	Mass (kg)	Weight (N)
North pole	90	9.83	1.00	9.83
New York, NY	42	9.80	1.00	9.80
Selinsgrove, PA	37	9.79	1.00	9.799
Equator	0	9.78	1.00	9.78

Table 4.2. The difference between mass and weight on different bodies in the Solar System.

	1 kg mass		Quarter-pounder burger	
	Mass	Weight	Mass	Weight
Earth	1 kg	9.8 N = 2.2 lbs	0.25 lbs	0.25 lbs $\approx \frac{1}{4}$ lbs
Moon	1 kg	1.66 N = 0.37 lbs	0.25 lbs	0.042 lbs $\approx \frac{1}{24}$ lbs
Jupiter	1 kg	24.8 N = 5.1 lbs	0.25 lbs	0.57 lbs $\approx \frac{1}{2}$ lbs

same 1 kg mass weighs 24.8 N on Jupiter due to its large g (≈ 26 m s^{-2})! Please note that even though these weights are different, the mass (= 1 kg) is the same in all places.

Example 4.8. Your quarter-pounder burger from Burger King does not weigh the same on the Earth and the Moon. Can you calculate how much it weighs on the Moon versus in Selinsgrove, PA. Once you have calculated the weights, you may check your answer with those provided in table 4.2.

Example 4.9. Write a play: weightlessness
You are the lucky winner of a raffle in which you get to go aboard the space shuttle. You are really excited to earn this opportunity. But when you get there, you realize you can't feel your weight. You call your mother and explain to her what you are experiencing. What do you think you will say to describe your feelings of weightlessness?
Solution:

4.6 Tension

Tension, usually represented by the symbol (T), is a pull force exerted by a string, cable, or a chain on an object of mass (m). The SI units of tension are newtons (N), and tension is always measured parallel to the string on which it applies. Suppose a mass is hung on the end of a rope and the rope is suspended from the ceiling by a hook, as shown in figure 4.7. A hanging mass experiences a pull towards the Earth due to the force of gravity (F_g), given by its weight, numerically equal to mg. To keep the mass hanging from the ceiling, the string must provide a force (tension) which is equal in magnitude but opposite in direction to the force of gravity. In this case, Newton's second law for the mass can be rewritten as

$$\Sigma F_y = T - mg. \tag{4.17}$$

There are two basic possibilities for systems of objects held by strings:
1. **Case 1: acceleration is zero.** When the acceleration in the system is zero, the system is in equilibrium and equation (4.17) may be rewritten as

$$\begin{aligned} \Sigma F_y &= T - mg \\ 0 &= T - mg \\ T &= mg. \end{aligned} \tag{4.18}$$

   In this case, the system is either at rest or moving at a constant velocity.
2. **Case 2: the system is accelerated.** When there is acceleration present in the system, for example, in the case of Atwood's machine, equation (4.17) becomes:

$$\begin{aligned} \Sigma F_y &= T - mg \\ ma &= T - mg \\ T &= ma + mg. \end{aligned} \tag{4.19}$$

Note that tensional force is always a pull force and that for our situations the string is always assumed to have negligible mass.

4.6.1 The tension in a guitar string

The string in a stringed instrument is fixed at two ends and held under a tension (T). To tune a string, the tension the string is held under is changed by turning the tuning pegs as shown in figure 4.10 until the string reaches the required pitch (frequency). For a stringed instrument, such as a guitar, each string vibrates at a natural frequency which is directly proportional to the tension in the string and inversely proportional to the diameter of the string. I briefly discuss this here; for a more detailed discussion, please refer to chapter 15.

4.6.2 Factors that influence the frequency or pitch of a guitar string

The factors that influence the frequency or pitch of a guitar string include:
1. The tension the guitar string is held under.

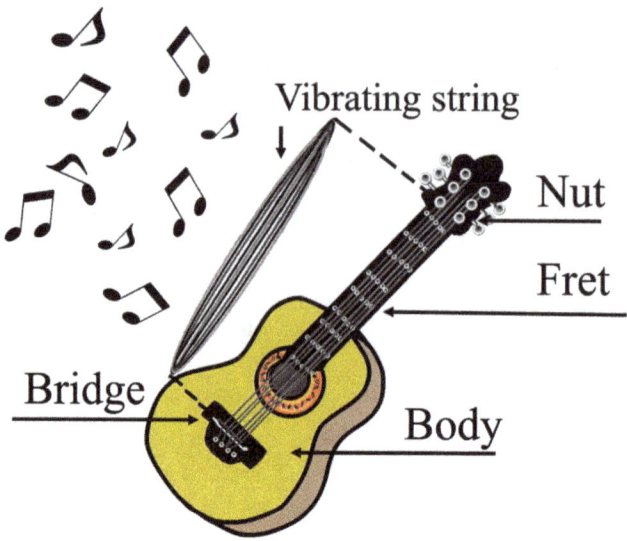

Figure 4.10. Tension in a guitar string.

2. The string material. Strings can be made up of materials such as steel, nickel, copper, and nylon.
3. The type of string, i.e. plain or wound. Plain strings are thinner and have a lower tension, which means that plain strings are quieter and don't last as long. Wound strings, constructed by tightly winding one material around a core material, are typically thicker and have a higher tension, which means they are louder and last longer.
4. The string gauge, which is the thickness or the diameter of a guitar string, is generally measured in thousandths of an inch. For example, a 10 gauge string is 0.010 inches in diameter. Thinner strings weigh less and can vibrate faster than thicker strings, resulting in higher-pitched sounds.
5. The guitar scale length, which is the length over which the string vibrates. It is the length of the open string between the guitar's nut and bridge. A longer scale length requires larger tension to pull a string tight to the required frequency (pitch), whereas the same string at a shorter scale length requires less tension to be tuned to the same frequency.

Higher frequencies are produced either by decreasing the diameter of the string and/or by increasing the tension in the string. Conversely, lower frequencies are obtained by increasing the diameter of the string and/or by decreasing the tension the string is held under. The desired string tension on an acoustic guitar, for which the typical string length is 25.5 inches, varies from about 70 newtons (16 pounds) for extra-light strings to around 180 newtons (40 pounds) for heavy-gauge strings. For example, to produce the note G3 at a frequency of 196 Hz on a typical full-size acoustic guitar, a plain string with a diameter of 0.475 mm must be held at a tension of 20 pounds. However, to produce the note G3 using a plain string with a diameter of 0.411 mm, a tension of 15 pounds is required. This changes if you have a wound string. For a wound string of, say, diameter 0.553 mm, a tension of 25 pounds produces a G3.

It is worth noting that at lower tensions, the strings are easier to play, although they do produce less sound, and at tensions somewhere below 14 pounds, the string may be too loose to produce acceptable sound quality. Higher string tension produces louder sounds, but at these tensions, the string is harder to play, and too much tension may damage the guitar body.

Review Questions 4.2.
1. The SI unit of weight is the newton. Do you remember what other quantity takes the newton for its unit?
2. Is additive weight ever useful? Have you ever tried to ride in a sports car in the snow?
3. Suppose you got to interview a fighter jet pilot about the forces and accelerations they feel during a flight. What kinds of questions do you think you should ask?
4. How does a car maintain its speed when cruise control is engaged?

Exercises 4.2.
1. Calculate your mass in kg and your weight in newtons.
2. All countries in the world measure in mass in kilograms (kg) instead of weight units such as the pound or newton. However, the scales used in grocery stores measure weight, which is a force, but give answers in kg. Calculate what a calibrated scale will give for 40 N of fresh strawberries?
3. Compare the weights of a 70 kg man when he is on Earth, on the Moon ($g = 1.6$ N kg^{-1}), on Venus ($g = 8.9$ N kg^{-1}), and on Mars ($g = 3.7$ N kg^{-1}).
4. Iron has a density of 7.87 gms cm^{-3}, and the mass of an iron atom is 9.27×10^{26} kg. Find the volume of an iron atom if the atoms are considered to be perfect spheres and are tightly packed together.

Further reading

- Ohanian H 1985 *Physics* (New York: W.W. Norton & Company)
- Resnick R, Halliday D and Krane K 2002 *Physics* vol 1 5th edn (New York: Wiley)
- Rossing T D, Moore R F and Wheeler P A 2002 *The Science of Sound* 3rd edn (Upper Saddle River, NJ: Pearson Education)

IOP Publishing

The Physics of Sound and Music, Volume 1
A complete course text (Textbook)
Samya Bano Zain

Chapter 5

Vibrating systems

5.1 Simple harmonic motion

Nature provides many examples of vibrating systems, for example, trees swaying in the wind and the motion of tides in the ocean are examples of vibrating systems. For all vibrating systems, some things are similar: the first is that the motion of the system repeats itself in equal intervals of time and the second is that a force, called the *restoring force*, constantly acts to restore the system to its original state. This leads us to conclude that there must be some essential requirements for any object to vibrate. These requirements include: (i) there must be a particular position, called the *equilibrium position*, at which no net force acts on the particle and the object could remain at rest indefinitely; (ii) when the object oscillates, it moves away from the equilibrium position (the distance between the oscillating object and its equilibrium position at any time is called *displacement*); (iii) there is *inertia*. Inertia, as we saw in section 4.3.1, is the inherent property of any object to oppose a change in its motion. This means that an oscillating pendulum does not stop once it reaches the equilibrium point but overshoots and continues on its path to the other side. At this stage in the motion, the *restoring force* acts to bring the oscillating pendulum back from the other side. This back-and-forth process continues until it stops due to drag forces.

5.1.1 Restoring forces

A force is called a *restoring force* when it acts on an object in such a way that it causes the object to accelerate towards the equilibrium position of the object. For example, in a simple pendulum made up of a mass hanging from a string, as shown in figure 5.1(a), the equilibrium position of the pendulum is reached when the mass is hanging straight down. When the mass is displaced, gravity provides the restoring force which works to bring the system back to its equilibrium position.

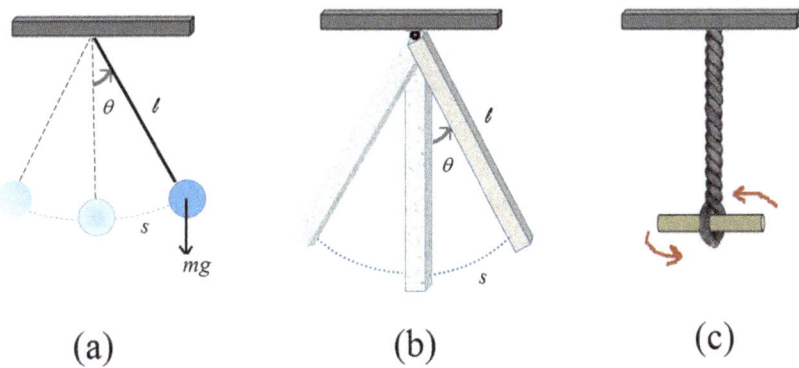

Figure 5.1. Types of pendulum: (a) a simple pendulum; (b) a physical pendulum; and (c) a torsional pendulum.

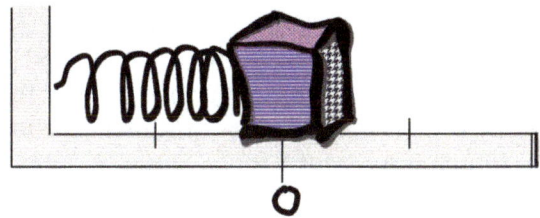

Figure 5.2. A mass attached to a spring.

The restoring force, usually represented by the symbol ($\vec{F}_R$), is always directed toward the equilibrium position of the system and is a function of the physical position of its mass only. Another way to think about this is that the restoring force is the force with which an object resists deformation. The applied force and the restoring force are equal in magnitude but opposite in direction, forming an action–reaction pair, as introduced in section 4.3. Under static conditions, the restoring force by which an object opposes applied forces obeys **Hooke's law**, which states:

'*The magnitude of the restoring force is directly proportional to the deformation.*'

For a mass attached to a spring as shown in figure 5.2, provided that the deformation is small, Hooke's Law is mathematically given (to a good approximation) by

$$\vec{F}_R = -kx, \tag{5.1}$$

where k is the spring constant and x is the displacement away from the equilibrium position.

The motion set up under these conditions is called *periodic motion*. Periodic motion occurs when an object moves back and forth over the same path. Periodic motion is also called *oscillatory* or *vibratory* motion. The period of harmonic motion is defined as the time required for a vibrating body to complete one complete round trip, also called one *oscillation* or one *cycle*. As you walk, your arms may be thought of as acting like pendulums. The arm is not quite a simple pendulum, since not all of its mass is attached to the end; it is more like a physical pendulum, like the one shown in figure 5.1(b).

5.1.2 The simple harmonic oscillator

Simple harmonic motion (SHM) is a special kind of periodic motion in which the restoring force is proportional to the displacement away from the equilibrium point. An object that experiences SHM is known as a *simple harmonic oscillator*. There are many examples of SHM in everyday life. Any spring or rubber band with a weight attached that is free to vibrate can exhibit SHM. Similarly, a pendulum, including a child on a swing set, exhibits SHM and is itself a simple harmonic oscillator. Indeed, the regular motion of a pendulum is the reason that it is used as a timing device for grandfather clocks.

For a motion to be called SHM, the object undergoing SHM must have a restoring force (given by Hooke's law) that keeps the object in SHM. The magnitude of the restoring force must be proportional to the displacement from the equilibrium point, and the frequency and the period of the SHM oscillator must be independent of the amplitude (maximum displacement).

Not all periodic vibrations are examples of SHM. An electrocardiogram (or EKG) which traces the periodic pattern of a beating heart is not an example of SHM, since the trace needle does not follow the sine wave path of SHM.

5.1.3 Examples of simple harmonic motion

In this section, we talk about objects undergoing SHM.

1. **A simple pendulum:** the simplest example of an object displaying SHM is a mass hanging from a long string that is given a gentle push. An example of a simple pendulum is seen in figure 5.1(a). If the amplitude of the oscillations is kept small, the period (T_p), i.e. the time taken to complete one oscillation, of a simple pendulum of length (l) is mathematically given by

$$T_p = 2\pi \sqrt{\frac{l}{g}}, \qquad (5.2)$$

   where g is the acceleration due to gravity.

2. **A physical pendulum:** a physical pendulum is any rigid object, such as a bar or a meter stick, that is free to oscillate about some fixed axis, as shown in figure 5.1(b). The period of oscillation of a physical pendulum is not necessarily the same as that of a simple pendulum. Since it does not directly relate to our discussion in this text, I will not go into detail about the mathematics of the physical pendulum.

3. **A torsional pendulum:** a torsional pendulum is a special kind of a pendulum which is made by suspending a mass, for example, a bar attached to a string, and twisting it by an angle (θ). In a torsional pendulum, gravity provides the

restoring force and the twisting motion as the string unwinds provides the oscillation. The string shortens slightly as it twists and lengthens as it untwists.

4. **A mass attached to a spring:** a mass (m) attached to a spring with a spring constant of (k) exhibits SHM when it is displaced from its equilibrium position. If the spring is not stretched too much, the displacement of the mass is proportional to the applied force and is given by Hooke's law (equation (5.1)). The period (T_m) of a mass–spring system, as shown in figure 5.2, is mathematically given by

$$T_m = 2\pi \sqrt{\frac{m}{k}}. \tag{5.3}$$

Example 5.1. Suppose a certain spring that is hanging from a hook in the ceiling stretches by 10 cm when it is loaded with a 1 kg mass.

1. Find the spring constant (k).
   Solution:
   start from Newton's force equation for a mass hanging from a ceiling,

   $$\Sigma F_y = F_{\text{spring}} - W.$$

   Since the system is not moving, $\Sigma F = 0$, and the above equation becomes

   $$F_{\text{spring}} = W.$$

   We know that $F_s = kx$ and $W = mg$; hence, we can find k as follows:

   $$k = \frac{mg}{x} = \frac{(1 \text{ kg})(9.8 \text{ m s}^{-2})}{0.1 \text{ m}} = 98 \text{ N m}^{-1}.$$

2. Find the period of vibration (T_m).
   Solution:
   use equation (5.3):

   $$T_m = 2\pi \sqrt{\frac{m}{k}} = 2\pi \sqrt{\frac{1}{98}} = 0.63 \text{ s}.$$

3. Find the frequency of vibration (f_m).
   Solution:
   the frequency of vibration is the inverse of the period of vibration,

   $$f_m = \frac{1}{T_m} = \frac{1}{0.63} = 1.57 \text{ Hz}.$$

5.1.4 'Seeing' sound waves

As sound waves pass a certain point in space, the sound pressure in the area rises and falls. Microphones are used to capture this rise and fall in the sound pressure. However, to measure sound waves passing through a certain point, a graph called a *waveform of sound* is used. A waveform is a graph of sound pressure as a function of time. To display the waveform, we connect the microphone to a computer or an oscilloscope, as shown in figure 5.3.

5.1.5 An energy breakdown of an object undergoing simple harmonic motion

The to-and-fro motion of any object that repeats itself at equal intervals of time is SHM. This means that at the maximum amplitude of the motion, the object comes to rest for an instant before turning around and coming back towards the equilibrium position. At the maximum amplitude, the velocity of the object has to be zero, which implies that the kinetic energy of the object is zero and hence, at this instant, all the energy of the system is potential energy. Similarly, the speed of the object is maximized as it passes through its equilibrium point. Thus, the kinetic energy (KE = $\frac{1}{2}mv^2$) of the object is at a maximum as it passes through the equilibrium point and its potential energy (PE = $\frac{1}{2}kx^2$) is at a minimum.

Conservation of energy, defined in section 4.1.4, states that the total mechanical energy of a system (assuming it is a closed system) undergoing SHM remains constant. For a mass–spring system, such as the one shown in figure 5.2, this is mathematically written as

$$E_{\text{total}} = \text{KE} + \text{PE} = \frac{1}{2}mv^2 + \frac{1}{2}kx^2 = \text{constant}. \quad (5.4)$$

5.1.6 Simple harmonic and circular motion

Imagine that you have a mass attached to a string and you whirl it in a circle. The projection of this uniform circular motion onto a wall is also an example of SHM.

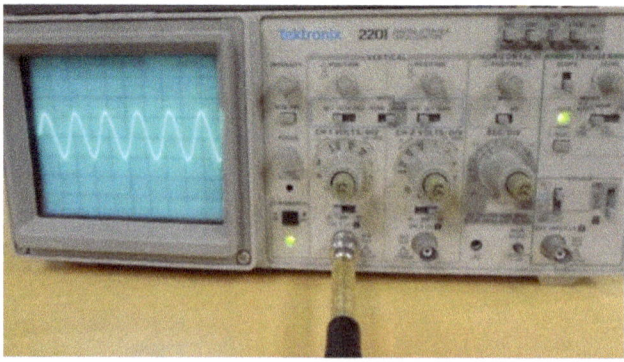

Figure 5.3. 'Seeing' sound waves.

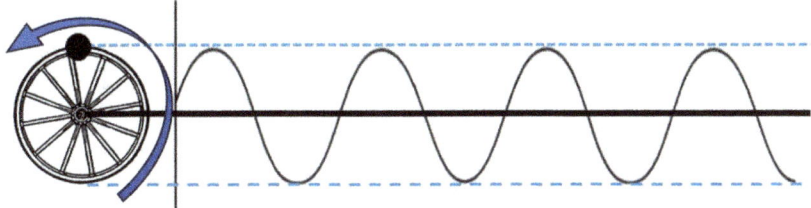

Figure 5.4. A projection of circular motion yields SHM.

As the mass rotates, its projection moves to and fro along a straight line. This to-and-fro motion can be plotted on a graph, and when it is traced on the graph, we see that it traces a sinusoidal wave pattern like the one seen in figure 5.4.

Example 5.2. Find the approximate spring constant for a mass–spring system for which the mass ($m = 0.50$ kg), and which experiences a velocity of 5 m s^{-1} when it is displaced by 50 cm.
Solution:
We know that the kinetic and potential energies for a mass–spring system are given by

$$KE = \frac{1}{2}mv^2$$

$$PE = \frac{1}{2}kx^2$$

and from the conservation of energy, equation (5.4),

$$E_f = E_i$$

$$KE_{final} + PE_{final} = KE_{initial} + PE_{spring}$$

$$\frac{1}{2}mv^2 + 0 = 0 + \frac{1}{2}kx^2.$$

After a bit of algebra, we find that the value of the spring constant (k) is

$$k = \frac{mv^2}{x^2} = \frac{(0.50 \text{ kg})(5 \text{ m s})^{-2}}{(0.50 \text{ m})^2} = 50 \text{ N m}^{-1}.$$

5.2 Standing waves

Standing waves are set up when an incoming wave is reflected at a boundary (see section 5.3 for details of reflection at a boundary). The reflected waves interfere with the incident waves, and the resultant waves appear not to propagate. This may occur

Figure 5.5. A visualization of standing waves.

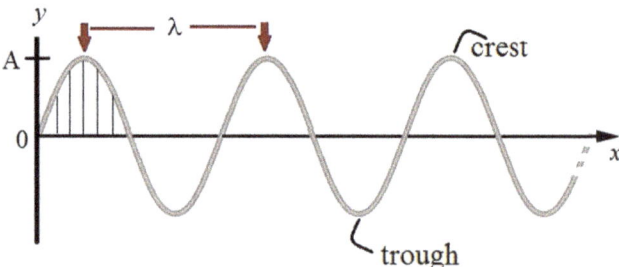

Figure 5.6. A string wave traveling along the positive *x*-axis. 'A' represents the amplitude of the wave and λ is the wavelength of the wave.

in two ways: either as a result of wave interference (discussed in section 5.5) between two waves traveling in opposite directions in a stationary medium, or it may occur when the medium itself moves in a direction opposite to that of the wave. In the first case, if the two waves of equal amplitude travel in opposing directions, then on average there is no net propagation of energy in the standing wave. Hence, standing waves are also called *stationary* waves.

In a standing wave on a string, as seen in figure 5.5, every point moves in SHM with the same frequency and reaches its maximum distance from equilibrium simultaneously.

5.2.1 Properties of standing waves

All standing waves share some common wave properties. I will discuss them in terms of waves in a string, as seen in figure 5.6.

1. **Equilibrium point**: the equilibrium point is the position of the string when no disturbance moves through it.
2. **Wavelength**: the distance between any point on a wave and the equivalent point on the next wave is called the 'wavelength.' Literally, 'wavelength' means 'the length of the wave.' The SI unit of wavelength is the meter, and its usual symbol is (λ). In other words, wavelength is defined as 'the distance between two consecutive cycles,' as seen in figure 5.6.
3. **Amplitude**: the strength or power of a wave signal is called its amplitude (A). It is interpreted as the *height* of a wave when viewed as a graph. The amplitude of a wave is the maximum displacement of a particle from its equilibrium position. For sound-producing devices, higher amplitudes represent higher volumes, hence the name *amplifier* is used for a device which increases the amplitude (volume) of sound waves. We measure the maximum displacement in both positive and negative directions. Both these points are equally distant from the equilibrium position, so for a standing wave, the amplitude is defined as the distance from rest to crest or from rest to trough.
4. **Crest**: the crest of a wave is the point that exhibits the maximum positive or upward displacement from the equilibrium position.
5. **Trough**: the trough of a wave is the point that exhibits the maximum negative or downward displacement from the equilibrium position.
6. **Period**: the period (denoted by T_P) of a wave is the time a wave takes to complete one complete vibrational cycle. The SI unit for the period is the second.
7. **Frequency**: the number of wavelengths that occur in one second is called the frequency (f) of the system. The SI unit of frequency is the hertz (Hz), defined as the number of cycles per second.

$$f = \frac{1}{T_P} \tag{5.5}$$

where T is the period of oscillation. Faster vibrations of sound sources imply higher frequencies of sound, which mean higher pitch. For example, singing in a high-pitched tone is achieved by forcing the vocal chords to vibrate quickly.

8. **Nodes**: an interesting feature of standing waves that results from superposition is that there are certain points along the string during interference that have zero displacements at all times. These zero displacement points are called nodes. It is worth noting here that for traveling waves (chapter 7.1), there may be some points of zero displacement as well; however, in traveling waves, the zero amplitude points change over time, so these points are not called nodes.
9. **Antinodes**: there are places between two nodes on a string at which the displacement oscillates at its maximum at all times. These points are called antinodes. The alternating patterns of the nodes and antinodes together set up a standing wave pattern.

5.2.2 The speed of standing waves

In section 1.7, we saw that speed of an object is mathematically given by

$$\text{speed} = \frac{\text{distance traveled}}{\text{time taken}}. \tag{5.6}$$

For a wave, we know that distance is given as a wavelength (λ), the time taken to complete one cycle is the period (T_P), and the relation between the frequency and the period of a wave is $f = 1/T_P$. For a wave, equation (5.6) may then be rewritten as

$$\text{speed of wave} = \frac{\text{wavelength}}{\text{period}} = \frac{\lambda}{T_P} = \lambda f. \tag{5.7}$$

5.2.3 The propagation of energy in standing waves

There is no net transfer of energy from one end to the other in a standing wave, and points vibrate with different amplitudes ranging from zero (at nodes) to a maximum amplitude (at antinodes). Since the energy cannot flow past the nodes in the string towards the right or the left, no net energy transfer occurs in a standing wave. So even though the energy alternates between vibrational kinetic energy and elastic potential energy, the energy can be said to be *standing* in the string.

In figure 5.7, at times $t = 0$ (figure 5.7 (top)) at the antinodes, the energy is nearly entirely the elastic potential energy associated with the stretching of the string, and

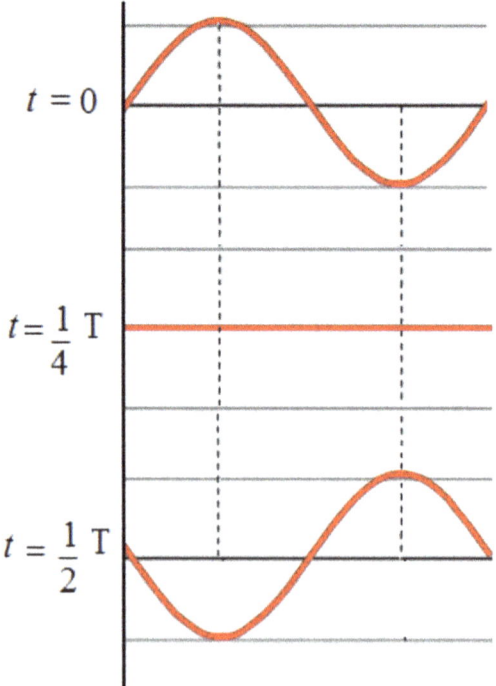

Figure 5.7. Energy propagation with respect to time.

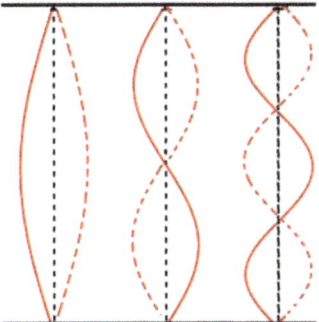 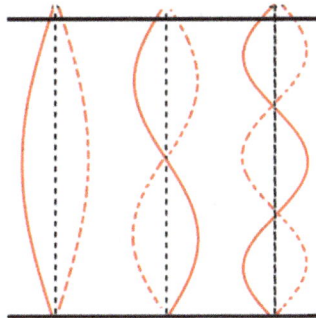

Figure 5.8. Standing waves in a string: (left) allowed states; (right) forbidden states.

at time $t = \frac{1}{4}T_P$ (figure 5.7 (middle)), when all parts of the string pass through the equilibrium position, nearly all of the energy is kinetic energy. Here, T_P is defined as the period of the system.

5.2.4 Allowed and forbidden standing wave conditions

A string held tightly at both ends is only allowed to vibrate at particular frequencies and wavelengths. At the fundamental frequency (f_0), the whole string vibrates up and down. The string is also able to vibrate at harmonics of the fundamental; for example, it may vibrate with a node at the middle of the string (called the first harmonic) at a frequency of ($f_1 = 2f_0$), at a frequency of ($f_2 = 3f_0$) (called the second harmonic), etc.

The fundamental and all harmonics always have nodes at each end. Any wavelength that does not have a node at each end does not form a standing wave and is hence forbidden. The allowed states of standing waves and those that are forbidden are given in figure 5.8(left) and figure 5.8(right), respectively.

5.2.5 Modes of vibrations

Natural objects in the world do not only have one mode of vibration but rather several modes of vibrations. A hammered piano string actually vibrates at several different rates at once. There are many examples of this kind of phenomenon in which several types of motions occur all at once. Imagine you get into your car and close the door; you feel the car vibrating at a particular frequency due to the door closing but this motion dissipates quickly, and the car remains stationary afterwards. However, on a windy day, you feel the car rocking back and forth with the speed of the wind. Once you start the engine, you feel a different kind of vibration through your seat. This motion is due to the oscillations of the engine and its parts. Next, when the car starts to move, you experience another vibration due to motion. At this point, you are experiencing three different kinds of motions: one due to the wind, one due to the engine, and one due to the motion of the car.

Similarly, when a sound is generated in a string, say by bowing, plucking, or hitting, it produces many modes of vibrations that are set up simultaneously. When you listen to a single note on an instrument, you are actually hearing many many pitches at once and not necessarily a single frequency. Most of us are not aware of this; however, musicians train themselves to hear these variations. The lowest pitch corresponds to the slowest vibration rate and is called the *fundamental frequency*, while all others are collectively called *overtones*.

In this section, we briefly define some important terms utilized to define modes of vibrations:

1. **The fundamental frequency**: each vibrating mode has a fundamental mode, which is the lowest possible mode in which it can vibrate. In terms of its frequency, the fundamental frequency of a periodic waveform is the reciprocal of its period.
2. **Harmonics**: the term 'harmonics' refers to modes of vibrations in a system that are whole-number multiples or nearly whole-number multiples of the fundamental mode. 'Nearly whole numbers' may mean 2.004 times the fundamental rather than two times the fundamental. In a guitar string, modes are usually so close to being whole-number multiples that we call them harmonics. The lowest common factor in a series of harmonic partials is called the fundamental, and it is also called the first harmonic, which can lead to confusion. When solving problems, please take care that terms 'fundamental frequency' and 'first harmonic' are sometimes used interchangeably.
3. **Overtones**: for oscillators that have modes that are not whole-number multiples of the fundamental mode, the term 'overtones' is used to describe their higher modes. Basically, 'overtones' is used for all higher modes of the fundamental mode, whereas 'harmonics' is used only for modes that are whole-number multiples of the fundamental mode. In addition, note that the term 'overtone' does not include the fundamental mode, whereas the term 'harmonics' includes the fundamental mode by default.
4. **Partials**: another term, 'partials,' is used to include all modes or components of vibrations of a system. This includes the fundamental mode as well as all overtones, whether they are harmonics or not.
5. **Inharmonicity**: inharmonicity is the degree to which the frequencies of overtones (known as partials, partial tones, or harmonics) depart from whole multiples of the fundamental frequency.

5.2.6 Combining various modes of vibrations

The actual motion of a vibrating system is the combination of its various modes of vibrations. The way in which these modes or partials combine is given by the *spectrum of the vibration*. The spectrum of vibration specifies the relative amplitudes of the partials.

For sound, the spectrum of sound specifies the amplitudes of its partials and is used in the Fourier analysis of complex tones. The Fourier analysis of complex tones, named after Joseph Fourier (1768–1830 CE), the mathematician who formulated it, is used to determine the component tones that make up a complex tone or waveform. Any periodic vibration, however complicated, can be built up from a series of simple vibrations whose frequencies are harmonics of a fundamental frequency by choosing the appropriate amplitudes and phases of harmonics. Electronic instruments called *spectrum analyzers*, shown in figure 5.9(middle) enable us to do this easily in the laboratory.

5.3 The reflection of waves

The word reflection comes from the Latin word, *reflexion*, which means *bending back*. The most common example of reflection is reflection in a mirror, which is caused by the bending of light that occurs when an incident light wave encounters an

Figure 5.9. Fourier analysis that gives the spectrum of vibration.

Review Questions 5.1.
1. What is a standing wave? What is the difference between a node and an antinode?
2. What is SHM?
3. What is a restoring force? How does SHM depend on a restoring force?
4. What do we mean by the period of vibration? How is this period related to the frequency?
5. Is there a motion that is periodic motion but is not simple harmonic? Explain your reasoning.
6. The movement of the windshield wipers on your car is an example of periodic motion. Can you approximate the periods of motion for the different settings of the windshield wipers on your car?

Exercises 5.1.
1. A simple pendulum undergoing SHM takes 0.5 s to travel from the highest point of zero velocity to the next such point (on the other side). The distance traveled between maximum amplitude points is 50 cm. Calculate the:
   (a) Period
   (b) Frequency
   (c) Amplitude of motion
2. Consider an object undergoing SHM. During this motion, in terms of amplitude, when is:
   (a) The kinetic energy at a maximum?
   (b) The potential energy at a maximum?
3. For a mass–spring system undergoing SHM, at what displacement, in terms of the amplitude, is the energy of the system half kinetic energy and half potential energy?
4. A pendulum consists of a mass connected to the ceiling by a string of length 50 cm. Find the period of oscillation.
5. An oscillating mass–spring system whose mass $m = 2$ kg takes 0.25 s to begin repeating its motion. Calculate:
   (a) The period of oscillation
   (b) The spring constant of the spring

abrupt boundary. When the incoming ray, which is also called the 'incident pulse,' reaches the boundary, the energy carried by the pulse is reflected and returns back to the medium that it started in. The disturbance that returns after bouncing off the boundary is known as the *reflected wave*.

The perpendicular to the surface, called *the normal to the surface* plays a crucial role in the laws of reflection. The angle that the incident ray makes with the perpendicular to the surface is called the angle of incidence (θ_i), and the angle that the reflected ray makes with the perpendicular to the surface is called the angle of reflection (θ_r), as shown in figure 5.10.

5.3.1 The laws of reflection

There are two **laws of reflection**:
1. The angle of incidence (θ_i) equals the angle of reflection (θ_r).
2. The reflected rays lie in the same plane as the incident rays. The two rays are on opposite sides of the normal, but at the same angle from it.

Example 5.3. A light ray strikes a reflective plane surface at an angle of 53° with the surface. Calculate:

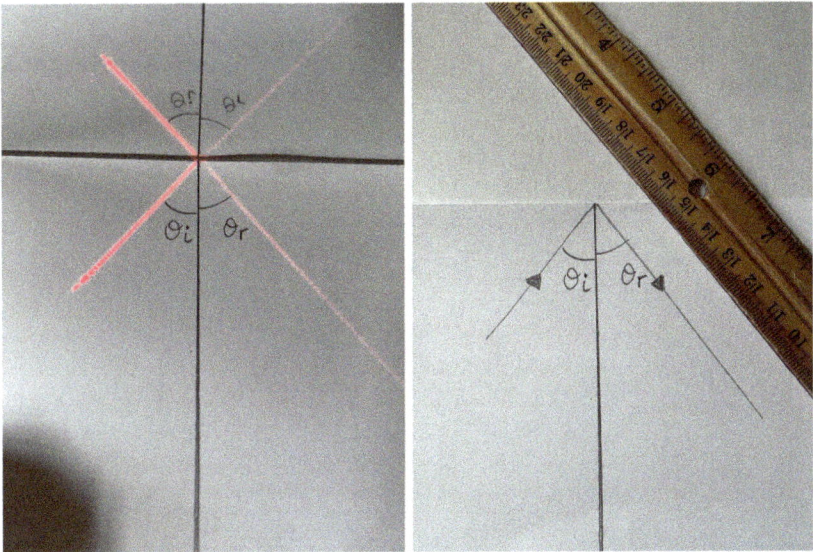

Figure 5.10. The reflection of light: θ_i is the angle of incidence and θ_r is the angle of reflection. The normal is perpendicular to the surface.

1. The angle of incidence (θ_i)
   Solution:
   The angle of incidence is the angle between the ray and the normal to the surface. Hence,
   $$\theta_i = 90° - 53° = 37°.$$

2. The angle of reflection (θ_r)
   Solution:
   From the law of reflection, we know that $\theta_i = \theta_r$, hence $\theta_r = 37°$.
3. The angle made by the reflected ray and the surface (θ)
   Solution:
   $$\theta = 90° - 37° = 53°.$$

5.3.2 Properties of a reflected wave

The speed, frequency, and wavelength of the reflected rays are the same as the speed, frequency, and wavelength of the incident rays. In general, the amplitude of the reflected pulse is less than the amplitude of the incident pulse. The type of reflection depends on the incoming wave.
   1. **Case 1: a plane wave incident on a plain barrier:** if the incident pulse is a plane wave, it is reflected back as shown in figure 5.11.

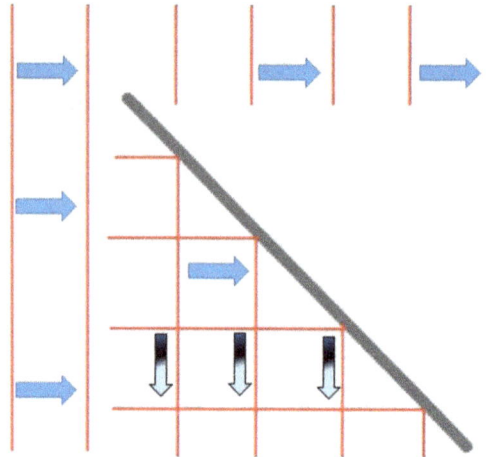

Figure 5.11. The reflection of a plane wave from a plane barrier.

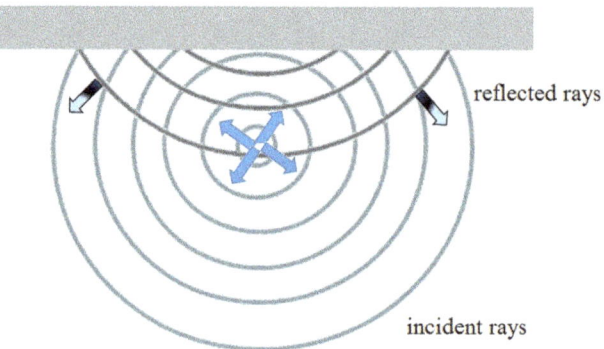

Figure 5.12. The reflection of a spherical wave from a point source.

2. **Case 2: spherical waves incident on a plane barrier:** spherical waves, also called *spreading waves* are waves that are emitted from point sources. When a spherical wave reaches a plane barrier, it is reflected back as shown in figure 5.12. As seen in figure 5.12, reflected waves are also spherical in shape and they seem to originate from a source on the other side (also called the *virtual side*) of the barrier. This is called an *image* source, and it is equally distant from the barrier on the virtual side as the point source is in front of the barrier on the so-called *real side*.

For sound waves, when a barrier obstructs the direct sound from reaching your ear, it appears as if the entire sound is coming from an *image* position behind the barrier.

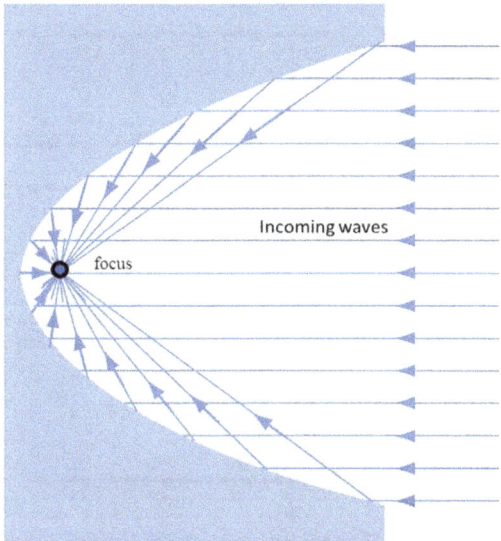

Figure 5.13. The reflection of waves from a concave barrier.

3. **Case 3: a plane wave incident on a parabolic barrier:** when a plane wave is incident on a concave surface, the concave surface focuses the wave and reflects it to a point, as shown in figure 5.13. In the case of sound waves, this produces a 'hot spot' where the sound waves are focused, which is highly undesirable when planning sound systems, for example, in rooms, theaters, and auditoriums.
4. **Case 4: a spherical wave incident on a spherical barrier:** suppose we have a point source located at the *focus*; it the waves it produces are incident on a spherical barrier, the result is that the reflected waves are plane waves. Theaters in ancient Greece (around 500 BCE) had curved shapes cut into hillsides that utilized this principle for maximum sound propagation.

5.3.3 Reflections of sound waves

Like all other waves, sound waves are also reflected. In this section, we discuss some ways in which sound waves are reflected.

1. **The reflection of a sound wave from an open end**: when an incident pulse (compression or a positive pulse) of a sound wave approaches an open end, the pressure of the sound wave drops to zero and the pulse is reflected back in the form of a rarefaction (or a negative pulse), as seen in figure 5.14.
2. **The reflection of a sound wave from a closed end**: when an incident pulse (compression or a positive pulse) of a sound wave approaches a closed end, the pressure of the sound wave builds up and the pulse is reflected back as in the form of compression or a positive pulse, as seen in figure 5.15.

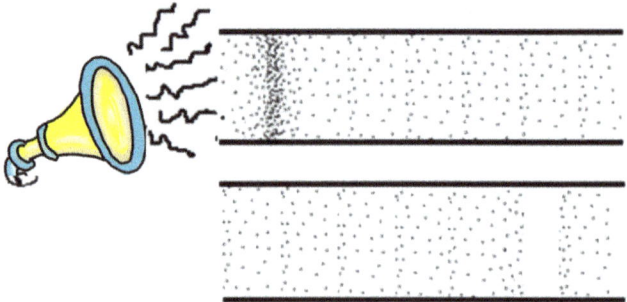

Figure 5.14. The reflection of a sound wave from an open end.

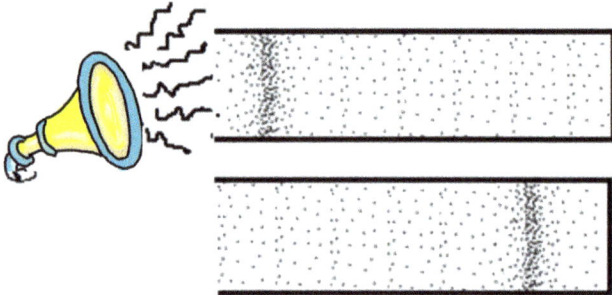

Figure 5.15. The reflection of a sound wave from a closed end.

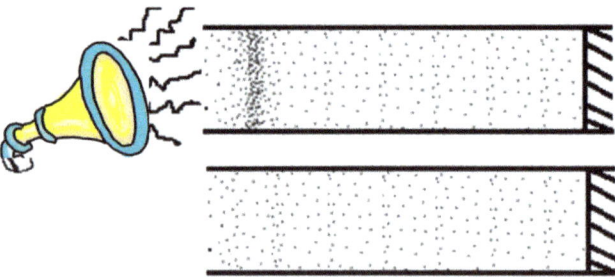

Figure 5.16. The reflection of a sound wave from an absorbing end.

3. **Reflection from an absorbing end**: when an incident pulse of sound wave approaches an end with a sound absorber attached to it, the sound absorber absorbs all of the incident pulse and virtually none of the pulse is reflected. This state is called *anechoic*, which translates to 'no echos' or *neither having nor producing echoes*, as seen in figure 5.16.

An anechoic chamber is a special room designed to completely absorb reflections of sound waves; it is also insulated from external noises. The term 'anechoic chamber' was first used in the 1940s by Leo Beranek to reference the minimization of reflections within rooms. Anechoic chambers may range in size from small compartments the size of household microwave ovens to large ones the size of aircraft hangers.

5.4 Waves in stringed instruments

Most waves observed in musical instruments are the vibrations set up in stringed instruments. A stringed instrument, as the name implies, is a musical instrument that uses vibrating strings to produce sound. Common stringed instruments are guitars (both acoustic and electric), the violin, the viola, the cello, the double bass, the banjo, the mandolin, the ukulele, and the harp. I will briefly talk about stringed instruments in this section and will leave a detailed discussion until chapter 15.

The history of stringed instruments is very old, and a great deal of work has been done over the years towards the development of mathematical analysis for the music made by stringed instruments. One of the earliest known forms of stringed instruments is the harp, whose existence can be traced back to 3000 BCE because of the funereal practices of the ancient Egyptians such as tomb paintings and grave relics. When discussing waves in strings, we must first define the different types of strings. Broadly speaking, strings can be divided into two main categories:

1. **Ideal strings:** ideal strings are made from circularly symmetric materials, which means that they are perfectly uniform and perfectly flexible. This also means that they are uniformly dense and hence require a uniform tension to move them from any point on the string. Ideal strings also have zero stiffness, which means they offer no resistance to bending; hence, the restoring forces must arise entirely from the externally applied tension and/or air resistance.
2. **Real strings:** real strings deviate from the ideal behavior due to the presence of unintentional damping forces that introduce a level of inharmonicity into the system under vibration. Inharmonicity is the degree to which an overtone departs from the actual overtone. Inharmonicity depends on the string itself and may be small or large. Thin strings approximate the ideal behavior, whereas very thick strings behave more like cylinders that have natural resonances that are not whole-number multiples of the fundamental frequency.

5.4.1 Allowed frequencies of the standing waves in strings

Figure 5.17 shows various standing wave patterns set up as a result of plucking a string of length L. Each pattern corresponds to a different wavelength and a corresponding frequency.

For standing waves set up in strings, we can find the frequency of all the modes using the formula

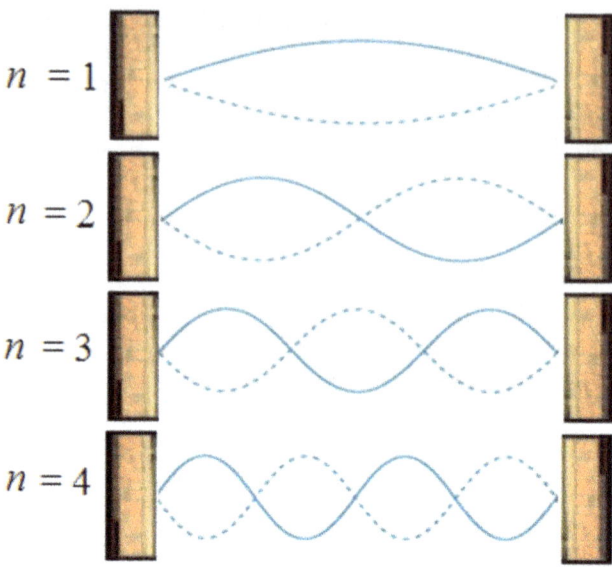

Figure 5.17. The fundamental and the first three harmonics of standing waves in a string.

$$f_n = n\frac{v}{2L}, \qquad (5.8)$$

where v is the speed of the wave, L is the length of the string, and n represents the mode under consideration. However, it has been experimentally observed that higher frequencies tend to damp out quickly, leaving only the lowest possible frequency, called the fundamental frequency (corresponding to $n = 1$). This fundamental frequency is obtained by plucking the string in the middle and is mathematically found using the formula:

$$f_1 = n\frac{v}{2L} = \frac{v}{2L}, \qquad (5.9)$$

where v is the speed of the wave, L is the length of the string, and n represents the mode.

The first harmonic (which corresponds to $n = 2$) is set up by lightly holding the middle of the string (creating a node at this point) and plucking the string about a quarter of the way from one of the sides. In this case, the wavelength ($\lambda = L$) and allowed frequency are

$$f_2 = n\frac{v}{2L} = 2\frac{v}{2L} = \frac{v}{L}. \qquad (5.10)$$

Similarly, the other patterns corresponding to $n = 3, 4, 5...$ can be set up by suitable damping and plucking, as seen in figure 5.17. I have summarized the results in table 5.1 for a quick review.

Table 5.1. The frequencies of modes for standing waves set up in strings.

Mode	Wavelength	Frequency
Fundamental ($n = 1$)	$2L$	$v/2L$
First harmonic ($n = 2$)	L	v/L
Second harmonic ($n = 3$)	$2L/3$	$3v/2L$
Third harmonic ($n = 4$)	$L/2$	$2v/L$

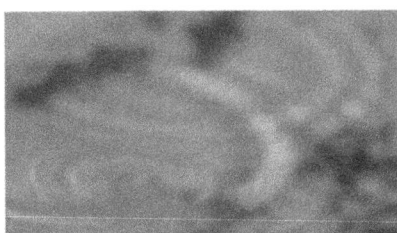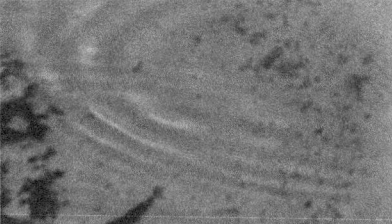

Figure 5.18. Interference in water waves.

Example 5.4. What is the third harmonic above the fundamental frequency of a standing wave set up in a string of length 30 cm with speed 40 m s^{-1}.

Solution:
The third harmonic corresponds to mode $n = 4$. Use equation (5.8) and convert the length of the string to meters, L = 30 cm = 0.30 m, to calculate the frequency as follows:

$$f_4 = n\frac{v}{2L} = 4\left(\frac{40 \text{ m s}^{-1}}{2(0.30 \text{ m})}\right) = 266.6 \text{ Hz}.$$

5.5 Wave interaction: superposition and interference

When two basketballs are thrown at each other, common experience tells us that they will collide and deflect away from each other. Basketballs are made of matter particles, so they are deflected as the result of a collision, but waves under the same conditions just pass through one another. This means that when two waves are thrown at each other at the same time, they 'superimpose' on each other for an instant before continuing on their original paths. 'Superposition' is the overlap or addition of two waves, whereas 'interference' is the effect of that overlap. In other words, interference is a pattern that you see when you observe a superimposed wave. The interference of waves set up in water is shown in figure 5.18.

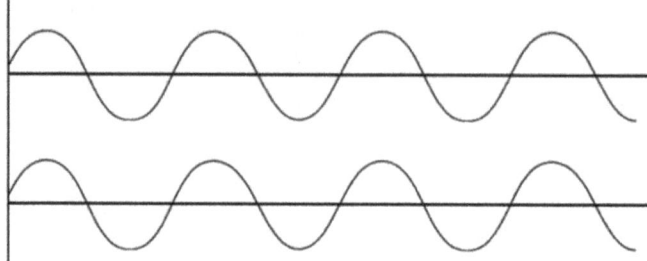

Figure 5.19. Waves in phase with each other.

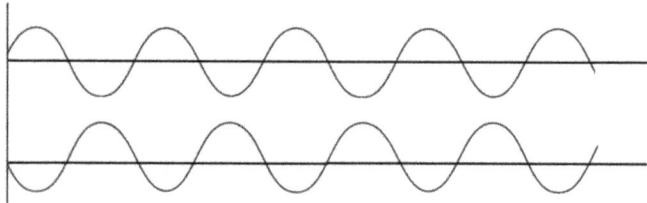

Figure 5.20. Waves out of phase with each other.

5.5.1 The phases of a wave

Before we talk about superposition and interference in waves, let us quickly discuss the phases of a wave. When a wave passes through a medium, it causes the particles in the medium to oscillate. A phase represents the particular stage in the repeating cycle of the vibrating particles of the medium. Two incoming waves can either be in phase or out of phase with each other.

1. **In phase:** the two waves shown in figure 5.19 are in phase with each other, which means that both waves do the same thing at same time and there is no phase difference. In terms of points of a wave, the in-phase points are located at the same position in the wave cycle.
2. **Out of phase:** the two waves shown in figure 5.20 are completely out of phase with each other, which means that both waves do opposite things at the same time. In terms of points of a wave, out-of-phase points are points that are not located at the same position in the wave cycle.

5.5.2 The principle of wave superposition

When two or more waves combine at a single point in a given medium, the displacement experienced by particles of the medium is given by a simple linear sum of the individual disturbances of the waves acting on the particle. This behavior of waves is called the *principle of superposition*, and when the amplitudes are not too large, it can formally be stated as follows:

'*When two or more waves overlap, the net displacement at a point in the given medium is the linear sum of the individual disturbances due to each wave.*'

Suppose two waves Ψ_1 and Ψ_2 that have amplitudes of A_1 and A_2, respectively, pass through a single point in a medium, say air, at the same time. The displacement that a molecule of air undergoes due to the presence of these two waves is given as the linear sum $\Psi_{(1+2)}$, mathematically written as

$$\Psi_{(1+2)} = A_1\Psi_1 + A_2\Psi_2. \tag{5.11}$$

There are two important aspects of wave superposition:
1. Each wave maintains its own identity and is unaffected by any other waves in the same space. In other words, the incoming waves pass through one another and emerge on the other side of the interaction nearly unchanged. This effect allows us to distinguish two people speaking in a room at the same time, since the sound waves pass through each other nearly unaffected.
2. When two or more waves are in the same medium, the overall amplitude at any point on the medium is simply the sum of the individual wave amplitudes at that point.

Figure 5.21 illustrates both of these aspects. In figures 5.21(a) and (b), two incident wave pulses move towards each other. Figure 5.21(c) shows the two pulses reaching the same spot in the medium. Here, the combined amplitude is just the plain algebraic sum of the two waves. The wave shape is also purely just the superposition of the individual waves. Figure 5.21(d) shows the two wave pulses moving away from each other after the interaction, clearly unchanged by their meeting. Both the outgoing waves here have the same shapes and amplitudes as they had before the interaction.

5.5.3 Types of wave interference

The pattern observed as a result of the superposition of waves is called wave *interference*. There are three ways in which waves can interfere with each other and hence three main types of interference patterns that may be set up: constructive, destructive, and intermediate.

1. **Constructive interference**: constructive interference occurs when both incoming waves are exactly in phase; such waves add together to produce a stronger wave. The amplitudes of the individual waves add, and the resultant wave is the sum of the amplitudes given by $|A_1 + A_2|$, as shown in figure 5.22 (left). In sound waves, constructive inference results in a louder sound.
2. **Destructive interference**: destructive interference occurs when both incoming waves are exactly inverted, or are 180 degrees out of phase. The amplitudes of the individual waves subtract, and the resultant wave is mathematically given by $|A_1 - A_2| = 0$, as shown in figure 5.22 (middle). In sound waves,

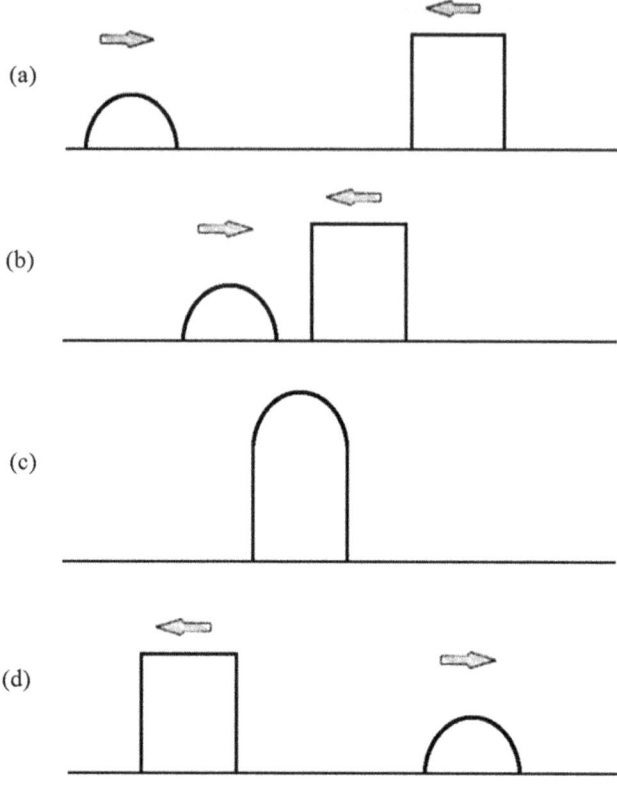

Figure 5.21. The superposition of waves.

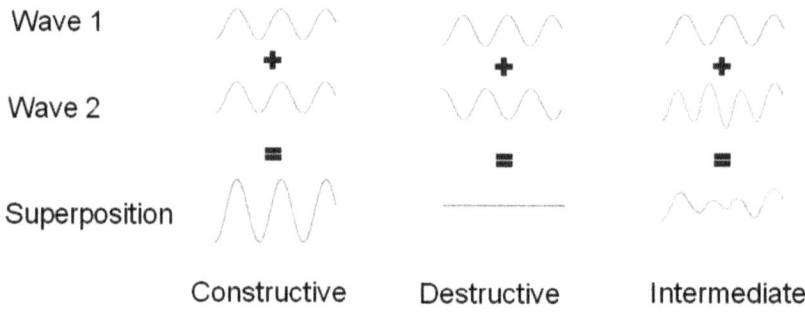

Figure 5.22. Types of wave interference.

destructive inference results in absolute silence, and this is how many noise cancellation devices work. Active noise cancellation works by taking external sound signals and canceling them out by using the headphones to output destructively interfering sound just in time. This technique relies on the speed of light, which is much greater than the speed of sound. Hence, the

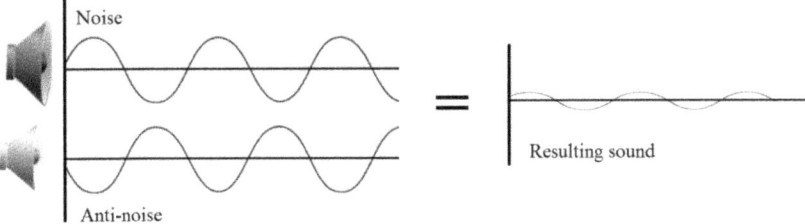

Figure 5.23. Sound interference in an active noise cancellation system.

headphone output occurs much faster than the incoming external sound. This allows both sounds to reach our ear at about the same time, which effectively cancels the external noise. At a very simple level, this technique is shown in figure 5.23.

3. **Intermediate interference**: intermediate Interference occurs when the incoming waves have a varying phase relationship. The superposition of these waves has a resultant amplitude somewhere between a maximum of $|A_1 + A_2|$ and a minimum of $|A_1 - A_2|$, as shown in figure 5.22 (right).

5.5.4 Interference in music

The idea of interference is exceptionally important in music. Musical sounds are often constant frequencies, called pitches, that are held for a sustained period. Sound waves interfere with other sound waves in much the same way as other waves do, and when the sound waves are musical sounds, the resulting superposition can sound either pleasant or unpleasant to the ear. When the sound is pleasant to hear, it is called *consonant*, and when it is unpleasant, it called *dissonant*.

The concept of interference of sound waves is essential when musicians develop musical scales. Musical scales consist of notes (pitches/frequencies), and musicians take great care to develop musical scales which consist of notes which sound consonant when played together and try to avoid notes that sound dissonant.

5.5.5 Interference between sound waves from two different sound sources

Suppose that two loudspeakers that produce sound waves of amplitudes A_1 and A_2, respectively, with the same frequency (f) are placed as shown in figure 5.24. If d_1 and d_2 are the distances between loudspeakers 1 and 2 and the observer, respectively, the path length is the extra distance traveled by wave 2 compared to the distance traveled by wave 1. The path length is mathematically given by

$$\Delta d = d_2 - d_1. \quad (5.12)$$

The principle of superposition says that for every point along the line connecting both speakers, the net sound pressure is the linear sum of the pressures from loudspeaker 1 and loudspeaker 2. The conditions for constructive and destructive interferences are then:

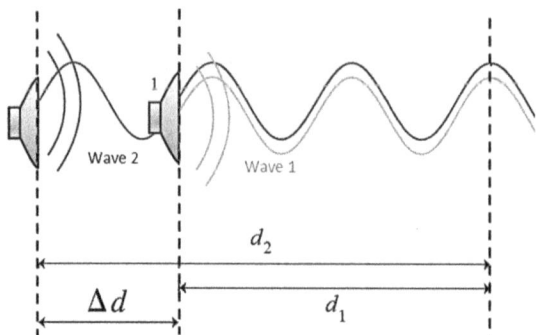

Figure 5.24. Constructive interference.

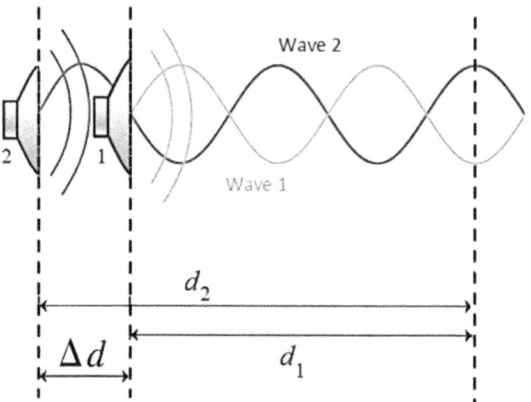

Figure 5.25. Destructive interference.

1. **Constructive interference**: when the speakers are placed one wavelength apart, the crests and troughs of both waves align, the two waves are in phase with each other, and they interfere constructively. The superposition of these waves results in a wave with larger amplitude ($|A_1 + A_2|$). This also means that at all times, the path length (Δd) is a whole number, say n, of wavelengths (λ), mathematically given by

$$\Delta d = n\lambda, \qquad (5.13)$$

where $n = 0, 1, 2, 3\ldots$.

2. **Destructive interference**: when the speakers are separated by half a wavelength, the two incoming waves are out of phase, as seen in figure 5.25. The superposition results in a wave with zero amplitude, which is the condition for destructive interference. The path length (Δd) is mathematically given by

$$\Delta d = \left(n + \frac{1}{2}\right)\lambda, \tag{5.14}$$

where $n = 0, 1, 2, 3\ldots$

Example 5.5. Suppose Sami stands in front of two loudspeakers that are in line with each other and producing sound at 343 Hz. Speaker 1 is 4 m away and speaker 2 is 5 m away from him. Assuming the speed of sound in air is 343 m s^{-1}, find:

(a). The path length

Solution:
The path lengths of the two waves reaching Sami from the speakers can be found using equation (5.12) as follows:

$$\Delta d = d_2 - d_1 = 5 - 4 = 1 \text{ m}.$$

(b). The wavelength corresponding to constructive interference

Solution:
At 340 Hz, this path length results in constructive interference; the wavelength corresponding to constructive interference is

$$\lambda = \frac{\text{speed}}{\text{frequency}} = \frac{343 \text{ m s}^{-1}}{343 \text{ Hz}} = 1 \text{ m}.$$

(c). The value of n that corresponds to constructive interference

Solution:
Use equation (5.13) to find the integer that corresponds to constructive interference, as follows:

$$n = \frac{\Delta d}{\lambda} = \frac{1 \text{ m}}{1 \text{ m}} = 1.$$

This means that for constructive interference, one wave fits in a path length of 1 m.

(d). Next, suppose that the frequency of source 1 is slowly increased until Sami no longer hears any sound. What is the wavelength for this destructive interference?

Solution:
The destructive interference (found using equation (5.14)) first occurs when the wavelength is decreased to

$$\lambda = \frac{\Delta d}{n + \frac{1}{2}} = \frac{\Delta d}{1 + \frac{1}{2}} = \frac{1 \text{ m}}{1.5} = 0.66 \text{ m}.$$

(e). What is the frequency of this destructive interference?

Solution:
This destructive interference corresponds to a frequency of

$$f = \frac{v}{\lambda} = \frac{343 \text{ m s}^{-1}}{0.66 \text{ m}} = 520 \text{ Hz}.$$

This means that destructive interference first occurs at 520 Hz, which is an increase of 180 Hz from the original 340 Hz.

5.5.6 Beats

When two sound waves of different but close enough frequencies approach your ear, the alternating constructive and destructive interference causes the sound to be alternately soft and loud, a phenomenon called *beating* or the production of beats.

5.5.6.1 The beat frequency (Δf)

The beat frequency, also called the *rate of throbbing*, is equal to the absolute value of the difference in frequency (Δf) of the two incoming waves. Consider two waves with frequencies f_1 and f_2 that interfere with each other. During this interference, the two waves either reinforce each other, causing constructive interference, or subtract from each other, causing the waves to interfere destructively. The beat frequency of the two waves is defined as:

$$f_{\text{beat}} = \Delta f = |f_2 - f_1| \tag{5.15}$$

Consider two incoming waves shown in figure 5.26 in red and purple colors. Both waves have the same amplitude A, but have different frequencies f_1 and f_2. Suppose

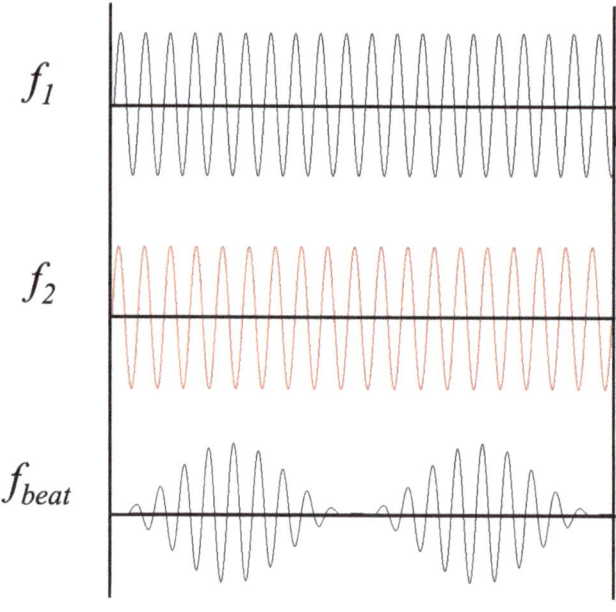

Figure 5.26. Two incoming waves with equal amplitude but slightly different frequencies and the resulting beat pattern. The resultant wave frequency is the average value of the two original frequencies; its amplitude varies between zero and the sum of the amplitudes of the original waves.

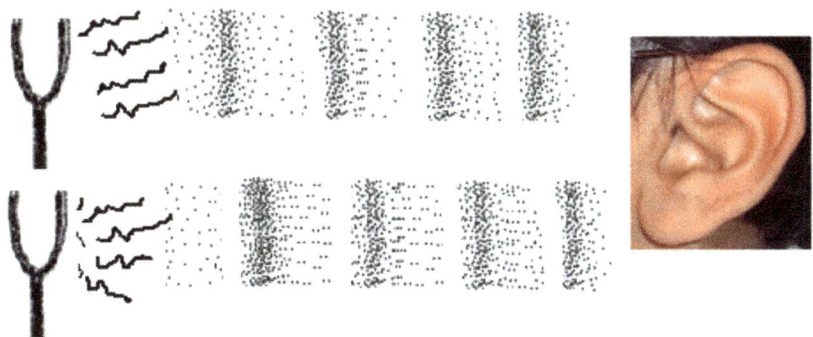

Figure 5.27. Compressions and rarefactions traveling to your ears.

f_1 and f_2 are relatively close to each other but that the red wave (f_1) has a slightly higher frequency than the purple wave (f_2). Next, suppose both waves start out in phase with each other, so that they initially add to give constructive interference.

Since the red wave (f_1) has a slightly higher frequency than the purple wave (f_2), it gradually gains on the purple wave until they are both eventually one half-cycle out of phase. At this point, the two waves cancel out and interfere destructively and the amplitude of the blue curve is zero. After an equal interval of time, they get back in phase with each other and again interfere constructively. This pattern of alternately constructive and destructive interferences results in a throbbing curve like the one shown in black in figure 5.26, and the resulting heard frequency is called a *beat frequency*. The compressions and rarefactions traveling to your ears are shown in figure 5.27.

As long as the frequency difference Δf is less than about 10 Hz, the beats are easily perceived. When Δf becomes greater than 15 Hz, beating is replaced by a distinctive roughness. As Δf increases further, the fused tones disappear completely and two distinct pitch signals are heard.

5.5.6.2 The perceived frequency
When two tones with similar frequencies, f_1 and f_2, occur in the same space, their interference causes beats, which result in an increase and a decrease in the perceived sound intensity. The perceived frequency is the average of the two frequencies, given mathematically by

$$f_{\text{perceived}} = \frac{(f_1 + f_2)}{2}. \qquad (5.16)$$

Please note that $f_{\text{perceived}}$ determines the pitch, whereas f_{beat} determines the frequency of the modulation.

Example 5.6. Two tuning forks with slightly different frequencies are struck simultaneously and the sound waves travel to our ear. The first tuning fork produces

a note at 512 Hz and the second tuning fork produces a note at 504 Hz. What do we hear?
Solution:
We know from equation (5.15) that the beat frequency is

$$f_{beat} = |f_2 - f_1| = |512 - 504| = 8 \text{ Hz}.$$

This means that we hear eight beats per second. The tone we hear is

$$f_{perceived} = \frac{(f_1 + f_2)}{2} = \frac{(512 + 504)}{2} = 508 \text{ Hz}.$$

Therefore, a tone of 508 Hz is heard at eight beats per second.

5.5.6.3 The uses of beats
Arising from simple interference, the applications of beats are far-ranging, and some are given in this section:
1. *Tuning musical instruments*: one way to tune a guitar is to compare the frequencies produced by the guitar to those of a known guitar. Using this technique, the difference between the frequencies of two strings is easily recognizable by all humans, not just musicians. It is clear that as the two frequencies get closer to each other, the beat frequency decreases, and when they are identical, the beat frequency disappears completely. At this point, the two frequencies are completely *in tune* with each other.
2. *Rotational speeds*: a twin-propeller airplane whose engines have slightly different rotational speeds produces a beating effect which is extremely uncomfortable for the passengers. In order to *tune* the two engines, engineers utilize the beat phenomenon. When the beating disappears, the two engines are perfectly *in tune* with each other.

Review Questions 5.2.
1. What happens to sound if there is no reflection of sound?
2. Does water reflect sound? What happens on a calm lake? on a choppy sea?
3. What are the effects of constructive and destructive interference in sound?
4. What are 'dead spots' in sound?
5. How do you calculate beats?
6. What is the difference between interference and beats?

Exercises 5.2.
1. A light ray strikes a reflective plane surface at an angle of 36° with the surface. Calculate:
   (a) The angle of incidence (θ_i)
   (b) The angle of reflection (θ_r)
2. What are the fundamental and the first harmonic of a standing wave moving at a speed of 100 m s^{-1} in a string 10 cm long?
3. A violin string that should be tuned to concert pitch (440 Hz) is slightly out of tune and we hear four beats per second. What are the possibilities for the frequency of the violin string? Hint: there are two possibilities.
4. If two people stand near each other and whistle, one at a frequency of 204 Hz and the other at a frequency of 214 Hz, what do people near them actually hear?

Further reading

- French A P 1971 *Vibrations and Waves* (Boca Raton, FL: Chapman and Hall)
- Arya A 1990 *Introduction to Classical Mechanics* 2nd edn (Upper Saddle River, NJ: Prentice-Hall)
- Symon K 1971 *Mechanics* 3rd edn (Reading, MA: Addison-Wesley)
- Resnick R, Halliday D and Krane K 2002 *Physics* vol 1 5th edn (New York: Wiley)
- Berg R E and Stork D G 1994 *The Physics of Sound* 2nd edn (Englewood Cliffs, NJ: Prentice-Hall)
- Rossing T D, Moore R F and Wheeler P A 2002 *The Science of Sound* 3rd edn (Upper Saddle River, NJ: Pearson Education)
- Emanuel D C and Letowski T 2009 *Hearing Science* (Philadelphia, PA: Wolters Kluwer)
- Hall D E 1991 *Musical Acoustics* 2nd edn (Pacific Grove, CA: Brooks/Cole)

IOP Publishing

The Physics of Sound and Music, Volume 1
A complete course text (Textbook)
Samya Bano Zain

Chapter 6

Damping and resonance in musical instruments

Most physical systems do not have only one frequency of vibration but rather can be set to vibrate at multiple frequencies. When a physical system is made to oscillate at its *natural frequency* by the application of an external force, a special phenomenon called *resonance* occurs. At resonance, the physical system oscillates at its maximum amplitude. The lowest possible frequency at which a system can be set into vibration is called the *fundamental frequency* or the *fundamental mode*, while all higher frequencies are called *harmonics* or *overtones*.

In this chapter, we discuss the phenomena of damping and resonance that take place in physical systems, including musical instruments. Let us begin by understanding what happens in real physical systems.

6.1 Damping in oscillators

A vibrating system is created by the addition of some external energy into a physical system, say, by pushing or pulling on an elastic object. For example, a vibrating system or a harmonic oscillator is set up when we pull on a mass in a mass–spring system, as discussed in section 5.1.3 and shown in figure 5.2, or when we pluck a guitar string. It is a common observation that without a continuous driving force, the vibrating system uses up the initial energy provided by the push or pull in overcoming resistive forces, such as drag or friction. This results in a decrease in the energy of the system over time, which means that the amplitude of the vibrating system also reduces over time, as shown in figure 6.1. Under such conditions, we say that this system experiences or undergoes *damping*.

If no additional external force is provided, the amplitude of the system eventually reaches zero and the oscillating system comes to rest in its equilibrium state. This kind of vibratory system is called a *damped system*, and its damping depends on the resistive forces it experiences. When the resistive forces are small, the damping is small and the decay is gradual, as shown in figure 6.1.

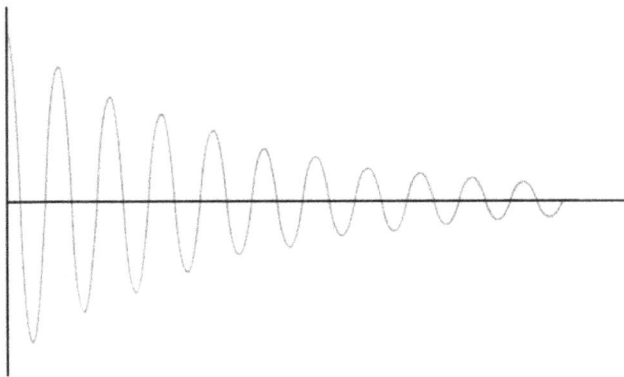

Figure 6.1. Damped simple harmonic motion.

6.1.1 Damped harmonic oscillators

Damped harmonic oscillators have non-conservative forces[1] that dissipate their energy. In this case, the energy of the system is not conserved, and therefore the law of conservation of energy does not hold. In general, the energy removed by non-conservative forces is given mathematically by

$$\Delta E = \Delta(\text{KE} + \text{PE}) = W_{\text{nc}} = W_{\text{friction}} = \vec{F}_{\text{f}} \cdot \vec{d}, \tag{6.1}$$

where W_{nc} is the work done by non-conservative forces, which in this case is W_{friction} given mathematically by the dot product of the force of friction ($\vec{F}_{\text{f}}$) and $\vec{d}$, which is the displacement through which the object moves. Please note that for a damped harmonic oscillator, W_{nc} is negative because it removes mechanical energy (kinetic energy (KE) + potential energy (PE)) from the system.

6.1.2 Energy and damping in vibrating systems

A real vibrating system always has dissipative forces, such as friction or drag, and the system tends to lose mechanical energy because of these forces. Unless the energy is renewed somehow, the amplitude of oscillation decreases over time and the system eventually comes to a stop. These are called *damped systems* or *damped harmonic oscillators*. The curve in figure 6.2, called the *envelope* or *decay curve*, indicates the change in amplitude with respect to time, and the rate of decrease in the amplitude is called the *damping constant*.

Damping in an oscillating system causes the amplitude and hence the energy of the system to decrease. However, the frequency and the period of the vibration do not change. For a real system, the rate of damping depends on the relationship between the coefficient of friction and the mass of the vibrating system. It is mathematically given by

---

[1] Non-conservative forces remove mechanical energy from the system. Examples include air resistance and kinetic friction.

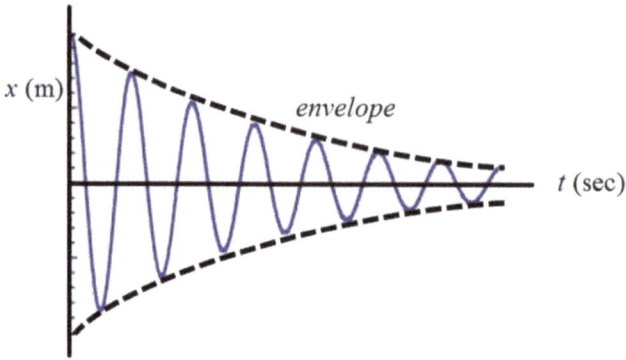

Figure 6.2. Damped simple harmonic motion with a decay curve or envelope.

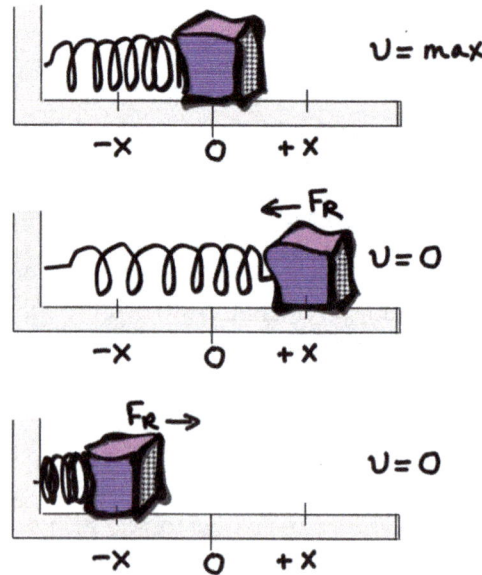

Figure 6.3. Example: a mass–spring system.

$$\text{damping coefficient } (a) = \frac{\mu}{2m}, \qquad (6.2)$$

where μ is the coefficient of friction and m is the mass of the object under consideration.

Example 6.1. Suppose a 0.2 kg object connected to a spring that has a force constant of $k = 50$ N m^{-1} rests on the top of a table, as shown in figure 6.3. Assuming the coefficient of friction between the mass and table, μ_k, is 0.1, calculate:
  (a). The frictional force between the mass and the table

Solution:
The frictional force (F_f) between the mass and the table:

$$F_f = \mu_k mg = (0.1)(0.2 \text{ kg})(9.8 \text{ m s}^{-2}) = 0.196 \text{ N}.$$

(b). The total distance the mass travels when it is released from rest 0.1 m from equilibrium

Solution:
Because the motion starts from rest, the initial energy in the system is the PE only. The total initial energy is

$$\text{PE}_i = \frac{1}{2}kx^2 = \frac{1}{2}(50 \text{ N m}^{-1})(0.1 \text{ m})^2 = 0.25 \text{ J}.$$

This potential energy is removed by friction, given by $W_{\text{friction}} = \vec{F}_f \cdot \vec{d}$, where $\vec{d}$ is the total distance traveled and $\vec{F}_f$ is the force of friction, given by step (a).

When the system comes to a stop, the frictional force balances the force exerted by the spring ($F_{\text{spring}} \equiv kx_f$), where x_f is the final position:

$$F_{\text{spring}} = F_f$$
$$kx_f = \mu_k mg$$
$$x_f = \frac{\mu_k mg}{k} = \frac{0.196 \text{ N}}{50 \text{ N m}^{-1}} = 0.004 \text{ m},$$

which corresponds to the final PE_f:

$$\text{PE}_f = \frac{1}{2}kx_f^2 = \frac{1}{2}(50 \text{ N m}^{-1})(0.004 \text{ m})^2 = 0.0004 \text{ J}.$$

This implies that the work done (W_{nc}) is equal to the loss of potential energy during one cycle:

$$W_{\text{nc}} = \text{PE}_f - \text{PE}_i = (0.25 \text{ J} - 0.0004 \text{ J}) = 0.25 \text{ J}.$$

The total distance that the object travels:

$$W_{\text{nc}} = F_f \, d.$$

Substituting in the values, we get

$$0.25 = 0.196 \, d$$
$$\Rightarrow d = 1.27 \text{ m}.$$

6.1.3 Classes of damped oscillators

Damped oscillators can be broadly grouped into three types:
1. **Underdamped:** oscillators are considered to be underdamped when the resistive forces opposing their motion are small. This is also referred to as the weak-damping condition. Oscillating systems under weak-damping conditions undergo many oscillations before coming to rest. A pendulum

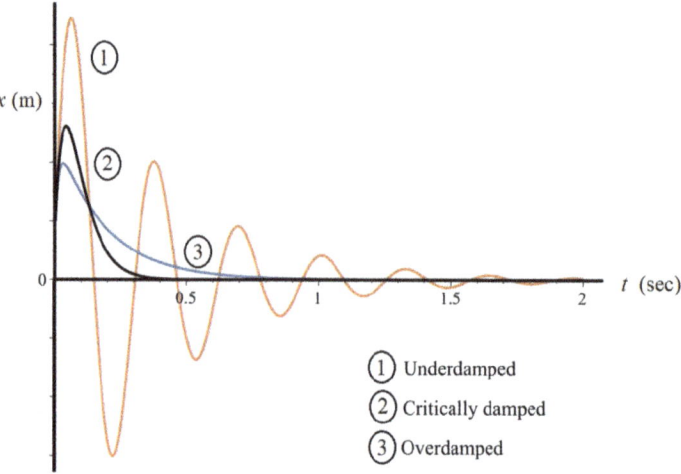

Figure 6.4. A graphical description of the various types of damping.

moving through air is underdamped. Most bells are also underdamped and hence they tend to vibrate for a long time before coming to rest. The underdamped condition is represented by curve (1) in figure 6.4.

2. **Critically damped:** critically damped systems experience higher resistive forces than underdamped oscillators. The displacement falls rather quickly in critically damped systems, and the system may pass through the equilibrium point once before slowly coming to rest. This is represented by curve (2) in figure 6.4. The suspension of a car should be critically damped to absorb bumps on the road. New cars usually have good damping, but over time, due to wear and tear of the shock absorbers, the suspension system of the car produces less damping and the car starts to bounce more during the ride, which makes the ride uncomfortable.

3. **Overdamped:** frictional forces are quite large in overdamped oscillators. This is also called a large damping condition. An overdamped system released from rest usually moves slowly to the equilibrium point without going much past it. This is represented as curve (3) in figure 6.4. A common example of an overdamped system is a pendulum moving through a very thick fluid, such as oil or honey.

6.1.4 Forces and frequencies in a vibrating motion

The forces in a vibrating motion are a function of time; that is, their magnitudes change with respect to time, hence we can use equation (6.3) to find the shape, frequency, envelope, etc. of the system. The three forces of inertia (F_I), elasticity (F_e), and friction (F_f) are essential in the study of vibrating systems. In the motion of a free system (one in which the initial energy is provided to the system and then the external force is removed) the net force is given mathematically by the vector sum of the characteristic forces:

$$\Sigma \vec{F} = \vec{F}_I + \vec{F}_e + \vec{F}_f. \tag{6.3}$$

The frequency of an underdamped real system (f_{ud}) can be found using the following equation:

$$f_{ud} = \frac{1}{2\pi}\sqrt{4\pi^2 f_R^2 - a^2}, \tag{6.4}$$

where f_R is the resonant frequency under ideal conditions and a is the coefficient of damping, calculated using equation (6.2). When the system is critically damped, f_{ud} is equal to zero, and when the system is overdamped, the value under the square root is negative and hence cannot be solved. This means that for an overdamped system, f_{ud} does not have a real value. The inverse of the coefficient of damping is called the *time constant*. The time constant specifies a time after which the amplitude of vibration in a system decreases by about 63%.

Example 6.2. Suppose a 2 kg mass resting on a table is attached to a spring. The coefficient of friction between the mass and the table is 0.1 and the spring constant is 20 N m^{-1}. Find the period of an underdamped real pendulum if the oscillations die out after about three seconds.

Solution:

We can find a period of an underdamped real pendulum using equation (6.4). In order to do that, we must find values for the resonant frequency (f_R) under ideal conditions and the coefficient of damping (a).

The resonant frequency (f_R) under ideal conditions is found by combining equation (5.3) and $f = 1/T_m$:

$$f_R = \frac{1}{2\pi}\sqrt{\frac{k}{m}} = \frac{1}{2\pi}\sqrt{\frac{20}{2}} = 0.50 \text{ Hz},$$

and the coefficient of damping (a) is found using equation (6.2):

$$a = \frac{\mu}{2m} = \frac{0.1}{2(2)} = 0.025 \text{ Hz}.$$

Hence, the period of an underdamped real pendulum

$$f_{ud} = \frac{1}{2\pi}\sqrt{4\pi^2 f_R^2 - a^2} = \frac{1}{2\pi}\sqrt{4\pi^2 (0.50)^2 - (0.025)^2} = 0.49 \text{ Hz}.$$

6.1.5 Main types of damping

There are two main types of damping:
1. **External damping**: external damping is the loss of energy due to external frictional forces, for example, air resistance or friction.

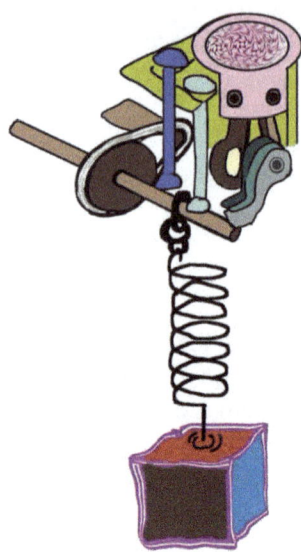

Figure 6.5. An example of a forced oscillation setup.

2. **Internal damping**: internal damping is the loss of energy due to extensions and compressions of the molecules within the system.

It is very difficult to completely eliminate all frictional forces from a periodic motion; they are always present in one form or another. However, the effects of frictional forces can be limited by the addition of forced oscillation into the system.

6.1.6 Forced oscillation in a mass–spring system

We saw above that oscillators left to their own devices eventually come to rest due to damping forces inherently present in the system. Thus, to keep a system vibrating, we have to continuously supply it with an external force. The resulting oscillations due to the external force are called *forced oscillations* or *driven oscillations* and are very important in physics. The effect of the forced oscillation is to make the system oscillate at the frequency of the applied external force, which may or may not be the same as the natural frequency of the oscillator. An example of a forced oscillation setup is given in figure 6.5.

6.2 Resonance

In the absence of frictional forces, an external force continuously applied at the correct frequency, even when it is very small, has the effect of increasing the amplitude of the system. However, if there is friction in the system, the amount of energy added in each cycle must be larger than the energy lost to overcome friction and increase the amplitude. If we provide the system with a large enough, albeit still small force, the overall effect is that the amplitude of the swing increases over time. This phenomenon is called *resonance*.

The word 'resonance' comes from Latin, and it means to *resound* or 'to sound out together with a loud sound.' Resonance is one of the most intuitively understood concepts. Very young children figure out how to play on rocking horses by setting them rocking using the phenomenon called *resonance*. Please note that it is more than likely that a toddler has never heard of the word resonance, but that does not stop them from employing it to perfection!

Similarly, a child on a swing almost always figures out how to maximize the oscillation of the swing. Let us discuss this a bit more. Before the child gets on the swing, the swing is at rest. This is called the *natural state* of the swing and is the equilibrium point for the swinging motion. As they get on the swing, they set the swing into motion, and it acquires some displacement. But like all things, the swing would prefer to remain in its natural state, so it starts towards its equilibrium point. However, when it gets there, inertia (section 4.3.1) causes it to overshoot its natural state. This overshoot continues until all the energy is dissipated and the swing stops for an instant before swinging back towards the equilibrium point. This process is repeated over and over, and a vibratory motion is set up. The amplitude of the swing can be maintained or increased by applying just a small force at its natural frequency during each vibration.

6.2.1 The effects of resonance

Let us briefly discuss some of the effects of resonance here:

1. While pedaling a bike, you can make the bike go faster if you apply a relatively small force to the pedal at the right instant. For example, Eman shown in figure 6.6 on her bike needs to apply just a little push during each cycle to get the bike moving faster and faster.

Figure 6.6. An example of resonance. On her bike, Eman needs to apply just a little force to get the bike moving faster and faster.

2. Soldiers march in step during a parade; however, they are required to break step when marching over a bridge. This is because if the marching soldiers match the natural frequency of the bridge, it sets up intense vibrations in the bridge and the bridge may collapse.
3. Similarly, during an earthquake, buildings vibrate at the frequency of the seismic wave. If the frequency of the seismic wave matches the natural frequency of the building, the building resonates at a greater amplitude, which can cause it to collapse. To avoid this, all buildings are built with resonance buffers at their bases, especially in earthquake-prone regions. The same is true for very tall skyscrapers that are subject to high wind conditions.
4. In music, resonance is used to increase the loudness of the sound made by musical instruments. The sound produced by just plucking guitar strings is very soft; however, these vibrations cause sympathetic vibrations to be set up in the soundboard, which in turn causes the column of air inside the guitar to resonate at the same frequency. The combination of all these greatly increases the loudness of the original sound produced by plucking the guitar strings.
5. Resonance also applies to the human voice. In humans, speech and/or singing is achieved by setting the vocal cords into vibration. This by itself does not produce enough sound, so it is further reinforced by resonant vibrations set up in the oral cavity and nasal passages, discussed in more detail in chapter 10.

6.2.2 Seashells and resonance

When you hold a seashell to your ear, it sounds as though you are listening to the sea. Have you ever wondered why? What you are hearing is not the sea, but resonances present in the room. There are always sound waves everywhere, even in the quietest of rooms. These sounds are mostly inaudible to humans and are called *background noises*.

Background noises fill all parts of our environments, including the seashell, which causes vibrations within the seashell. Each seashell, like everything else in our environment, has a set of natural frequencies at which it vibrates. When one of the frequencies present in the background noise causes the air within the seashell to vibrate at the seashell's natural frequency, resonance occurs. Resonance causes amplification of sounds at the resonant frequency and the previously inaudible sound becomes audible.

Next time you hear the sea in a seashell, remember that you are not hearing the sea in a seashell but are rather hearing the amplification of one of the many background frequencies that already exists in the room!

Example 6.3.
1. Suppose a vibrating system is vibrating at its natural frequency. What is the result of a forced vibration that acts on the system at the same frequency?
    (a) Damping
    (b) Vibration
    (c) Resonance.

Solution:

(a). A forced vibration that has the same frequency as a vibrating system acts on that system to increase its amplitude, which is called 'resonance.'
2. When does a vibrating system experience damping?
   (a) When its amplitude remains constant
   (b) When its period decreases
   (c) When its frequency decreases
   (d) When its amplitude decreases.

   Solution:

   (d). A vibrating system experiences damping when its amplitude decreases.

6.3 Ways to drive a string at one of its resonances

A stretched string can be driven at resonance in the following ways:
1. **A string attached to a tuning fork**. Tuning forks made from elastic metals, such as steel, are resonators with two prongs in a U-shape. The prongs of tuning forks vibrate at a specific constant frequency when struck. The frequency of the tuning fork depends on the length of the prongs. If you attach a string to the prongs of a tuning fork, it sets the string vibrating at the same frequency as the natural frequency of the tuning fork, as seen in figure 6.7.
2. **A violin bow moving over a string**. A violin bow moving over a taut string causes vibrations to be set up in the string, as seen in figure 6.7 (middle), because of slip-and-stick action, which is discussed further in chapter 15.
3. **Electromagnets**. Electromagnets are most important when talking about electric violins, electric bass guitars, or electric guitars. Electromagnets, shown in figure 6.7 (right), produce standing waves in steel strings and other magnetic materials.

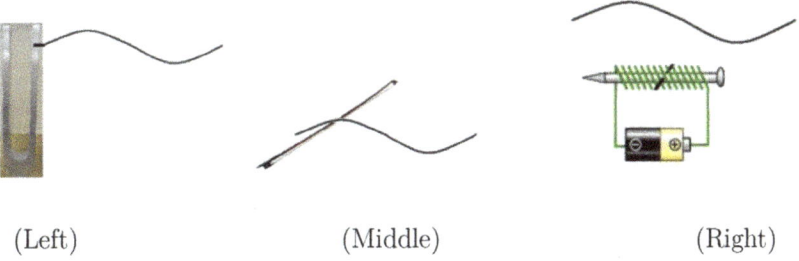

(Left)　　　　　　　　(Middle)　　　　　　　　(Right)

Figure 6.7. Waves in strings: (left) a string attached to a tuning fork; (middle) string vibration due to bowing; (right) a string attached to an electromagnet.

Review Questions 6.1.
1. What is resonance? Can you give some examples from musical and non-musical fields?
2. Sami is pushing Eman on a swing. If Sami pushes her with the same force in each cycle, does the amplitude of the swing increase by the same amount in each cycle?
3. A vibrating system experiences resonance when:
   (a) Its amplitude remains constant
   (b) Its period increases
   (c) Its frequency increases
   (d) Its amplitude increases.

Exercises 6.1.
1. How would a car bounce after going over a speed bump under each of these conditions?
   (a) Underdamping
   (b) Critical damping
   (c) Overdamping.
2. The amplitude of a particular oscillating mass–spring system halves in amplitude every minute. What is its amplitude after five minutes? after ten minutes? Hint: $\left(\frac{1}{2}\right)^2 = \left(\frac{1}{2\times 2}\right) = \frac{1}{4}$.
3. Suppose the initial amplitude of the mass–spring system in question 2 is 50 cm. Will the oscillations have ceased at the end of ten minutes? Explain in your own words.
4. Suppose a 5 kg mass attached to a spring is placed on a table. The coefficient of friction between the mass and table is 0.2 and the spring constant is 20 N m^{-1}. Find the period of an underdamped real pendulum if the oscillations die out after about five seconds.

6.4 Understanding resonance

Resonance occurs when the frequency (f) of a forced oscillation matches the natural frequency (f_0) of the system. This results in the maximum transfer of energy to the oscillator, and the amplitude of the system steadily grows until it reaches a maximum (A_{max}). The numerical value of A_{max} depends on the damping forces affecting the oscillator.

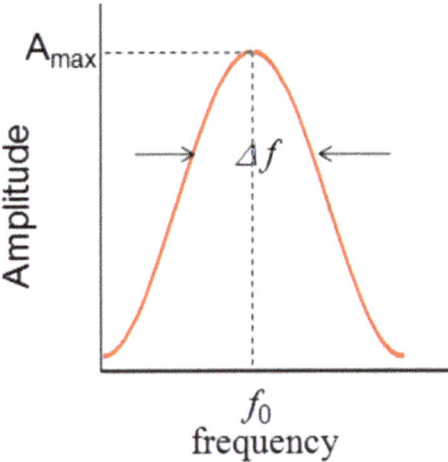

Figure 6.8. The amplitude of oscillation as a function of frequency. The amplitude is maximized at the resonant frequency. Here, f_0 is the natural frequency of the system and Δf represents the bandwidth.

The plot of amplitude versus frequency shown in figure 6.8 is almost symmetric around its peak. The peak value of this graph corresponds to the numerical value of the maximum amplitude ($A_{\max}$). The width of this curve, called the *bandwidth* or *line width* (Δf), is given by $\frac{A_{\max}}{\sqrt{2}}$, which is numerically equal to 71% of $A_{\max}$. Since $A_{\max}$ depends on the friction of the system, the *bandwidth* (Δf) also depends on the friction of the system.

6.4.1 The Q-factor

Physicists and engineers use a dimensionless parameter called the *quality factor* or *Q-factor* to describe the energy damping in oscillatory systems. In terms of energy, the Q-factor is given by

$$Q - \text{factor} = 2\pi \left(\frac{\text{average energy stored}}{\text{energy lost per cycle}} \right) \qquad (6.5)$$

The Q-factor compares the time constant for the decay of an oscillating system's amplitude to the oscillation period of the system. In this case, the Q-factor is mathematically given by

$$Q - \text{factor} = \frac{f_0}{\Delta f}, \qquad (6.6)$$

where Δf is the *bandwidth* or *line width*.

6.4.2 Damping and the Q-factor

The concept of the Q-factor originated in electrical engineering as a measure of 'good quality' in a circuit. A higher Q-value represents a lower rate of energy dissipation relative to the oscillation frequency, so in circuits with high Q-values, the oscillations die out slowly, as shown in figure 6.9 (left). For example, a pendulum oscillating in air has a high Q-value, while a pendulum immersed in oil has a low Q-value.

As the Q-factor increases in a system, the oscillations produced in the system are sustained for longer. In addition, higher Q-factor values produce cleaner results with lower levels of noise in the signal. It is worth noting that oscillators with higher Q-values exhibit more abrupt transitions from *in-phase* to *out-of-phase* states, whereas for low-Q-value oscillators, these transitions are more gradual. In other words, a highly damped vibrator exhibits a gradual change of phase, whereas a lightly damped one has rapid changes in phase.

In lightly damped systems, Δf is small and it produces a large A_{max}. In heavily damped systems, Δf is large, hence A_{max} is small. We can summarize the various situations for oscillating systems as follows:

1. **When the Q-factor $>1/2$**: this means that for the oscillating system, $\Delta f < 2f_0$ and the Q-factor is just over a half. We call this kind of system an underdamped system. Underdamped systems may oscillate once or twice before the oscillation dies away.
2. **When the Q-factor $<1/2$**: this means for the oscillating system, $\Delta f > 2f_0$ and the Q-factor is less than a half. We call this kind of oscillating system an overdamped system. These systems decay exponentially; they have high losses and no overshoot.
3. **When the Q-factor $=1/2$**: ($\Delta f = 2f_0$) and the system is called critically damped. The output does not oscillate in critically damped systems and these systems do not overshoot.

Example 6.4. A particular oscillator has a resonant frequency of 400 Hz and a Q-factor of 20. What is the bandwidth of its resonance curve?
Solution:
Using equation (6.6), and knowing that $f_0 = 400$ Hz:

$$Q\text{-factor} = \frac{f_0}{\Delta f}$$

$$20 = \frac{400 \text{ Hz}}{\Delta f}$$

$$\Rightarrow \Delta f = 20 \text{ Hz}.$$

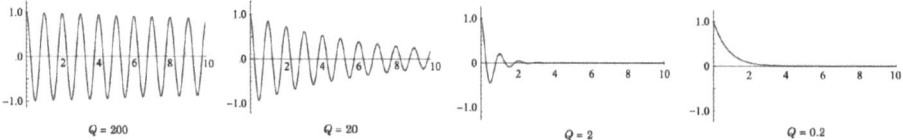

Figure 6.9. Damping as a function of the Q-factor.

Figure 6.10. Sympathetic vibrations in tuning forks.

6.5 Sympathetic vibrations

Sympathetic vibrations are vibrations that are set up in a material when another vibrating material is brought close to it without direct contact between the two materials. A traditional example that illustrates the principle of sympathetic vibration is two tuning forks of the same frequency, as seen in (figure 6.10). Strike the first tuning fork to produce sound and then bring it close to the other tuning fork, taking care not to physically make contact between the two tuning forks. You will then observe the second fork (which was not struck) also start to vibrate. The vibrations set up in the second tuning fork are called *sympathetic vibrations*.

Another example of a situation that puts the principle of sympathetic vibration to good use is the soundboard of a musical instrument. Musical instruments use vibrating strings that are either plucked or struck. The energy produced and transmitted by the plucked string is quite small, so all musical instruments use soundboards to enhance the sound produced by the instrument. When the string vibrates, it forces the soundboard in the instrument to vibrate at exactly at the same frequency, and the soundboard is able to transform the rather small sound produced by the string into a louder sound by moving a greater volume of air. In other words, the resonant properties of the soundboard increase the loudness of the instrument compared to the loudness of the string alone.

6.6 Resonances in musical instruments

Most musical instruments are played by either striking, hitting, plucking, or somehow disturbing the instrument. These methods cause a vibrational motion to be set up in the instrument. The physical structure of the particular instrument is an essential component in sound production. We need the instrument itself to be set into a standing wave condition in which all parts of the instrument work together to make the sound.

In a brass instrument, the sound comes from a vibrating column of air inside the instrument. The player makes this column of air vibrate by buzzing the lips while blowing air through a cup- or funnel-shaped mouthpiece. In a trumpet, the trumpet player creates sound waves by buzzing their lips against the mouthpiece. The buzzing lips cause a burst of air to enter the trumpet. The air then travels down the length of the tube and reaches the bell, at which point it is mostly reflected back into the trumpet. The reflected burst of air is reinforced by a new burst of air from the player's lips. As the player adds bursts of air at the correct frequency, the amplitude of the sound increases. Eventually, the amplitude becomes large enough that a small portion of the standing wave escaping the trumpet becomes audible. This is called the *resonant* condition.

6.6.1 The Helmholtz resonator

The Helmholtz resonator is a device created in the 1850s by Hermann von Helmholtz, and it is used to identify various frequencies present in music and other complex sounds.

The Helmholtz resonator, as shown in figure 6.11 (left) is a simple spherical cavity with a wide mouth. Some resonators have a small opening on the side opposite to the mouth, while others do not. When a burst of air is blown into the cavity through the mouth of the cavity, as shown in figure 6.11 (right), it compresses the air inside the cavity. In response, the air expands and rushes out, which creates a low-pressure condition inside the cavity. The low-pressure condition causes air to flow back into the cavity at a pressure greater than the atmospheric pressure. This process repeats multiple times, and the pressure in the cavity quickly rises and then drops below the normal equilibrium atmospheric pressure. This sets up a resonant condition in the Helmholtz resonator.

 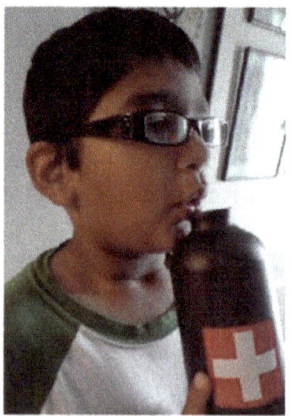

Figure 6.11. Helmholtz resonators: (left) a typical Helmholtz resonator; (right) Sami blowing air across a simple Helmholtz resonator.

The biggest advantages of the Helmholtz resonator are its simplicity of construction and the fact that it possesses a single isolated resonant frequency which can be mathematically determined by the size and the shape of the resonator. If we say that the air in the neck of a Helmholtz resonator has an effective length (L) and cross-sectional area (A), the frequency of the Helmholtz resonator is mathematically given by

$$f_H = \frac{v}{2\pi}\sqrt{\frac{A}{VL}}, \tag{6.7}$$

where v is the speed of sound in air and V is the volume of the spherical cavity.

Helmholtz used many of these resonators in the 19th century to verify the existence of harmonics in complex tones using the knowledge that smaller resonators have higher frequencies. Helmholz was able to hear an increase in the amplitude of any frequency present in the harmonic structure of the instrument by holding multiple resonators to his ears as a musical note was played. From this, he could roughly determine the spectrum of the note being played.

Historically, Helmholtz resonators were used in amphitheaters to enhance or attenuate sounds, and they are still used in the design and construction of theaters and churches. In addition, noises generated by artificial systems can be reduced using Helmholtz resonators. The classic example is car exhaust systems, in which a Helmholtz resonator (commonly known as a muffler) is added to the end of the exhaust system to reduce the noise of the engine. A muffler reflects the sound of the engine back to the engine, and in the process it muffles the engine noise.

Example 6.5. Suppose we have a 1 liter bottle with an effective neck length (L) of 5 cm and a cross-sectional area (A) of 3 cm³. Find the resonant frequency.
  Solution:
  The given values include, the volume of the bottle = 1 liter = 0.001 m³, the length of the neck (L) = 5 cm = 0.05 m; and the cross-sectional area (A) of 3 cm² = 3×10^{-4} m³.
  We use equation (6.7) together with 344 m s⁻¹ as the speed of sound at room temperature:

$$f_H = \frac{v}{2\pi}\sqrt{\frac{A}{VL}} = \frac{344}{2\pi}\sqrt{\frac{3 \times 10^{-4}}{(0.001)(0.05)}} = 134 \text{ Hz}.$$

Hence, the resonant frequency is 134 Hz.

6.6.2 Chladni patterns

Ernst Friedrich Chladni, born in Germany on November 30, 1756, was a physicist and a musician. He studied meteorites, researched vibrating plates, and calculated the speed of sound in different gases. Due to his contributions to the field of acoustics, he is called the *father of acoustics*. Chladni is best known in acoustics for the invention and development of a technique used to experimentally show the modes of vibration in rigid surfaces.

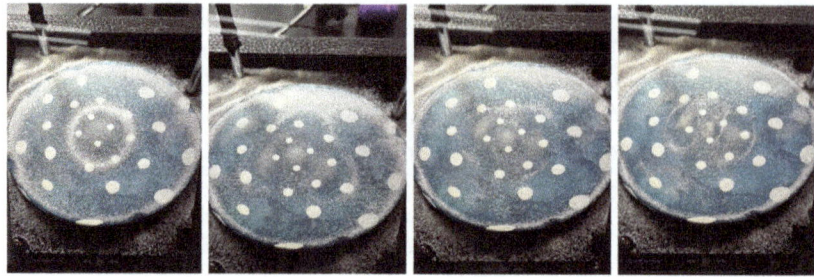

Figure 6.12. Chladni patterns created in the laboratory. Student project (Amanda M, Jocelyn M, Samuel H). The frequency increases from left to right.

In 1680 CE, Robert Hooke of Oxford University had previously observed nodal patterns emerge when he ran a violin bow along the edge of a plate covered with flour. Nodal lines are the places where the amplitude of the vibrating surface is zero and the surface does not vibrate. Chladni developed this technique and used a piece of flat metal covered lightly with sand and drew a bow over its side to show that bowing sets up vibrations in the plate until the resonant condition is reached. At resonance, the vibration causes the sand to collect along the nodal lines where the surface is at rest. This process outlines the nodal lines of the metal surface. Chladni's technique was first published in 1787 CE in his book *Entdeckungen über die Theorie des Klanges* ('Discoveries in the Theory of Sound'), and these patterns are now called *Chladni figures* or *Chladni patterns*. There are many ways of forming Chladni patterns, but the three main methods that have been used throughout history are:

1. A plate can be bowed with a violin bow. This requires a lot of skill and practice but does not require anything mechanical and can be seen in figure 16.4.
2. A plate can be excited mechanically at the frequency of the desired mode.
3. The plate can be made to resonate by another powerful sound wave tuned to the frequency of the desired node.

Chladni patterns created in the laboratory at Susquehanna University are shown in figure 6.12.

6.6.3 Singing rods

Any rod can be made to vibrate along its length at its resonant frequency, thus making it *sing*. The rod most commonly used for this purpose is an aluminum rod, because aluminum a has relatively low density for a metal, which makes it a good conductor of sound. The modal frequencies (f_S) for a singing rod are given mathematically by

$$f_S = \frac{n}{2L}\sqrt{\frac{Y}{\rho}}, \tag{6.8}$$

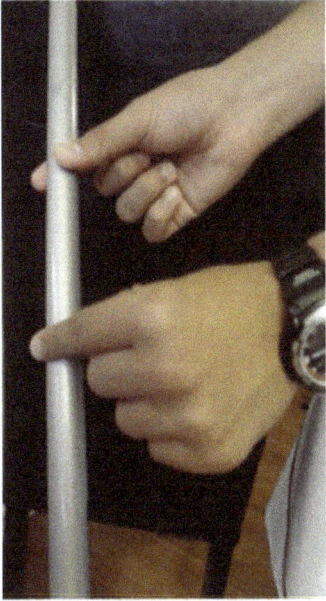

Figure 6.13. Hareem playing a singing rod.

where L is the length of the rod, Y is Young's elastic modulus (discussed in chapter 7, section 7.3), and ρ is the density of the rod material. It is important to note that the wave speed does not depend on frequency or the diameter of the rod. To make the rod sing, Hareem has to hold the rod at its midpoint with the thumb and index finger of his right hand, called the *nodal hand*, as shown in figure 6.13. To find the exact middle, or the center of mass, he must start with his hands at the ends of the rod. Now he must slowly bring his arms together while allowing his fingers to slide beneath the rod. Since the rod is uniform in shape, his fingers will automatically come together at the center of mass[2] of the rod. Next, to make the rod sing, he should scrape the thumb and pointer finger of the left hand (call it the *scraping hand*) along the length of the rod, as seen in figure 6.13. This process causes a sound to be produced in the rod, and we thus call it a *singing rod*.

The reason for using an aluminum metal rod for this purpose is that aluminum is an elastic solid which elongates when pulled. Therefore, during its motion, the scraping hand actually pulls on the rod and stretches it slightly. This stretching causes the rod to gain potential energy. In addition, this pulling motion simultaneously causes both ends of the rod to vibrate longitudinally along their lengths at its resonant (fundamental) frequency (f_0). These longitudinal vibrations travel to the end of the rod, where they are transmitted to the surrounding air molecules.

---

[2] The center of mass of a system of particles, such as a baseball bat, is defined so that we can predict the possible motion of the system. It was first introduced by the Greek philosopher Archimedes, and is defined as the point in the system that moves as though all of the system's mass were concentrated there and all external forces were applied there.

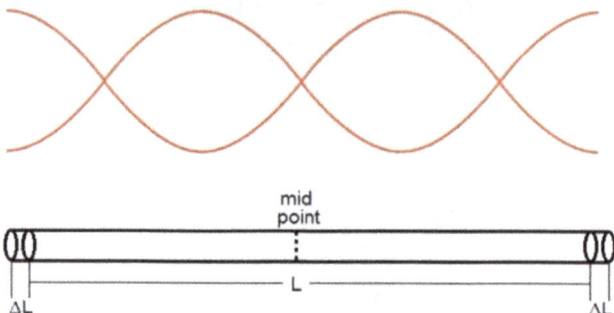

Figure 6.14. A schematic of a singing rod with nodes and antinodes.

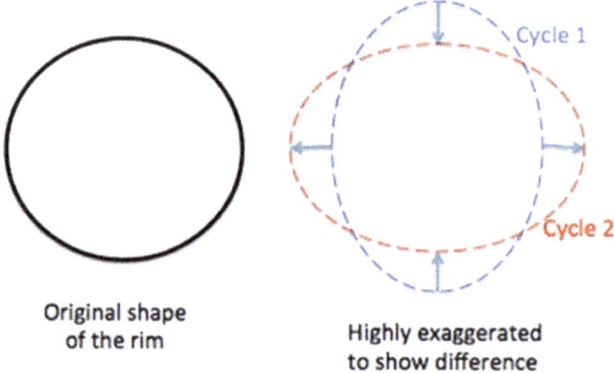

Figure 6.15. Vibrating wine glasses: highly exaggerated overhead view.

These then travel to the ear and are interpreted as sounds or *songs*. Some vibrations travel toward the end of the aluminum rod and are reflected back from the end of the rod. These reflected waves head back and collide with waves still being created by one's scraping hand. The two waves interfere with each other and create regions of constructive and destructive interferences. Where the interference is constructive, antinodes are formed and the amplitude increases and you hear a louder sound. Where the interference is destructive, the amplitude goes to zero (nodes), as shown in figure 6.14.

6.6.4 Singing wineglasses

If you tap the rim of a glass, you hear a sound. This is because tapping sets the glass into vibration, generally at its resonant frequency. The pitch of the sound depends on several factors, such as the size and the shape of the glass. The rim and sides of the glass vibrate together. When viewed from above, it is observed that the rim varies between two elliptical shapes shown in the highly exaggerated figure 6.15. When the points at the top and bottom of the (cycle 1) circle move inwards, the points on the left and right of the same (cycle 1) circle move outward, and vice versa

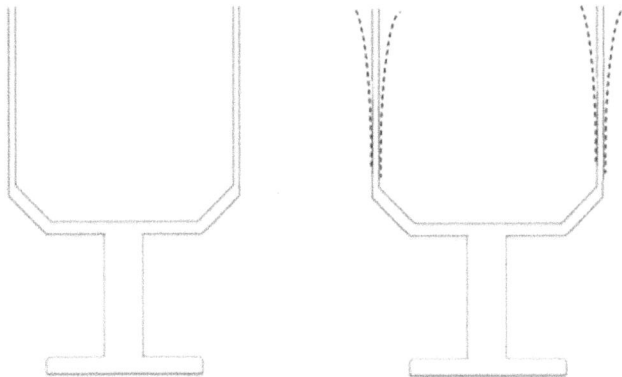

Figure 6.16. Vibrating wine glasses: highly exaggerated side view.

half a cycle later. The sides of the glass also vibrate, and the shape of the glass changes, often several hundred times per second, as shown in the figure 6.16. As the sides of the glass vibrate, they push air back and forth, creating sound vibrations in the air that travel to your ears.

6.6.4.1 Ways to make the glass vibrate
Ways to make the glass vibrate include:
1. **Rubbing your finger around the rim of the glass.** As your finger moves along the rim, it alternately slips and sticks to the rim, just like a violin bow slips and sticks as it moves across a violin string. This is called *slip–stick motion*, detailed in section 15.6.2. This slip–stick motion produces a vibration pattern at the natural frequency of the glass, which in turn results in the production of a sound. Please note that the sound can be made louder or softer by changing the speed and/or applied pressure, but the vibration frequency remains the same as the resonant frequency (f_0) of the glass.
2. **Wetting your finger.** Wetting your finger makes the slip–stick motion easier to produce. This is analogous to rubbing rosin on a violin bow, which prepares the bow and string for slip–stick motion, making the production of sound easier.
3. **Adjusting the pitch.** The pitch of a glass can be adjusted by simply adding or removing water from the glass. This happens because the water absorbs the vibrations and reduces its frequency. Adding water to the glass lowers the pitch and removing the water from the glass makes the pitch higher. A whole series of different musical notes may be produced by adding different amounts of water to different glasses. This is principle behind 'glass harmonicas' used to play songs.

Review Questions 6.2.
1. While playing the singing rod, why do you have to hold it in the middle? What happens if you do not hold the rod at its center of mass?
2. Can you break a wineglass by singing? Theoretically? Practically?
3. What is the Q-factor of an oscillatory system?
4. Why are drums not made out of just a stretched membrane that you beat to make sound?
5. Why do guitars or violins need bodies? Why can they not just be a bunch of strings?

Exercises 6.2.
1. We calculate that the bandwidth of the resonant curve of a particular oscillating pendulum is 5 Hz. If the Q-factor of the pendulum is ten, find its resonant frequency.
2. If the resonant frequency (f_o) of a Helmholtz resonator is 500 Hz and the neck has a cross-sectional area of 0.01 m^2 and a length of 0.3 cm, what is the volume of the enclosure in liters? Take the speed of sound at room temperature = 344 m s^{-1}.
3. Find the fundamental modal frequency (f_S) of an aluminum singing rod 1 meter long. Assume that the density of aluminum is 2.71 gm cm^{-3} and the value of its Young's modulus is 60 GPa.

Further reading

- Rossing T D, Moore R F and Wheeler P A 2002 *The Science of Sound* 3rd edn (Upper Saddle River, NJ: Pearson Education)
- Rossing T D and Fletcher N H 2004 *Principles of Vibration and Sound* (Berlin: Springer)
- Fletcher N H and Rossing T D 1998 *The Physics of Musical Instruments* (Berlin: Springer)
- French A P 1971 *Vibrations and Waves* (Boca Raton, FL: Chapman and Hall)
- Chababy R and Sherwood B 2011 *Matter & Interactions* 3rd edn (New York: Wiley)

Part III

Sound propagation

IOP Publishing

The Physics of Sound and Music, Volume 1
A complete course text (Textbook)
Samya Bano Zain

Chapter 7

Sound propagation

How do you move something from one place to another? The simple answer is that you take it there yourself. But how do you transfer energy from one place to another? One way would be to take it there physically yourself, or you could transfer it in some chunk of matter. For example, if you want to break a piece of glass, you do not have to physically make contact with it yourself. You could throw a rock at the glass; by your action of throwing the rock, you inevitably put some energy into it. This energy travels with the rock to the glass and shatters it upon impact.

There is another way to transfer energy that does not involve the transfer of matter; rather, it involves using a wave. A wave is a transfer of energy without a transfer of mass. If you and a friend hold the ends of a rope, you can transfer energy over to her simply by moving the rope back and forth. Although the rope has a back-and-forth motion, the rope itself does not move toward or away from you or from her. Only the energy is transferred. In this chapter, we will talk about sound propagation that utilizes waves.

A wave is a disturbance induced in a medium which travels away from its source. Particles in the medium a wave travels through are disturbed from their equilibrium position as the wave passes through; however, these disturbed particles immediately return to their equilibrium position once the wave has passed. Waves can transport energy and information from one place to another without moving any material particles from one place to another. It is almost like saying that we are transporting passengers from New York City to Selinsgrove, PA without a bus! Two of our five basic senses are wave detectors; the ear is a sound wave detector and the eyes are light wave detectors.

At the microscopic scale, forces between atoms are responsible for the propagation of waves. In solids, all atoms can be assumed to be connected to neighboring atoms through a lattice of springs, like those we saw in the mass–spring system (figure 5.2). Atoms are fixed in place in the lattice, but are free to vibrate back and forth. When one atom is displaced from its equilibrium position, say by a wave passing by, it collides with the next atom in its path and energy is transmitted to the next atom. Once the wave passes by the original atom, it returns to its initial state. These collisions keep moving forward and the wave is transmitted through the medium. There is no net displacement of the particles as the wave passes; the individual atoms only oscillate around their equilibrium positions. The maximum amplitude of the oscillations is proportional to the initial disturbance.

7.1 Traveling waves

Traveling waves, as their name implies, are waves that travel. They occur when a wave is not confined to a given space in a medium. Traveling waves transport energy from one place to another, and for an undamped wave, all points through which the wave travels oscillate with the same amplitude. Traveling waves are also called *progressive waves*; an example of a traveling wave is given in figure 7.1.

7.1.1 Setting up a traveling wave

A wave can be set up in a stretched rope by attaching it on one side to a wall and moving the other end up and down in a vertical direction. If the end is moved upward f times per second, the wave that is set up oscillates with a frequency of f, as shown in figure 7.2. This wave travels with speed v, which is determined by the mass of the rope and the tension applied to it. For a string of length (L) and mass (m) held under a tension (T), the speed of the wave is given by

$$v = \sqrt{\frac{TL}{m}}. \qquad (7.1)$$

This equation shows that the speed of the wave is independent of the frequency of the wave.

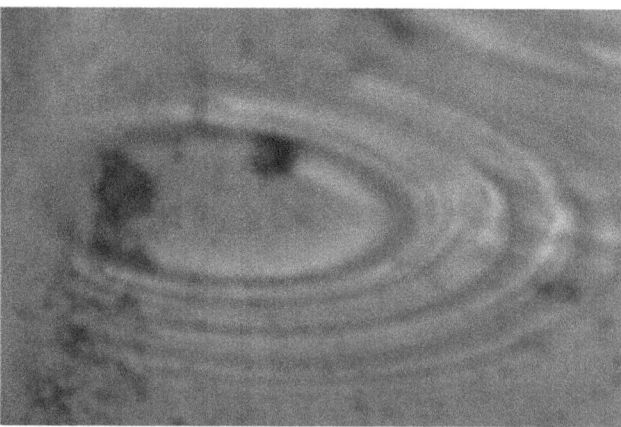

Figure 7.1. Traveling waves passing through water.

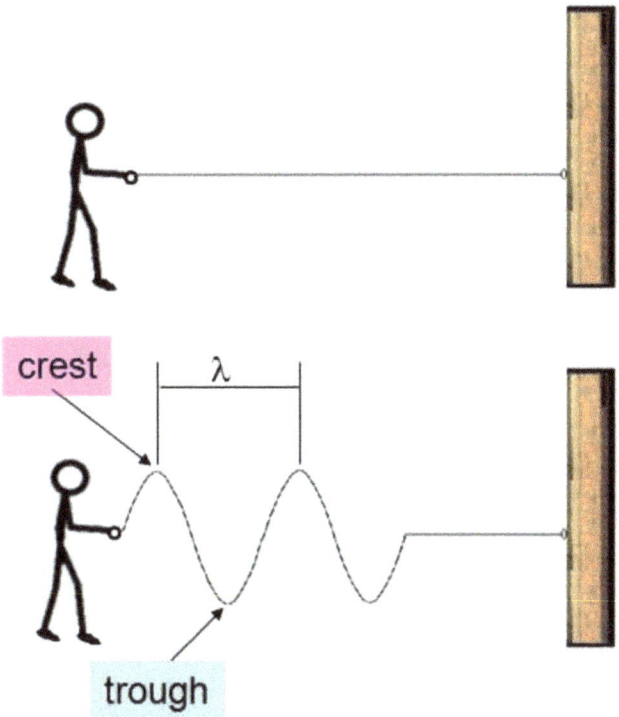

Figure 7.2. Setting up a traveling wave.

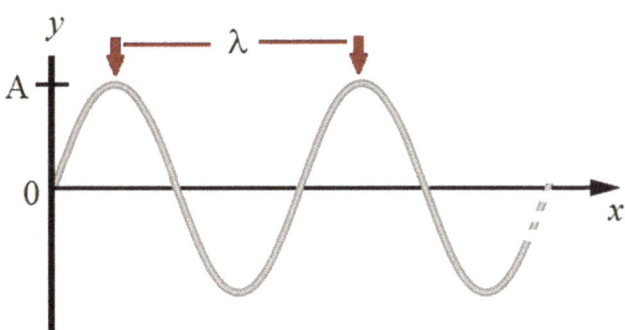

Figure 7.3. A string wave with amplitude (A) traveling along the positive x-axis. The diagram shows a typical wavelength (λ), measured from an arbitrary position.

7.1.2 The mathematical description of a wave in a string

Mathematically, a wave is represented as a change in some quantity, for example, displacement (y) or pressure (P), as a function of both the x-position and time. Suppose a sinusoidal wave like the one shown in figure 7.3 travels in the positive x-direction. As the wave travels forward, it moves through small elements of the string, and they oscillate parallel to the y-axis. At time (t), the displacement along the y-axis is called Ψ and is mathematically given by

$$\Psi(x, t) = A \sin(kx - \omega t), \quad (7.2)$$

where A is the amplitude of the wave, k is the wave number ($k = 2\pi/\lambda$), x is the position of a point in the direction of wave propagation, t is time, and ω is the angular frequency ($\omega = 2\pi f$, where f is frequency). The minus sign in equation (7.2) applies to waves moving in the positive x-direction; for waves moving in the negative x-direction, a plus sign is added to the equation instead. Please note that equation (7.2), which is expressed in terms of position x, is used to find the displacements of all elements in the string as a function of time. Therefore, equation (7.2) determines the shape of the wave at all times.

7.1.3 Properties of traveling waves

Some properties of traveling waves that are essential to our understanding in this text are as follows:

1. **Amplitude:** the amplitude (A) of a wave, as shown in figure 7.3, is the magnitude of the maximum displacement of the string elements from their equilibrium positions as the wave passes through them. Since amplitude is a magnitude, it is always positive, even when it is measured downward instead of upward.
2. **Wavelength:** the wavelength (λ) of a wave is the distance between repetitions of the wave shape.
3. **Period:** the period of oscillation (T_P) of a wave is defined as the time any string element takes to move through one complete cycle or oscillation.
4. **Frequency:** the frequency (f) of a wave is defined as the inverse of the period of a wave, ($1/T_P$).
5. **Angular frequency:** $\omega = 2\pi f$ is called the angular frequency of the wave and its SI units are radians[1] per second. The relationship between period, frequency, and angular frequency is,

$$f = \frac{1}{T_P} = \frac{\omega}{2\pi}. \quad (7.3)$$

The maximum speed and maximum acceleration of a point on the string depend on the amplitude and the *angular frequency* of the wave.

Example 7.1. A transverse wave traveling along a string on the x-axis has the form:

$$\Psi(x, t) = y(x, t) = (3 \text{ m}) \sin (72(\text{rad m}^{-1})x + 2(\text{rad s}^{-1})t).$$

---

[1] A radian is an angle whose corresponding arc in a circle is equal to the radius of the circle. There are 2π radians in a full circle, hence 2π rad = 360°.

What are the frequency, angular frequency, wave number, period, and amplitude of this wave?

Solution:

We can find the values by comparing the given equation to equation (7.2).
1. The amplitude of the wave, $A = 3$ m.
2. The angular frequency of the wave, $\omega = 2$ rad s^{-1}.
3. The wave number, $k = 72$ rad m^{-1}.
4. The frequency of the wave, $f = \frac{\omega}{2\pi} = 0.32$ Hz.
5. The period of the wave $= T_P = \frac{1}{f} = 3.14$ s.

7.1.4 The speed of a transverse wave in a string

As we saw in section 7.1.1, the speed of a wave in a string depends on the properties of the string. For a string of length (L) and mass (m) held under a tension (T), the speed of the wave is given by equation (7.1):

$$v = \sqrt{\frac{TL}{m}}.$$

However, in reality, the length and the mass of the string are not independent, so we have to take into account the mass variation due to the length of the string, which is very difficult to do. So instead, writing the speed of the wave as

$$v = \sqrt{\frac{T}{\mu_l}}, \qquad (7.4)$$

where μ_l is a quantity called the **linear mass density** (mass per unit length), gives us certain advantages. If we write the speed of a wave as shown in equation (7.4), we have made the amount of medium an independent variable. We only care about the mass per unit length, not the total mass of the string. In other words, when trying to find the wave speed at a point x on a string, we only have to look at the immediate surroundings of point x and do not have to worry about the length of the entire string. Only the properties of the string around point x can determine how fast the wave travels as it passes through it.

According to equation (7.4), as the tension (T) increases, the wave speed also increases, and as the linear mass density increases, the wave speed decreases. In other words, a larger restoring force makes the wave go faster and more inertia makes the waves go slower. In addition, please keep in mind that the wave speed is not the same as the speed of the particle in the medium. For a transverse wave, the particle moves (up and down) along the y-axis, whereas the wave moves (forward) along the x-axis. The particle movement gives the amplitude (A), whereas the movement of the wave along the x-axis gives the speed (v) of the wave.

7.1.5 The energy and power of a wave traveling along a string

When a wave is set up in a stretched string, our hand provides energy for the motion of the string. As the wave moves away in the string, it transports that energy as both kinetic and elastic potential energies. We have previously discussed the concepts of kinetic and potential energies in general in section 5.1.5, but here we will limit our discussion to energies in a wave traveling along a string.

1. **Kinetic energy:** kinetic energy is associated with the motion of an object. It is mathematically given by

$$\text{KE} = \frac{1}{2}mv^2, \tag{7.5}$$

   where m is the mass and v is the velocity of the object. When a transverse wave passes through a string, the string moves up and down in response to the propagation of the wave. Consider an element of the string of mass (dm) and length (dx), as shown in figure 7.4. When the wave passes through the string, the element (dm) moves up and down. When element (dm) passes through position $y = 0$, its transverse velocity (and hence its kinetic energy) is at a maximum. As the element (dm) moves away from the equilibrium point, it slows down and its kinetic energy decreases. When the element (dm) is at its maximum amplitude, it briefly comes to rest before turning back; at this point, its kinetic energy is zero.

2. **Elastic potential energy:** in order to travel through a string, the wave must stretch the string. Consider a small length of the string (dx); this small length (dx) must increase and decrease periodically in response to the motion of the transverse wave. Elastic potential energy is hence associated with these changes in length, as shown in figure 7.4. At the maximum amplitude position, the length of the string element (dx) has its normal undisturbed value, so its elastic potential energy is zero. However, when the element is rushing through its equilibrium position, it reaches its maximum stretch and thus has its maximum elastic potential energy.

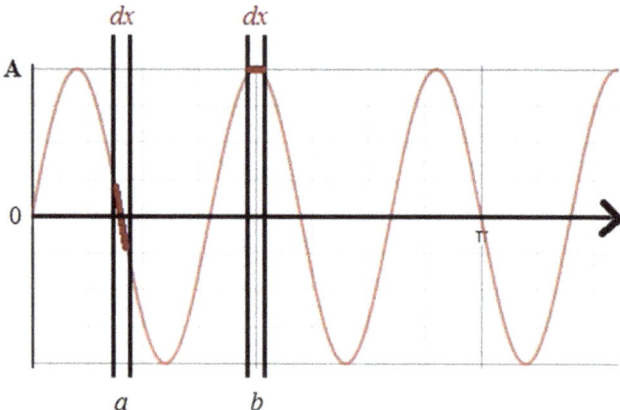

Figure 7.4. A wave traveling along a string. Element a is at an equilibrium point and element b is at maximum displacement. The kinetic energy of the string element at each position depends on the transverse velocity of the element. The potential energy depends on the amount by which the string element is stretched as the wave passes through it.

3. **Energy transport:** from the above discussion, we see that the string element (dm) has its maximum values of both kinetic and elastic potential energies when it passes through equilibrium ($y = 0$), whereas at maximum displacement (dm), it has zero energy. This means that as the wave travels along the string, the tension in the string continuously works to transfer energy between various regions of the string. Hence, the energy initially provided by our hand is transported through the string.
4. **Average power:** in an oscillating system, such as a mass–spring system (section 5.1.3), the average kinetic and the average potential energies are equal. The average power is the average rate at which energies of both kinds (kinetic and potential) are transmitted by the wave. The SI unit of power is the watt (W), and its mathematical expression is

$$P_{\text{avg}} = \frac{1}{2} A^2 v \mu_1 \omega^2 = \frac{1}{2} A^2 \sqrt{\mu_1 T} \omega^2, \qquad (7.6)$$

where μ_1 is the linear mass density, v is the speed, and T is the tension. The angular momentum (ω) and amplitude (A) depend on how the wave was initially generated, whereas the linear mass density and speed depend on the material the string is made from and the tension (T) it is held under.

Example 7.2. A transverse sinusoidal wave (with $f = 100$ Hz) travels along a string held under a tension of $T = 50$ N and has a linear density of $\mu_1 = 0.5$ kg m^{-1}. If the maximum amplitude is $A = 1$ cm, find the average rate of energy transport.

Solution:

The average rate of energy transport is the average power P_{avg}, as given by equation (7.6). In order to use this equation, we must first calculate the angular frequency (ω) and convert the amplitude to meters:

$$\omega = 2\pi f = 628 \text{ rad s}^{-1}.$$

Amplitude $= 1$ cm $= 0.01$ m.

$$P_{\text{avg}} = \frac{1}{2} A^2 \sqrt{\mu_1 T} \omega^2 = 98.5 \text{ W}.$$

7.2 Periodic waves

A wave in which the same pattern repeats at equal time intervals is called a *periodic wave*. The exact shape of the pattern is not important; however, a periodic wave repeats over and over in time, and each section of a periodic wave transports the energy that was originally used to generate it. A periodic wave can be easily observed in a taut string; if the struck plucked or shaken, it periodically oscillates with a small time delay. The generated wave travels down the string, and every point on the string oscillates with the same pattern in time. Musical sounds are mostly periodic, whereas noise is *aperiodic*. In human speech, vowels sung at a steady pitch (constant frequency) are mostly periodic, whereas most of the consonant sounds are aperiodic. Harmonic waves are special kinds of periodic waves in which the disturbance is sinusoidal. For a harmonic wave, every point on the string moves in a *simple harmonic motion* (SHM) with the same amplitude and frequency, even though all points may reach maximum displacements at different times.

Review questions 7.1.
1. Distinguish between transverse waves and longitudinal waves. Give examples.
2. Sketch a graph of a transverse wave as a function of position, identifying amplitude (A), wavelength (λ), and period (T) and places where the string element (dx) has a positive velocity, a negative velocity, and zero velocity. Also identify the points on the graph where its transverse speed is zero and where it is at a maximum.
3. What happens to the speed of a wave that passes through the string if the mass of the string is doubled?
4. What happens to the speed of a wave that passes through the string if the length of the string is doubled?

Exercises 7.1.
1. A sinusoidal wave has a frequency of 100 Hz and a wavelength of 2 m. Calculate:
   (a) The speed of the wave
   (b) The angular velocity
   (c) The angular wave number.
2. A sinusoidal wave that has a wavelength of 2 m travels along a string. The time taken for a particular point to move from its maximum displacement to the equilibrium point is 0.2 s. Find the following:
   (a) Period
   (b) Frequency
   (c) Wave speed.
Hint: the motion from maximum displacement to zero is one-quarter of a cycle.
3. What is the speed of a transverse wave that has a frequency of 2 Hz traveling in a rope of length 4 m and mass 0.060 kg held under a tension of 450 N?
4. For a transverse wave traveling along a string, the x-axis has the form:

$$\Psi(x, t) = y(x, t) = (5 \text{ m})\sin(4(\text{rad m}^{-1})x + 3(\text{rad s}^{-1})t).$$

What are the frequency, angular frequency, wave number, period, and amplitude of this wave?

7.3 The speed of sound waves

We know from chapter 1, section 1.7 that the speed of a particle is defined as

$$\text{speed} = \frac{\text{distance traveled}}{\text{time taken}} = \frac{\text{m}}{\text{s}}, \qquad (7.7)$$

and that the speed of a wave that has a wavelength of (λ), a period of (T), and a frequency of ($f = \frac{1}{T}$) can be rewritten as

$$v = \lambda f. \tag{7.8}$$

In air, sounds of all different wavelengths travel at the same speed. If this were not true, it would be very difficult to hear any performance with low and high frequency components at some distance away from the speaker. This is informally known as the '**equal-speed rule.**' It is worth noting here that not all types of waves obey the equal-speed rule. For example, surface water waves passing over deep water do not obey this rule; longer-wavelength waves from a distance arrive before shorter ones. This is because these waves propagate at a speed proportional to the square root of their wavelength, so at a distance from the source, the long wavelength arrives first.

Sound needs an elastic medium through which it can travel, and its speed is a property of the medium itself. However, its speed is independent of the amplitude and the frequency of the wave, which only depend on how the sound is generated. This applies to both periodic and non-periodic waves. Periodic sound waves are steady tones, and non-periodic sound waves are beeps or short clicks, as we saw in figure 7.5. For air at room temperature, the speed of sound is approximately 344 m s^{-1}. In general, the speed of sound is given by equation (7.8).

Sound does not travel in outer space because there are no molecules to transmit the sound energy. However, sounds can easily travel in Earth's atmosphere, since each cubic centimeter of air contains about 3×10^{19} air molecules. At sea level and at about 60 °F, the density of air is approximately 1.23 kg m^{-3}. However, the density of air does not remain the same throughout the atmosphere; in fact, the density of air decreases as you go higher in altitude, and at about 20 km above the Earth's surface (and at $-70°$ F), the density of air drops to about 0.1 kg m^{-3}. This means there are relatively fewer air molecules per centimeter of space than at sea level. This decrease in both air temperature and density is due to the decrease in atmospheric pressure, which is about 101 325 N m^{-2}, also called the pascal (Pa), which is the SI unit of pressure, at about room temperature.

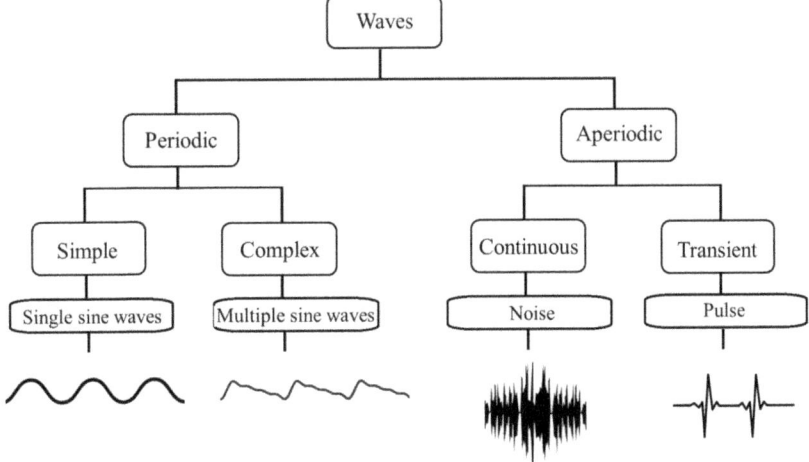

Figure 7.5. Types of waves.

7.3.1 The dependence of the speed of sound on temperature

The speed of sound changes with different local temperatures of the atmosphere around it. For instance, the speed of sound in air (v_o) at about sea-level at 0 °C is 331 m s^{-1}, and at room temperature ($T = 20$ °C), the speed of sound is about 344 m s^{-1}. Sound created by an aircraft flying at approximately 20 km above the Earth at an outside temperature of about −70 °F travels at approximately 295 meters per second or about 1062 km h^{-1}.

The simple explanation for the fact that the speed of sound depends on the air temperature is that sound, like heat, is a form of kinetic energy. Molecules at higher temperatures have more energy and therefore vibrate faster. Since the molecules vibrate faster, sound waves travel more quickly. Another equivalent explanation is that as the temperature increases, provided the pressure remains constant, the density of air decreases. As density of air decreases, molecules move closer to each other and the speed of sound increases. Conversely, as the temperature decreases, the density of air increases, the air molecules move away from each other, and the speed of sound decreases. This may seem a little counterintuitive, but think about a balloon filled with air. If you heat the air inside the balloon, the size of the balloon increases even though you have not added any more air molecules to the balloon. This size increase is due to a decrease in the air density (mass per volume); in other words, the same number of air molecules have taken up more space.

Over the range of temperatures we normally encounter, the speed of sound increases by about 0.606 m s^{-1} for each degree Celsius increase in temperature. Generalizing, the speed of sound in air can be found by the approximate formula

$$v = (331.5 + 0.606 T_C), \tag{7.9}$$

where T_C is the air temperature in Celsius[2]. The value 331.5 m s^{-1} used in this equation is the value of the speed of sound at $T = 0$ °C and pressure = 1 atm = 101 325 N m^{-2}. Please note that these temperature and pressure conditions are often used as references and are known as standard pressure and temperature (STP). This equation gives accurate speeds to better than 1% in the temperature range of [−66, 89] °C.

Example 7.3. What is the wavelength of a 1000 Hz tone at 20 °C?
Solution:
Given the frequency of 1000 Hz and knowing $v_{20°} = 344$ m s^{-1}, we can say from equation (7.8) that

$$\lambda = \frac{v_{20°}}{f} = \frac{344 \text{ m s}^{-1}}{1000 \text{ Hz}} = 0.344 \text{ m}.$$

---

[2] To convert temperature from Fahrenheit to Celsius, use the formula

$$T°C = \frac{5}{9}(T°F - 32).$$

Example 7.4. What is the wavelength of a 1000 Hz tone at 40 °C?
Solution:
Since the speed of sound changes with temperature, we now have to find the new speed of sound using equation (7.9):

$$v_{40°} = (331 + 0.606T_C) = 331 + 0.606(40) = 355 \text{ m s}^{-1}.$$

Given a frequency of 1000 Hz and knowing that $v_{40°} = 355$ m s^{-1}, we can say from equation (7.8) that

$$\lambda = \frac{v_{40°}}{f} = \frac{355 \text{ m s}^{-1}}{1000 \text{ Hz}} = 0.355 \text{ m}.$$

7.3.2 Other factors that influence the speed of sound

It is useful to recognize that the speed of sound does not only depend on temperature, but it also depends on the relative humidity of the air and the direction of the wind. The greater the relative air humidity, the greater the speed of sound. This is because moist air is less dense than dry air and because water vapor molecules have a lower molecular mass ($H_2O = 18$) than both nitrogen ($N_2 = 28$) and oxygen ($O_2 = 32$) molecules. However, the actual amount of water vapor in air is only a small fraction of the total air mass, even when the relative humidity reaches 100%. Therefore, changes in air humidity do not change the actual speed of sound in air by much. Similarly, the direction and the speed of the wind both have little influence on the speed of sound, so they can be practically ignored under normal conditions. However, even a slight wind can have a large effect on the range of sound propagation.

One property that does not affect the speed of sound is the frequency of the sound itself. All waves of various frequencies produced by the same sound source arrive at the listener at the same time. This is essential, otherwise low and high notes in a performance would reach us at different times, which would be disastrous for both the listener and the performer!

7.3.3 The speed of sound in solids

'Solid' refers to one of the three states of matter. Solids can hold their shape because their molecules are tightly packed together and have a definite shape that is not easy to change, as seen in figure 7.6. In solids, sound travels as longitudinal waves, in which the displacement of the molecules in the solid is parallel to the direction of propagation. To compress a solid, large forces are required, and the compressibility of a solid is directly proportional to the speed of sound in the solid. The speed of sound waves in a solid depends on the elastic properties of the solid. Sound travels faster in 'stiffer' materials.[3] In addition, sounds travel slower in denser materials, and for crystals, sound waves may even have different speeds in different directions because it may be easier to stretch or compress the bonds along one axis of the crystal than another.

---

[3] Stiffness defines the rigidity of a material and degree of difficulty in bending it.

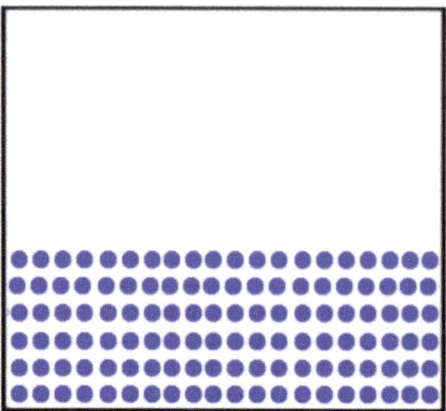

Figure 7.6. An example of the placement of molecules in a solid.

7.3.3.1 Waves traveling the length of a thin rod or bar

Rods are defined as materials whose diameter is shorter than a wavelength. The speed of sound waves traveling along the length of a thin rod is given by the formula

$$v = \sqrt{\frac{Y}{\rho}}, \tag{7.10}$$

where Y is the Young's modulus of the material, which is a measure of the stiffness of an elastic material; its SI units are N m^{-2}. Young's modulus is a quantity used to characterize and distinguish materials. Here, ρ is the density (mass/volume) of the material.

Suppose a force (called a tensile force) is applied to a solid rod or bar of a certain material that has a length of L_o. It is observed that the rod stretches by a small amount ΔL. The value of ΔL depends on the material itself; in other words, a softer material stretches more, and a stiffer material stretches less. The magnitude of the applied tensile force ($\vec{F}$) and the change in length ΔL are related as follows:

$$\frac{\vec{F}}{A} = Y\frac{\Delta L}{L_o}, \tag{7.11}$$

where A is the area of the bar (= length × width). The ratio F/A, which is the force applied per unit cross-sectional area is called *stress*, and the ratio $\frac{\Delta L}{L_0}$, which is the fractional change in length, is called *strain*. The units of Young's modulus (Y) are N m^{-2} but since $\frac{\Delta L}{L_0}$ is a ratio, it is dimensionless.

$$\frac{\vec{F}}{A} = Y\frac{\Delta L}{L_o} = \frac{N}{m^2} = \text{pressure} \tag{7.12}$$

Hence, equation (7.10). can be rewritten as:

$$v = \sqrt{\frac{\text{stiffness } (Y)}{\text{density } (\rho)}}. \qquad (7.13)$$

From equation (7.13), it is apparent that speed of sound is larger in stiffer materials. Since solids are stiffer than liquids and liquids are stiffer than gases, the speed of sound is highest in solids and lowest in gases. On the other hand, we see that the density (ρ = mass per unit volume) is in the denominator of equation (7.13). This means that as the density decreases, the speed of sound increases. Hence, when the medium becomes 'lighter,' the speed of sound increases. Aluminum has a relatively low density for a metal, which makes aluminum a good conductor of sound (as we saw for singing rods in section 6.6.3), whereas lead has a very high density, which means that the speed of sound is low in lead.

Example 7.5. A copper alloy has a Young's modulus of 1.1×10^{11} Pa and a density of 8.92×10^3 kg m^{-3}. What is the speed of sound in a thin rod made of this alloy?
Solution:
We can find the speed of sound using equation (7.10). Substituting the given values yields:

$$v = \sqrt{\frac{Y}{\rho}} = \sqrt{\frac{1.1 \times 10^{11}}{8.92 \times 10^3}} = 1.23 \times 10^7 \text{ m s}^{-1}.$$

To ensure that we are using the correct units, we know that pressure is force per unit area and can be expressed as,

$$\text{Pa} = \frac{\text{N}}{\text{m}^2} = \frac{\left(\text{kg } \frac{\text{m}}{\text{s}^2}\right)}{\text{m}^2} = \frac{\text{kg}}{\text{m s}^2}.$$

Then, using equation (7.10), we see that

$$v = \sqrt{\frac{\frac{\text{kg}}{\text{m s}^2}}{\frac{\text{kg}}{\text{m}^3}}} = \sqrt{\frac{\text{kg m}^3}{\text{m s}^2 \text{ kg}}} = \sqrt{\frac{\text{m}^2}{\text{s}^2}} = \frac{\text{m}}{\text{s}}.$$

Hence, the speed of sound is numerically equal to 1.23×10^7 m s^{-1}.

7.3.4 The speed of sound in liquids

Molecules in liquid, as shown in figure 7.7, are not fixed in a lattice structure like those in a solid, and the bonds between the molecules in a liquid continuously break and reform. Applying a little force increases the pressure on the molecules in a liquid, which causes the molecules in the region of applied force to be compressed. Under this compression, the liquid molecules move away from their initial spots.

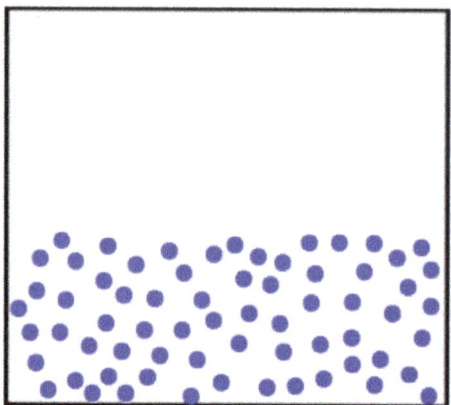

Figure 7.7. An example of the placement of molecules in a liquid.

In addition, inertia causes these molecules to go farther than they would have gone under just the effect of the applied force. The pressure in the region where the compression initially occurred now becomes lower than the average pressure in the liquid, which pulls the molecules back to their initial state. This process repeats until all the energy is carried away as sound waves. This is why multiple waves spread out from the place where a rock is dropped in a pond, as seen in figure 7.1.

In fresh water, sound travels at about 1497 m s^{-1} at 25 °C, whereas in salt water that is free of air bubbles or suspended sediment, sound travels at about 1560 m s^{-1}. The speed of sound in seawater depends on pressure (and hence on depth), temperature (a change in temperature of 1 °C changes the speed of sound by about 4 m s^{-1}), and salinity (a change in salinity of 1% changes the speed of sound by about 1 m s^{-1}). For longitudinal sound waves in a fluid, the change in pressure (ΔP) is proportional to the fractional volume change ($\Delta V/V$):

$$\Delta P \propto -\frac{\Delta V}{V}. \tag{7.14}$$

The negative sign in equation (7.14) indicates that a positive pressure results in a negative value of ΔV; it also indicates that larger pressures result in larger volume changes. The constant of proportionality for this equation is called the '**bulk modulus**' (B). The bulk modulus measures a material's resistance to uniform compression; its SI unit is the pascal (Pa). The bulk modulus is mathematically defined as 'the ratio of infinitesimal pressure increase to the resulting *relative decrease* of the volume of the material.' We can then rewrite equation (7.14) for a fluid as follows:

$$\Delta P = -B \frac{\Delta V}{V_o}. \tag{7.15}$$

The speed of sound in a fluid is mathematically written as

$$v = \sqrt{\frac{\text{bulk modulus } (B)}{\text{density } (\rho)}}. \tag{7.16}$$

Note that a larger restoring force makes the waves faster, and conversely, larger inertia results in slower waves. This means that in media with a larger bulk modulus (i.e. media that are harder to compress), sound waves travel faster, whereas in media that have greater density (ρ = mass/volume), the wave speed is smaller.

7.3.5 The speed of sound in gases

The speed of sound in a gas depends on the composition of the gas and the temperature at which it is held. Gases are less dense than liquids, hence they are more compressible. Very little pressure is required to squeeze a gas into a smaller volume (figure 7.8).

From the above discussion, we know that sound travels slower in materials that can be easily compressed and it travels faster in less dense materials. The net effect of these two factors is that sound travels slower in a gas than it does in a liquid of the same substance because the change in compressibility generally has a much bigger effect than the change in density. Mathematically, the speed of sound in a gas has the same formula as that in a fluid and is given by

$$v = \sqrt{\frac{\text{bulk modulus } (B)}{\text{density } (\rho)}} \tag{7.17}$$

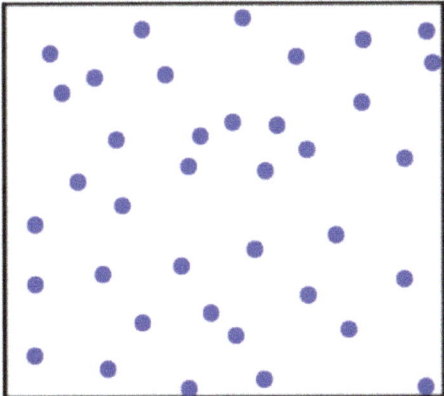

Figure 7.8. An example of the placement of molecules in a gas.

7.3.6 The speed of sound in an ideal gas

An ideal gas[4] is an approximation scientists make that allows them to deal with gases more easily. An ideal gas is a theoretical gas composed of a set of randomly moving, noninteracting point particles. Under normal conditions of temperature and pressure, most real gases, such as hydrogen, helium, etc. behave qualitatively like an ideal gas. The speed of sound in an ideal gas is given by the formula:

$$v = \sqrt{\frac{\gamma RT}{M}}, \quad (7.18)$$

where T is the absolute temperature in Kelvin (K)[5], M is the molecular weight of the gas, also called molar mass of the gas, and γ and R are constants for the gas: γ is the adiabatic index and R is the universal gas constant (=8.31 J mole^{-1}. K).

Example 7.6. The speed of sound in air: we can find the speed of sound in air if we know the values of the constants; the molecular weight of air = 0.0288 and γ = 1.4 for air.

In addition, $T_{\text{freezing point}}$ is given by $(t + 273) = 0\ °C + 273 = 273$ K and T_{roomtemp} is $(t + 273) = 21\ °C + 273 = 294$ K. Substituting these values into the above equation gives:

1. The speed of sound at 0 °C = 331 m s^{-1}.
2. The speed of sound at 21 °C = 344 m s^{-1}.

7.3.7 The speeds of sound in various media

The speed of sound varies from substance to substance and is generally highest in solids and liquids. It travels about 4.3 times faster in water (1484 m s^{-1}) and nearly 15 times faster in iron (5120 m s^{-1}) than in air at 20 °C (344 m s^{-1}). The speeds of sound in some solids, liquids, and gases are provided in table 7.1.

Example 7.7. Helium voice: have you ever inhaled some helium and then tried to speak? What do you hear? Have you ever wondered why?
  Solution:
  The answer is very simple: it is the result of the higher speed of sound in helium than in air. The speed of sound in helium is 972 m s^{-1}, whereas the speed of sound in air is 343 m s^{-1}. This means that the speed of sound in helium is about 2.8 times the speed of sound in air. Please note: the pitch of the voice does not change.

---

[4] By 'ideal gas,' we mean a collection of particles for which: 1. when the temperature, T, and number of particles, N, are fixed, the energy of the gas does not depend on the volume, V; 2. at fixed temperature, the number of states available to each particle is proportional to the volume (V).
[5] To convert temperature to absolute temperature (Kelvin), use the formula:

$$T_K = T°_C + 273.15.$$

Table 7.1. Speeds of sound in selected solids, liquids, and gases.

Solid	Young's modulus (GPa)	Density (kg m^{-3})	Speed (m s^{-1})
Aluminum	71	2700	5128
Granite	70	2600	6000
Steel	200	7700	5941
Wood (hard)	11	720	4267
Glass (Pyrex)	62	2300	5190
Lead	16	11 300	1219
Bone	19	1900	3160
Soft tissue (muscle)	2.5	1050	1545
Soft tissue (fat)	2	940	1460
Liquid	Bulk modulus (GPa)	Density (kg m^{-3})	Speed (m s^{-1})
Fresh water (20 °C)	2.2	1000	1485
Seawater (20 °C)	2.3	1026	1500
Mercury (20 °C)	28	13 600	1435
Gas	Bulk modulus (GPa)	Density (kg m^{-3})	Speed (m s^{-1})
Helium (20 °C)	0.000 19	0.18	1025
Air (20 °C)	0.000 142	1.21	344
Air (0 °C)	0.000 142	1.29	331

7.4 Sound absorption

A sound wave moving through a medium loses energy over time for two main reasons, namely, internal absorption and external absorption:

1. **Internal absorption:** internal absorption denotes the sound absorbed by a medium due to internal friction that exists within the medium. Internal friction transforms sound energy to heat energy as the sound passes through the medium. Internal absorption is the main reason for the decrease in sound by absorption in open spaces and large areas.
2. **External absorption:** external absorption denotes the sound that is absorbed at space boundaries as sound passes from one medium to another. External absorption is the dominant reason for the decrease in sound by sound absorption in covered spaces such as rooms, auditoriums, theaters, etc.

The ability of any medium and its boundaries to absorb sound is defined using a quantity called the *absorption coefficient*. The absorption coefficient of the medium is the ratio of the internal friction coefficient of the medium to the effective mass of the medium. The total sound absorbed by a medium during propagation depends on the frequency of the sound, the distance it travels, and the physical properties of the medium. The absorption of sound by the air depends heavily on the number of dust particles and the humidity present in the air. The average coefficient of absorption for air is twice the average coefficient of absorption of water; therefore, a high-intensity wave with a frequency of about 100 Hz is completely absorbed by air in less than 500 km, whereas the same wave may travel almost halfway around the world in water.

Review questions 7.2.
1. What is the speed of sound in air at 0 °C and at room temperature?
2. How does the speed of sound change with temperature?
3. How does the speed of sound change with atmospheric pressure? Compare the speed of sound at 1 atm and at 2 atm.
4. What does the term 'absolute temperature' mean?
5. Compare the speed of sound in two otherwise identical steel rods with the same mass, one has a diameter of 1 cm and the other has a diameter of 2 cm. Hint: the volume of a rod is ($\pi r^2 L$), where r is the radius and L is the length of the rod.
6. Compare the speeds of sound at room temperature in air, water, wood, and aluminum.

Exercises 7.2.
1. During a thunderstorm, you see a flash of lightning and then you count ten seconds before you hear thunder. Knowing the speed of sound, can you estimate how far you are from the storm in miles?
2. What are the speeds of sound in miles per hour at 0 °C, at room temperature (21°C), and at 45 °C.
3. If a supersonic plane is flying at the speed of sound, it is said to be traveling at Mach 1. Calculate the speeds if the supersonic plane is flying at Mach 1.5 and at Mach 2:
   (a) at 10 °C.
   (b) at room temperature.
4. What is the speed of a longitudinal wave in a brass bar, given that the density of brass is 8.4 g per cubic centimeter and the value of Young's modulus for brass is 1.10×10^{11} N m^{-2}?
5. If a brass bar is held under a tension of 2500 N, for what value of mass per unit length does the transverse wave speed equal the longitudinal wave speed of the brass wire given in Q4?
6. A lightning flash is seen in the sky and 8.2 seconds later the boom of thunder is heard. The temperature of the air is 12.0 °C.
   (a) What is the speed of sound in air at that temperature?
   (b) How far away is the lightning strike?
   (c) The speed of light is 3.0×10^5 km s^{-1}. How long does it take the light signal to reach the observer?

Further reading
- Rossing T D, Moore R F and Wheeler P A 2002 *The Science of Sound* 3rd edn (Upper Saddle River, NJ: Pearson Education)

- Rossing T D and Fletcher N H 2004 *Principles of Vibration and Sound* (Berlin: Springer)
- Fletcher N H and Rossing T D 1998 *The Physics of Musical Instruments* (Berlin: Springer)
- Berg R E and Stork D G 1994 *The Physics of Sound* 2nd edn (Englewood Cliffs, NJ: Prentice-Hall)
- Emanuel D C and Letowski T 2009 *Hearing Science* (Philadelphia, PA: Wolters Kluwer)
- French A P 1971 *Vibrations and Waves* (Boca Raton, FL: Chapman and Hall)

IOP Publishing

The Physics of Sound and Music, Volume 1
A complete course text (Textbook)
Samya Bano Zain

Chapter 8

Factors that impact sound propagation

In this chapter, we discuss some of the principles that govern wave behavior, starting with a phenomenon that might be considered fundamental to the way in which waves propagate, called Huygens' principle.

8.1 Huygen's principle

Huygens' principle, named after the Dutch physicist Christian Huygens (1629–95) states that every point on a propagating wave front at any instant (t) serves as a source of spherical secondary wavelets at a later time ($t + \Delta t$), as we see in figure 8.1. If the propagating wave has a speed of (v) and a frequency of (f), then the secondary wavelets have the same frequency (f) and the same speed (v). This seemingly simple principle has quite an impact, because it is used to derive both the law of reflection (see section 5.3) and the law of refraction, called Snell's law (discussed in section 8.2).

8.2 The refraction of waves

When an incident wave reaches a boundary and crosses over to the other side, it is called a *transmitted* ray, and the associated process is called 'transmission.' When an incident wave that is transmitted bends due to the change in its speed when passing between two different media, it is called a *refracted* wave. This process of bending of the incident wave is called 'refraction.'

It is commonly observed that a beam of light that is incident at an angle other than 90 degrees to the surface is bent from its original direction. This change in the angle of the beam's direction is called refraction. In addition, when a beam of light is transmitted from a less dense medium (say air) to a denser medium (say water), the light beam bends towards the perpendicular (called the normal) to the boundary between the two media, as shown in figure 8.2. When the same beam of light is transmitted from water into air, the beam bends away from the normal. The amount of bending (refraction) is given by Snell's law, mathematically written as

$$n_1 \sin(\theta_1) = n_2 \sin(\theta_2), \tag{8.1}$$

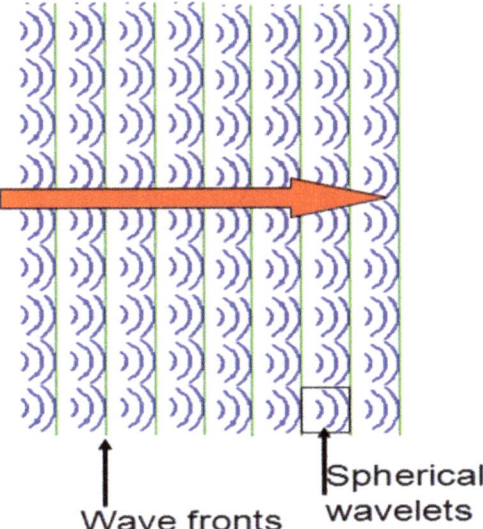

Figure 8.1. Wave fronts and wavelets.

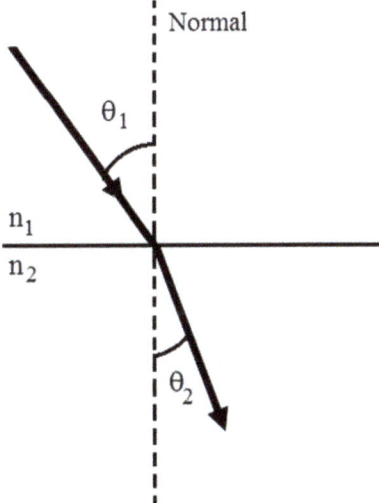

Figure 8.2. The refraction of a light wave at the boundary between air and water.

where n_1 is the refractive index of medium 1, n_2 is the refractive index of medium 2, θ_1 is the angle of incidence, and θ_2 is the angle of refraction. Both the angle of incidence and the angle of refraction are measured with respect to the normal to the surface, as shown in figure 8.2. If the light beam enters one side of a glass surface, say a glass slab or prism, as shown in figure 8.3, it is called an incident ray, and when it emerges from the

Figure 8.3. The refraction of a light wave at the boundary between air and glass.

other side of the glass after passing through the glass or prism, it is called the emergent ray. In refraction through a glass slab, the emergent ray is parallel to the incident ray. Please note that all the rays involved in this process, that is, the incident ray, the refracted ray, and the emergent ray, are measured with respect to the normal to the surfaces.

Example 8.1. A light ray travels from air (with $n_a = 1.0$) into water (with $n_w = 1.33$). The angle of incidence is 30°. What is the angle of refraction?
Solution:
Using equation (8.1),
$$n_a \sin \theta_i = n_w \sin \theta_r.$$
Substituting in the given values yields:
$$(1)\sin(30°) = (1.33)\sin \theta_r$$
$$\frac{1}{1.33} \sin(30°) = \sin \theta_r$$
$$0.374 = \sin \theta_r.$$
$$\sin^{-1}(0.375) = \theta_r$$
Solving the above expression, we find the angle of refraction, $\theta_r = 22°$.

8.2.1 The properties of a refracted wave

Some important properties of a refracted wave are given in this section:

1. The direction of a refracted ray is different from that of the incident ray.
2. The speed and wavelength of a refracted ray are different from the speed and wavelength of the incident ray.
3. The frequency of a refracted ray is the same as the frequency of the incident ray. If this were not true, waves would not preserve their colors.

8.2.2 Refraction in surface water waves in the ocean

In water, waves travel faster in deeper water than in shallower water. Water waves in the ocean are formed far away from the shore in deeper waters but they move quite slowly as they reach the shore; in fact, by the time the waves reach the shore, they are almost head-on to the shore. This happens because of the gradual change in the slope of the ocean bed as it approaches the shore. The gradual slope allows the waves to slow down and hence refract (bend) in such a way that water waves are almost parallel to the shoreline when they get there.

8.2.3 The refraction of sound under various conditions

The speed of a wave depends on the elastic and inertial properties of the medium. When a wave moves from one medium to another, the speed of the wave changes, which implies that the direction of the wave also changes. Hence, we observe the refraction or bending of the wave. Sudden refraction is frequently observed in the study of light rays, say, when a light ray encounters a glass or water surface and transmission occurs. Sound waves, on the other hand, mostly travel through air, and the properties of air luckily do not change quickly in the Earth's atmosphere. This means that the speed of sound waves and hence the direction of sound waves do not change quickly. In fact, the speed of sound, and hence the refraction of sound, mainly depends on changes in the local air temperature, as given by equation (7.9) ($v = 331.5 + 0.606 T_C$), where T_C is the air temperature in Celsius.

Let us briefly discuss the refraction of sound waves in the Earth's atmosphere under commonly encountered conditions. Please keep in mind that sound waves propagate faster in warm air than in cooler air as we go through this discussion.

1. **The refraction of sound waves during the daytime:** during a sunny and hot day, the Sun heats up the Earth's surface and it starts to radiate. The air just above the ground gets hot, whereas the air at higher elevations remains relatively cooler. The heated air just above the surface rises but as it rises it cools slowly, which causes a temperature gradient to be formed in which the atmospheric temperature decreases with elevation. Such a gradient is called the *adiabatic lapse rate*. We know that sound waves propagate faster in warm air than in cold air; hence, sound waves closer to the ground travel faster than sound waves in the adjacent higher layers and under these conditions sound refracts upwards and becomes 'lost.'

2. **The refraction of sound waves during the nighttime:** during the night, the opposite effect occurs: hot air rises and cold air moves to the surface of the Earth. This results in what is called an *atmospheric temperature inversion*, in which the temperature rises with elevation above the ground. Since the speed of sound is faster in warmer air, sound emitted by a source under the condition of atmospheric temperature inversion bends towards the Earth.

3. **The refraction of sound waves during storms:** the refraction of sound is also the reason why lightning is seen but the thunder not heard beyond about 14 miles from lightning originating at about 4 km in the atmosphere. When the storm starts, the cooler air is at higher elevations, while the warmer air is near the surface of Earth, which means that the speed of sound is greater

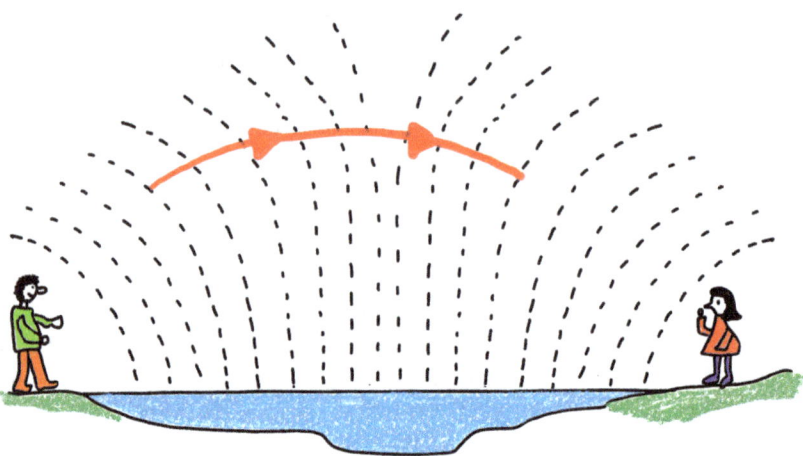

Figure 8.4. The refraction of a sound wave over a body of water.

near the ground. Hence, the sound of thunder traveling to the ground is refracted back up into the atmosphere. This creates a region called a *shadow region* in which the lightning can be seen but the thunder cannot be heard.

4. **The refraction of sound over water:** the air right above the water surface tends to be cooler than the air higher above the water surface. Since sound waves travel slower in cooler air than in warmer air, the part of the wave directly above the water travels slower than sound waves above it. This phenomenon results in a change in the direction of the sound wave, and the sound wave refracts downwards towards the water, as seen in figure 8.4. Hence, in order to get your message across a river, it might help to shout upwards.

5. **The refraction of sound into or against the wind:** the propagation of sound is directly dependent on the medium through which it moves. In the presence of a steady wind, air in the atmosphere tends to drag against the ground. This causes the wind speed to be slower near the ground, which means that the speed of sound is also slower near the ground. When the sound propagates in the same direction as the direction of the wind, the speed of sound also increases with height. So, when sounds are produced *with* the wind, they are refracted towards the ground and hence *carried on* the wind. On the other hand, if sounds are produced *against* the wind, the speed of sound decreases with elevation, the sounds are refracted upward (away from the ground), and the sound is *lost*.

6. **The refraction of sound under the ocean:** sound travels faster in solids than in gases; hence, it travels faster in water (approximately 1500 m s^{-1}) than in air (343 m s^{-1}). Under normal circumstances, the speed of sound changes with the temperature, salinity, and pressure of water. Among these, changes in temperature play the most important part in determining the speed of sound under water. Sound waves travels faster in warmer water than in cold water,

and we know that the temperature of the ocean decreases with depth. This means that sound waves slow down as we go deeper under water, which results in the downward refraction of sound waves that originate under water. For submarines submerged just under the surface of the water, this refraction creates regions called *shadow regions*. These shadow regions limit submerged submarines' ability to locate ships or other submarines at a distance. This refraction, on the other hand, is also believed to enhance the propagation of the sounds of marine mammals such as dolphins and whales, allowing them to communicate over large distances.

8.3 Diffraction

Diffraction is the bending of incoming waves around corners and the spreading of an incoming wave around an obstacle located in its path. The amount by which a wave diffracts depends on two things: the wavelength of the wave itself and the size of the obstacle. As a sound wave front propagates forward, it produces point sources which emit spherical waves, as we saw in figure 8.1. As long as all these point sources start at the same time and are allowed to emit waves without interruption, the wave front moves in the direction of propagation in a straight path. However, if we insert an obstruction into its path, some of the waves are blocked and cannot contribute further to the wave front. The result is that the wave front becomes curved, as seen for water waves in figure 8.5 (left). A similar phenomenon is observed when a wave encounters a barrier with a slit, as seen in figure 8.5 (right).

In terms of wave fronts, we can generalize a few scenarios for waves when they encounter obstacles.

1. **Case 1: The obstruction is much larger than the wavelength of the incoming wave.** If the obstruction or opening is much larger than the wavelength of the incoming wave, the part of the wave front that naturally encounters the opening propagates forward and the shape of the front remains unbent and straight, as seen in figure 8.6.

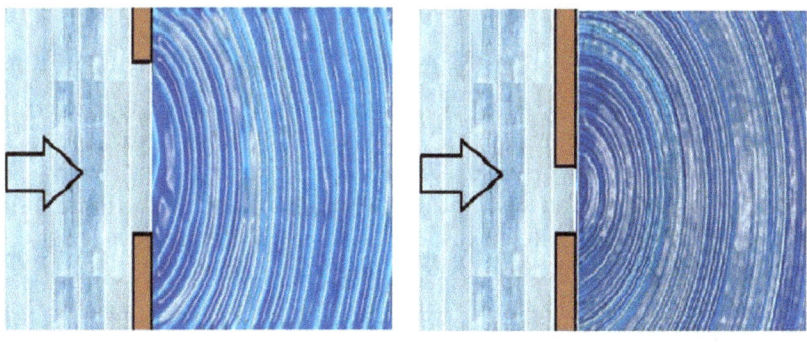

Diffraction through large opening   Diffraction through small opening

Figure 8.5. Diffraction in water waves: (left) diffraction through a large opening; (right) diffraction through a small opening.

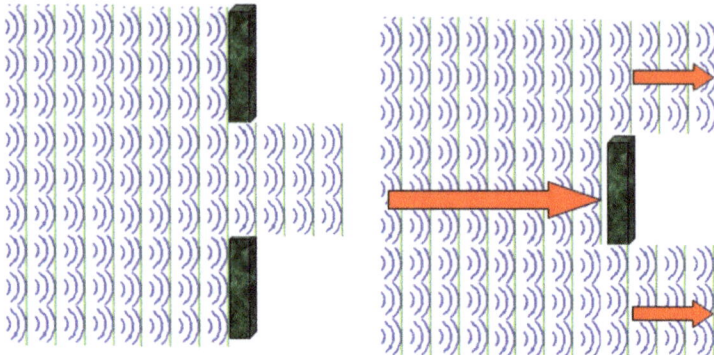

Figure 8.6. An obstruction that is much larger than the wavelength of the incoming wave.

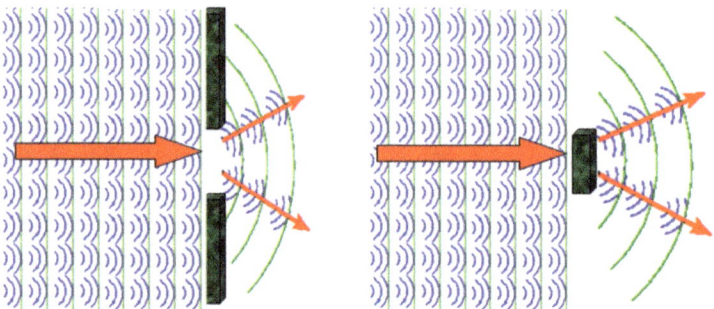

Figure 8.7. An obstruction that is comparable to the wavelength of the incoming wave.

2. **Case 2: The obstruction is comparable to the wavelength of the incoming wave.**
   If the size of the opening is comparable to the wavelength of the incoming wave, the diffracted waves 'bend through' or 'around' the opening or obstruction, as seen in figure 8.7.

8.3.1 Hearing around corners

Suppose you want to know who is in a group around the corner walking towards you that you cannot see yet. How do you find out without actually looking? This is where diffraction comes to the rescue. Adult men generally have low-frequency (long-wavelength) voices, whereas adult women have higher-frequency (shorter-wavelength) voices. High-frequency sounds tend to be more directional and hence diffract less, whereas low-frequency sounds are easily diffracted around corners. Hence, you should be able to pick out the male voices, which are easily diffracted, from around the corner much quicker than their female counterparts. This phenomenon is shown in figure 8.8.

Many animals use diffraction and utilize the fact that long-wavelength (low-pitched) sounds carry further than short-wavelength (high-pitched) sounds. A low-pitched (long-wavelength) adult lion roar can be heard up to five or six miles

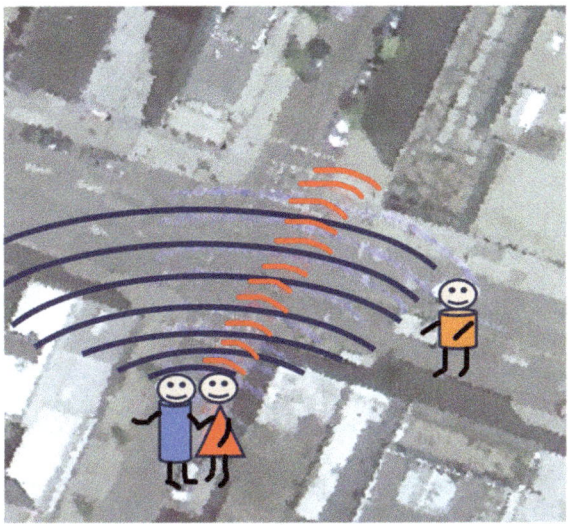

Figure 8.8. Hearing around corners. Male voices are represented by blue curves and female voices are represented by red curves.

away. Similarly, long-wavelength hoots made by owls can diffract around obstacles such as trees, which allows owls to communicate over long distances in the forest. On the other hand, songbirds' musical tweets are very localized because they generally have high-pitched (short-wavelength) tweets.

8.3.2 The diffraction equation

The relationship used to find the first minimum in sound intensity for a slit with uniform width (W) is

$$\sin\theta = \frac{\lambda}{W}, \tag{8.2}$$

where W is the width of the opening, λ is the wavelength of the sound wave, and θ is the diffraction angle. It is important to note that the diffraction equation used to determine the minimum for a circular opening is not the same and is, in fact, given mathematically as follows:

$$\sin\theta = 1.22\left(\frac{\lambda}{d}\right), \tag{8.3}$$

where d is the diameter of the hole, λ is the wavelength of the sound wave, and θ is the diffraction angle.

Example 8.2. A 1000 Hz sound and a 8000 Hz sound both emerge from a loudspeaker through a circular opening whose diameter is 45 cm. Assuming that the speed of sound in air is 343 m s^{-1}, find the diffraction angle (θ) for each sound.

Solution:
We know that $v = \lambda f$; from this, we can find the wavelengths of both incoming sounds:

$$\lambda_1 = \frac{344 \text{ m s}^{-1}}{1000 \text{ Hz}} = 0.344 \text{ m}$$

$$\lambda_2 = \frac{344 \text{ m s}^{-1}}{8000 \text{ Hz}} = 0.043 \text{ m}.$$

Using the circular-opening diffraction equation (8.3), we can now find the diffraction angles:

$$\sin \theta = 1.22 \left(\frac{\lambda}{d} \right).$$

For the 1500 Hz sound:

$$\sin \theta_1 = 1.22 \left(\frac{0.344 \text{ m}}{0.45 \text{ m}} \right) = 0.932$$

$$\theta_1 = \sin^{-1}(0.932) = 68.8°.$$

For the 8000 Hz sound:

$$\sin \theta_2 = 1.22 \left(\frac{0.043 \text{ m}}{0.45 \text{ m}} \right) = 0.116$$

$$\theta_2 = \sin^{-1}(0.116) = 6.7°.$$

It is clear that the θ values for the two frequencies are very different. The difference between the locations where these two sounds can be heard is illustrated in figure 8.9 and is the reason for having woofers and tweeters. The tweeters of a loudspeaker are specially designed to be shaped like a fan to obtain good dispersion of the high frequencies, since this does not occur naturally through diffraction.

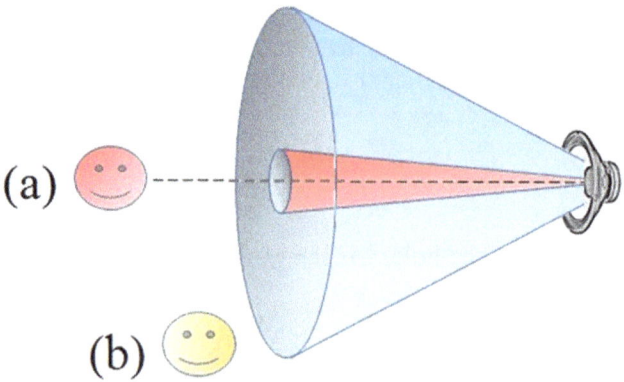

Figure 8.9. An example of diffraction: a person at point (a) hears both high and low frequencies, whereas a person at point (b) primarily hears only the low frequencies.

8.3.3 Acoustic diffraction in amphitheaters

Scientists and archaeologists have been equally baffled by how good the acoustics are in ancient Greek amphitheaters. Ancient Greek amphitheaters were large, semicircular, open-air venues constructed in hills and used primarily for performances such as plays and Greek dramas. The phenomena that made the acoustics of the ancient amphitheaters so remarkable were the chosen locations, the geometry of their design, and the materials used in their construction.

Scientists and archeologists have a number of theories about how the voices of actors and singers traveled so well without external amplification or noise reduction. What they found is that this is primarily due to a principle called *acoustic diffraction*. Acoustic diffraction occurs when a sound wave moves around an object whose dimensions are about equal or smaller than the wavelength of the sound. High-frequency sounds have short wavelengths and do not diffract around most obstacles, whereas low-frequency sounds that have longer wavelengths diffract relatively easily. High-frequency sounds are absorbed or reflected by obstacles, whereas low-frequency sounds just bend around the obstacle. Hence, because high-frequency sounds are absorbed or reflected by obstacles, they create *sound shadows* (discussed in section 11.1.1) behind the object. In the region of a sound shadow, higher-frequency sounds are attenuated. Please note that the acoustic diffraction in amphitheaters does not actually make the voices of the actors performing on stage louder; it just allows the voices to travel farther by damping the other noises present in the environment, such as spectators talking or birds chirping.

Review questions 8.1.
1. What is refraction?
2. Explain the difference between diffraction and refraction.
3. What is diffraction? How does acoustic diffraction help us in designing large performance stages?
4. How do you hear around corners? Why are adult men easier to hear around corners than their female counterparts?

Exercises 8.1.
1. What is the angle of incidence for a light ray traveling from water into flint glass ($n = 1.70$):
   (a) If the angle of refraction is 30°?
   (b) If the angle of refraction is 20°?
2. A speaker has a diameter of 25 cm.
   (a) Assuming that the speed of sound in air is 343 m s^{-1}, find the diffraction angle for a 2.0 kHz tone.
   (b) What speaker diameter should be used to generate a 6.0 kHz tone whose diffraction angle is as wide as that of the 2.0 kHz tone?

8.4 The Doppler effect

The Doppler Effect, named after the Austrian physicist Christian Doppler who proposed it in 1842 CE, is the effect produced by a moving wave source. The movement of the source produces a shift in the apparent frequency of the produced waves. This change in frequency is dependent on whether the source is moving towards or away from the observer. For a source such as an ambulance with its sirens turned on that is moving towards an observer, the received frequency is:
1. Higher compared to the actual frequency emitted from the source as the source approaches.
2. Identical to the frequency of the source at the instant it passes the observer.
3. Lower as the source recedes from the observer.

This is called the *Doppler effect* and is observed in all types of waves, for example, water waves, sound waves, and light waves. There is always an apparent upward shift in frequency when the observer and the source approach each other and an apparent downward shift in frequency when observer and source travel away from each other. The Doppler effect observed for a sound source and an observer is shown in figure 8.10.

8.4.1 The reason for the Doppler effect

The source of sound always emits sounds at the same frequency; however, the Doppler effect is observed when the source and/or the observer move with respect to each other. The reason is simple, and it relates to the equation for the speed of a wave:

$$\text{speed} = \frac{\text{distance}}{\text{time}} = \lambda f.$$

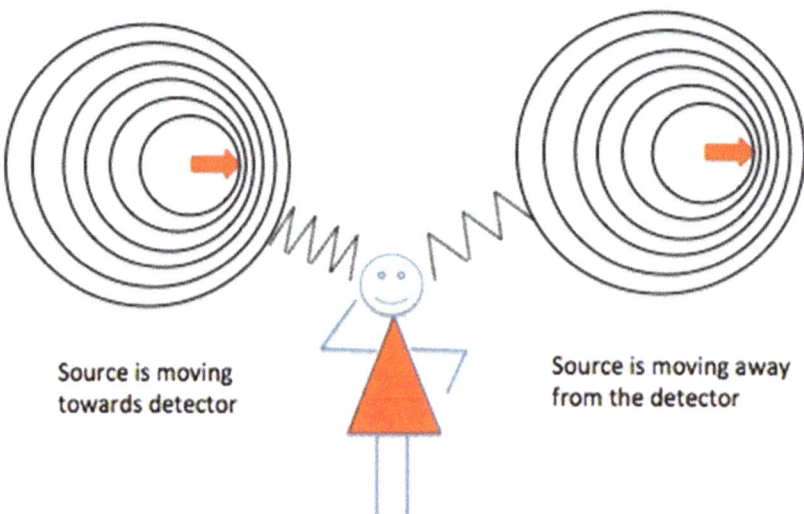

Figure 8.10. The Doppler effect: the apparent change in frequency of a wave in relation to an observer moving relative to the wave source.

Since the speed of the wave is constant but the source and/or the observer move with respect to each other, the distance constantly changes. This means that the perceived frequency must change accordingly in order to keep the speed the same. If the source and the observer move towards each other, the distance between them decreases but for the same period of time, the same number of waves must fit between the source and the observer; thus, the waves are compressed into smaller distance intervals. For these reasons, taking the speed and the frequency of the noise maker as (v_{noise}) and (f_{noise}), respectively, the speed and frequency of the observer/listener as (v_L) and (f_L), respectively, and the speed of sound as (v_{sound}), we can write:

1. **Case 1: the source and listener are both at rest.** When the source is at rest, the emitted wave fronts travel outwards from the source as shown in figure 8.11. The circular lines in the figure represent compressional wave fronts that travel away from the source in three dimensions. If the source and listener are both at rest, the speed of the listener relative to the source is zero; $v_L = v_{noise} = 0$; the listener perceives the sound waves at a frequency equal to the source frequency (f_{noise}), i.e.

$$f_L = f_{noise} \tag{8.4}$$

2. **Case 2: a moving source:** suppose the source starts to move and speeds up. As the source accelerates, we observe that the wave fronts ahead of the source start to bunch up, whereas those behind the source spread out. As the source speeds up, this bunching effect is enhanced, as can be seen in figure 8.12 between part (a) and part (b).

    A moving source leads to two possibilities: (i) when the source moves towards a stationary listener and (ii) when the source moves away from a

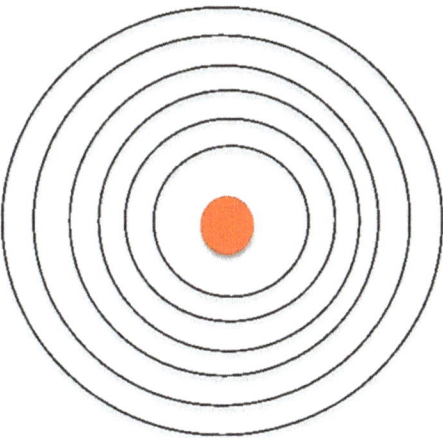

Figure 8.11. A sound source emitting while at rest.

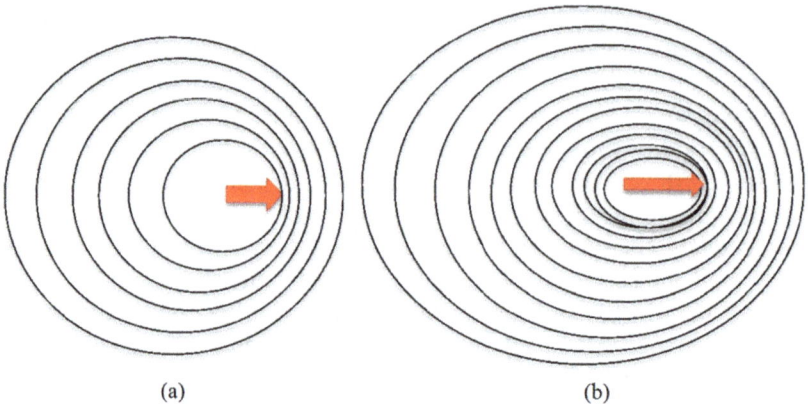

Figure 8.12. The source starts at rest and accelerates and speeds up from part (a) to part (b).

stationary listener. For simplicity's sake, I will only discuss cases here for which the source moves at a constant velocity.

(a) **Case 2a: the source moves towards a stationary listener.** When the source moves towards a stationary listener, the listener perceives sound waves reaching them at a faster rate (i.e. at a higher frequency/higher pitch) and hence f_L is greater than f_{noise} ($f_L > f_{noise}$). Mathematically, we can write the frequency perceived by the listener (f_L) in this case as

$$f_L = f_{noise}\left(\frac{v_{sound}}{v_{sound} - v_{noise}}\right), \qquad (8.5)$$

where v_{noise} is the speed of the source, f_{noise} is the frequency emitted by the source, and v_{sound} is the speed of sound in air.

(b) **Case 2b: the source moves away from a stationary listener.** When the source moves away from a stationary listener, the listener perceives sound waves reaching them at a lower rate (low frequency/low pitch), that is, f_L is less than f_{noise} ($f_L < f_{noise}$):

$$f_L = f_{noise}\left(\frac{v_{sound}}{v_{sound} + v_{noise}}\right). \qquad (8.6)$$

3. **Case 3: a moving listener.** When the listener moves instead of the source, it again leads to two different situations: (i) when the listener moves towards a stationary source and (ii) when the listener moves away from a stationary source:

(a) **Case 3a: the listener moves towards a stationary source.** Suppose a sound source radiates sound at a constant frequency (f_{sound}) and is at rest. A listener moves towards it with speed ($v_{listener}$). The wave fronts are produced at the same frequency (f_{sound}). However, since the listener is moving towards the source, the listener counts more waves as they *meet* the source. Therefore, the listener perceives the frequency reaching them (f_L) as

$$f_L = f_{noise}\left(\frac{v_{sound} + v_{listener}}{v_{sound}}\right), \quad (8.7)$$

where $v_{listener}$ is the speed of the observer, f_{noise} is the frequency emitted by the source, and v_{sound} is the speed of sound in air.

(b) **Case 3b: the listener moves away from a stationary source.** If the listener moves away from a stationary source, they count fewer waves as they *leave* the source, so the listener perceives the frequency of sound waves reaching them (f_L) as

$$f_L = f_{noise}\left(\frac{v_{sound} - v_{listener}}{v_{sound}}\right). \quad (8.8)$$

4. **Case 4: both source and listener move simultaneously.** When both source and listener move simultaneously, we have to combine equations (8.5), (8.6), (8.7), (8.8) together to determine the perceived sound frequency as follows:
   (a) When source and listener move towards each other:

   $$f_L = f_{noise}\left(\frac{v_{sound} + v_{listener}}{v_{sound} - v_{noise}}\right). \quad (8.9)$$

   (b) When source and listener move away from each other:

   $$f_L = f_{noise}\left(\frac{v_{sound} - v_{listener}}{v_{sound} + v_{noise}}\right). \quad (8.10)$$

Example 8.3. An ambulance producing a 3000 Hz tone moves toward you while you are stopped at a red light. Suppose the velocity of the ambulance is 50 m s^{-1}:
1. What sound frequency do you hear as the ambulance approaches you? What is the actual shift in frequency that you hear?
   Solution:
   The sound source (ambulance) is moving toward you (the listener), hence we can use equation (8.5):

   $$f_L = f_{noise}\left(\frac{v_{sound}}{v_{sound} - v_{noise}}\right) = 3000\left(\frac{344}{344 - 50}\right) = 3510 \text{ Hz}.$$

   The resulting shift in frequency is (3510–3000) = 510 Hz.

2. The ambulance now passes you and heads down the road at 50 m s^{-1}. What frequency of sound do you hear now? What is the frequency shift now?
   Solution:
   The sound source (ambulance) is moving away from you (the listener), hence we can use equation (8.6):

$$f_L = f_{noise}\left(\frac{v_{sound}}{v_{sound} + v_{noise}}\right) = 3000\left(\frac{344}{344 + 50}\right) = 2619 \text{ Hz}.$$

The resulting shift in frequency is $(2619 - 3000) = -318$ Hz.

Example 8.4. Suppose you are going down the road at 25 m s^{-1} toward a police cruiser producing a 2000 Hz tone and simultaneously the police cruiser is moving toward you at a velocity of 30 m s^{-1}. What is the perceived frequency just before the police cruiser passes you?
Solution:
The sound source (police cruiser) and you (the listener) are both moving towards each other, hence we can use equation (8.9):

$$f_L = f_{noise}\left(\frac{v_{sound} + v_{listener}}{v_{sound} - v_{noise}}\right) = 2000\left(\frac{344 + 25}{344 - 30}\right) = 2350 \text{ Hz}.$$

You perceive the frequency of the police cruiser just before it passes you to be 2350 Hz.

8.4.2 Some applications of the Doppler effect

Please note that the Doppler effect does not take place because of an actual change in the frequency of the source. The source emits the waves at a certain frequency and continues to do so; it is the observer who perceives a different frequency because of the relative motion between the source and the observer. Also note that the Doppler effect takes place whenever the speed of the source and/or observer are much slower than the speed of the wave. Some applications of the Doppler effect include but are not limited to:

1. The Doppler effect for light is used extensively in astronomy to determine new planets, binary stars, and changes in the universe. Astronomers often observe photons coming from distant galaxies that do not quite fit into the emission spectrum of the luminous matter of the Galaxy under observation. The difference between the expected and observed wavelengths allows astronomers to assign either *red-shift* or *blue-shift* values to these galaxies to determine whether those galaxies are moving towards us or away from us.
2. Echocardiography is a noninvasive medical test that uses high-frequency sound waves (called ultrasounds) to provide a detailed picture of the

structure of the heart. The image formed by echocardiography is called an echocardiogram (echo for short); it shows blood flow through the heart and heart valves. An echocardiogram can help to detect serious heart conditions such as blood clots, heart failure, aneurysms, heart valve disease, etc.
3. Doppler radar uses the Doppler effect for radio waves to detect changes in the atmosphere by measuring and detecting motion toward or away from the radar in order to predict weather conditions. Using Doppler radar, meteorologists have greatly improved our ability to predict dangerous weather patterns, which helps to save thousands of lives every year from thunderstorms, tornadoes, and hurricanes.

8.4.3 Shock waves

So far, we have talked about situations in which the velocity of the moving object is much lower than the speed of sound. But what happens in situations in which the source is moving at or above the speed of sound? If an airplane or any other source, for example, a bullet, moves faster than the speed of sound, it is said to be traveling at *supersonic* speed. At supersonic speeds, the source is always at the leading edge or ahead of the waves that it produces. This results in the bunching up of compressional wave fronts along the conical edge of the wave pattern, as seen in figure 8.13. This produces a zone of very high pressure, which results in a shock wave.

In a normal wave, high-pressure regions (compressions) are followed immediately by low-pressure regions (rarefactions) and these reach the listener one after another. However, for supersonic waves, all regions of high and low pressure reach the listener at the same time. This results in a very loud noise like a *boom*, which is commonly known as a *sonic boom*. It is a general misconception that we hear sonic booms only at the instant the source reaches and/or crosses the speed of sound. In reality, sonic booms are observed whenever a source traveling faster than the speed of sound passes an observer. Sonic booms are only an indication that the source is

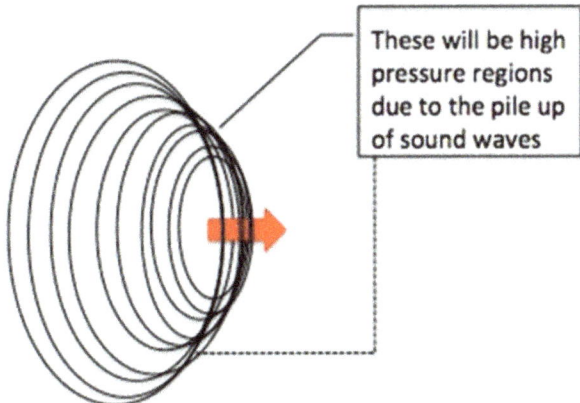

Figure 8.13. A source moving at, or faster than, the speed of sound.

traveling faster than the speed of sound and not that it has reached the speed of sound. However, the pilot in the source only hears the sonic boom as they reach the speed of sound.

A sonic boom shock wave created by an airplane has the shape of a cone which intersects the ground. At the point of intersection, a hyperbolic shape is formed on the ground, as seen in figure 8.14. The sonic boom hits every point at the same time on this hyperbolic curve, which means that people at all points along the curve hear the sonic boom at the same time.

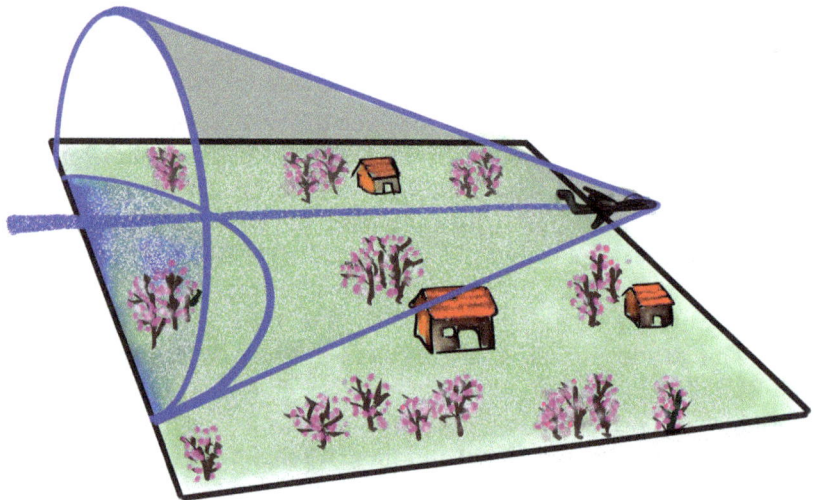

Figure 8.14. When a plane traveling faster than the speed of sound (supersonic) passes overhead, the compressions and rarefactions are heard at the same time, which results in a sonic boom. The shape of a sonic boom shock on the ground.

Review questions 8.2.
1. Explain the Doppler effect in your own words.
2. Describe the sound of an ambulance siren as an ambulance passes you standing at the roadside.
3. What causes a sonic boom?

Exercises 8.2.
1. An ambulance siren at the side of the road emits a tone at a frequency of 440 Hz. What is the apparent frequency when a car approaches the ambulance at 75 mph?
2. What is the apparent frequency when it recedes at the same speed?
3. Calculate the percentage frequency change for both cases.
4. A police officer at the side of the road notes that an 18-wheeler coming towards them is traveling at 60 mph in a 30 m.p.h zone. What is the shift in frequency in their radar gun for waves reflected from the speeding 18-wheeler? Assume that the radar gun transmits microwaves at 9700 MHz. Hint: microwaves are electromagnetic waves, so they travel at the speed of light.
5. What is the apparent frequency as the police cruiser passes you and moves away from you in example 8.4?

Further reading
- French A P 1971 *Vibrations and Waves* (Boca Raton, FL: Chapman and Hall)
- Resnick R, Halliday D and Krane K 2002 *Physics* vol 1 5th edn (New York: Wiley)
- Fowles G and Cassiday G 2005 *Analytical Mechanics* 7th edn (Belmont, CA: Thomson Learning)
- Rossing T D, Moore R F and Wheeler P A 2002 *The Science of Sound* 3rd edn (Upper Saddle River, NJ: Pearson Education)

Part IV

Sound reception

IOP Publishing

The Physics of Sound and Music, Volume 1
A complete course text (Textbook)
Samya Bano Zain

Chapter 9

Sound power and sound intensity

Before we talk about sound power (W)[1] and sound intensity (I), we have to restate a few formulas.

9.1 Power and pressure

9.1.1 Pressure

According to Newton's second law (section 4.3), we know that one of the effects of a net force acting on an object is to cause the object to accelerate in the direction of the applied force. This net force may be a single force acting on an object, or it could be the result of multiple forces acting on the object. Newton's second law does not distinguish between a single force or a distribution of forces, but rather it is concerned about the net effect of the forces. However, there are situations in which it is important to consider the distribution of forces in order to fully comprehend the situation. For example, when we walk on the ground, we push against the ground; according to Newton's third law, the ground pushes back at us with the same force, as seen in figure 9.1. This is true for all cases, but we see different effects depending on the condition of the ground. If the ground is hard, nothing happens. If the ground is soft, it gets squashed underfoot and we leave footprints in it. If there is snow on the ground, the snow gets squished, and our feet sink into the snow. To counteract the effect of snow on the ground, people use snowshoes, which distribute the force applied by our weight over a larger area and prevent us from sinking into the snow.

On the other hand, if a person were to walk across a wooden floor with spiked heels, they could severely damage the floor. The difference between snowshoes and spike-heeled shoes is the area over which same total force (weight) is applied. This implies that the distribution of the force per area is the important factor. Pressure is

---

[1] I have changed the symbol of power from the generally used (P) to the symbol (W) to avoid confusing it with the symbol for pressure (p).

Figure 9.1. The pressure applied to the ground when you walk.

defined as *the magnitude of the force acting perpendicular to a surface per unit cross-sectional area*. Mathematically,

$$\text{pressure} = \frac{\text{force}_{\text{perp}}}{\text{area}} \qquad (9.1)$$

where force$_{\text{perp}}$ is the magnitude of the force acting perpendicular to a surface and area is the cross-sectional area on which it acts. The SI unit of pressure is newtons per square meter, (N m^{-2}), which is also called the '**pascal**' (Pa). Pressure is a scalar quantity and hence does not have a direction. It is important to note that although pressure is produced by a force and forces are vector quantities, pressure itself does not have a direction.

Example 9.1. The pressure on our heels

Suppose a lady who weighs 100 lbs (445 N) stands on two feet so that her weight is distributed equally on the ground. Also assume that the underside of each foot has an area of about 0.01 m^2. Her weight is then distributed over both feet (area = 0.02 m^2) such that the pressure she applies to the bottom of each foot is approximately 22 250 Pa.

Now say that she lifts one foot in the air, what happens? Her weight is still 445 N but the area over which her weight is distributed is halved. This means the new pressure applied to the ground by her foot is 44 500 Pa. Is it hard to imagine why high heels make your feet and legs hurt?

9.1.1.1 The pressure exerted by fluids

Fluids, that is liquids and gases, exert forces on the walls of their containers and anything immersed in them. One important property of fluids at rest is that the pressure always acts perpendicular to all surfaces, i.e. the walls as well as anything immersed in the fluid. The pressure at any point in an open container of fluid is determined by the weight of the fluid above that point.

Figure 9.2. The effect of pressure on velocity.

Change in pressure with depth:
Pressure increases with depth, which means the deeper you go underwater, the higher the pressure becomes. Pressure is also directly proportional to velocity, which means the speed of water leaking from the bottom hole in figure 9.2 (right) is larger than the speeds of the leaks from the higher holes.

9.1.1.2 Atmospheric pressure
The standard pressure exerted by the Earth's atmosphere at sea level is called atmospheric pressure, and its symbol is 'atm'. Numerically, 1 atm = 1.01×10^5 Pa = 14.7 lb in^{-2}. It is interesting to note that the value of the atmospheric pressure, 1.01×10^5 Pa or 101 000 Pa, is a surprisingly large force for the areas that we normally consider. For example, the force due to atmospheric pressure on our heads, assuming our heads have an approximate diameter of 16 cm = 0.16 m and a surface area of approximately ($4\pi r^2$) = 0.08 m^2, is;

$$\text{pressure} = \frac{\text{force}}{\text{area}}$$
$$\Rightarrow \text{force} = \text{pressure} \times \text{area}$$
$$= 8119 \text{ N}$$

This is a lot of force! Actually, this force is equivalent to about nine people, each weighing 100 kg (about 220 lbs), standing on our head at all times. If this is the true value of atmospheric pressure, how are we able to walk, run, or function at all? The answer is simple, the air exerts this force equally on all sides of our heads, and hence the sum of the forces (the net force) on our head is zero. So, we are fine.

Some pressure values:
- The pressure at sea level = 1000 hPa = 1.01×10^5 N m^{-2} = 1 atm.
- The pressure at an altitude of 5.5 km = 510 hPa = $\frac{1}{2}$ atm sea level.
- The pressure at Mount Everest (8848 m) = 327 hPa = $\frac{1}{3}$ atm sea level.
- The pressure in Selinsgrove, PA (160 m) = 981 hPa ≈ atm at sea level.

Example 9.2. Anybody that scuba dives knows that the pressure increases as they dive deeper. This increasing water pressure limits the depth a submarine can reach.

Look at the bubbles coming out of the scuba gear in figure 9.3.
1. What do you observe?
2. Can you explain what is happening here?
3. Can this also help explain why submarines do not have large windows?

Figure 9.3. A scuba diver.

9.1.1.3 Blood pressure

Blood pressure is the pressure exerted on the walls of the arteries and veins in our bodies by the blood circulating through them. Blood flows from the heart to all parts of the body via blood vessels, and blood pressure results from the pumping action of the heart muscle. During each heartbeat, blood pressure varies between a maximum (systolic) and a minimum (diastolic) pressure. The rate of blood flow is related to the resistance presented by the different blood vessels it flows through (figure 9.4).

The mean blood pressure drop is a function of the distance from the heart, that is, the blood pressure value becomes lower as you move farther away from the heart. However, while standing, gravity plays a role; because of gravity, the blood pressure in your feet while standing is generally greater than the blood pressure in your head, even though your feet are further away from your heart than your head.

9.1.2 Work and energy

Energy and work have various meanings in everyday language. However, in physics, they have very particular definitions. Energy is the central idea of all sciences, and it takes many forms. Most natural and human activities depend on the efficient

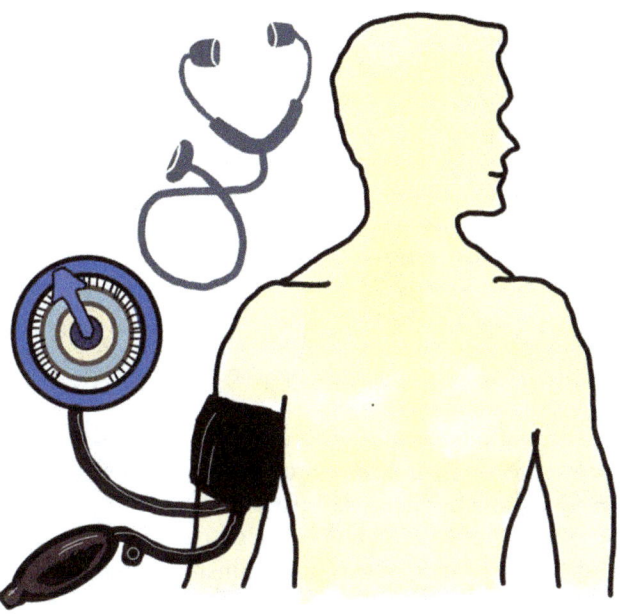

Figure 9.4. Blood pressure.

conversion of one form of energy to another, accompanied by the least possible loss in initial energy. Mechanical energy is closely related to work, and work is defined as the transfer of energy. In acoustics, our main concern is the conversion of mechanical energy (produced by a vibrating object) into sound energy, which is carried by the movement of molecules in a medium as a sound wave. Both work and energy have the same SI units, called 'joules' (J).

The work–energy relation: *the net work done on an object by an applied force is equal to the change in the kinetic energy in the of the object.*

Suppose you are a supervisor on a job, and you want to send a heavy box of construction materials from the ground to the workers on the 40th floor. By definition, the work done is,

$$\text{work} = \text{force} \cdot \text{displacement} = \vec{F} \cdot \vec{d} = mg \cdot h, \tag{9.2}$$

where $\vec{F}$ is the force and $\vec{d}$ is the displacement. Please note that even though force and displacement are both vector quantities, work is a scalar. Once the construction material is taken to the 40th floor, this work is stored as the *potential energy* of the system. Potential energy is defined as *the capability of an object to do work, in this case by the virtue of its position in a gravitational field*. There are some interesting features of work as defined in physics that I will briefly discuss here:

1. The work done is the product of the force applied to an object times the displacement parallel to the applied force. It stands to reason that cars in the Indy 500 race do no work, since they start and end at the same point!

2. Work is not an intrinsic property of an object; you cannot say that a body gains or loses a certain amount of work. Work is simply a force that acts on a body and moves it through a certain distance.
3. The work done in moving a body of mass m to height h against the force of gravity was found to be mgh in equation (9.2). This is interesting because it is the same as the mathematical description of potential energy (PE = mgh).
4. If the work done is zero, this means that either:
    (a) $\vec{F} = 0$, or
    (b) $\vec{d} = 0$.
5. The work done depends on the applied angle. Mathematically,

$$\text{work} = \text{force} \cdot \text{displacement} = \vec{F} \cdot \vec{d} = Fd \cos \theta, \qquad (9.3)$$

where θ is the angle between the force ($\vec{F}$) and the displacement ($\vec{d}$). When:
    (a) $\theta = 0°$, the work done is (force × displacement)
    (b) $\theta = 90°$, the work done is zero
    (c) $\theta = 180°$, the work done is negative (force × displacement), for example, raising a mass up against gravitational force through a distance h

9.1.3 Power

As a supervisor, you could devise multiple ways of sending the materials up at a construction site. However, for the manager, time is money and hence you have to send the materials up in as short a time as possible. This is where power comes in the picture.

Power is defined as the rate at which work is done. So if you are a construction manager, the term that matters is power. You want the work completed in the shortest time, hence you have to provide more power. In physics, power is defined as,

$$\text{power (watt)} = \frac{\text{work}}{\text{time interval}} = \frac{W}{\Delta t} = \frac{\text{joule}}{s} \qquad (9.4)$$

The SI unit for power is the watt (W), which is equivalent to one joule per second (J s^{-1}).

A regular light bulb that is rated at 100 watts consumes 100 joules of energy every second. In other words, this is the amount of energy the bulb uses (and what you get billed for). The bulb 'outputs' energy in the form of visible light and heat.

For historical reasons, horsepower is used to describe the power delivered by a machine. One horsepower is approximately equal to the power that a 'standard' horse is able to provide. Mathematically, one horsepower is ≈ 750 watts.

Example 9.3. Jack and Jill went up a hill to fetch a pail of water. Jack and Jill weigh about the same, but Jill goes up the hill in half the time. Calculate:
1. Who did the most work?
    Solution:
    Jack and Jill weigh the same and both covered the same distance, and we know that work is defined as force times distance. Hence, both Jack and Jill did the same work.

2. Who delivered the most power?
   Solution:
   Jack and Jill did the same amount of work; however, Jill does the work in half the time. Power is work divided by time, hence Jill delivers twice the power that Jack does.

9.1.4 Stress and strain

In chapter 4, we said that the effect of an applied force on an object is to cause it to accelerate in the direction of the applied force. However, another effect of this applied force is to change the shape of the object at the contact surface. This change is referred to as *deformation*. The deformation can be permanent or temporary. If the change is not permanent, we call the object an *elastic object*. Hence, an elastic object is one that returns to its original size and shape after the contact forces have been removed. However, if the forces acting on the object are too large, even an elastic object can be permanently deformed.

If a tensile force[2] is applied to an object, it causes the object to stretch or elongate by an amount proportional to its original length. Suppose that you have two wires of different lengths L_1 and L_2, and a force ($\vec{F}$) is applied to these wires. Both wires stretch by different amounts ΔL_1 and ΔL_2 with respect to their original lengths L_1 and L_2. This fractional change in length is called strain and is mathematically defined as

$$\text{strain} = \frac{\Delta L}{L}, \qquad (9.5)$$

where L is the original length of the object under consideration and ΔL is the change in length due to the external force. Note: strain is a ratio of lengths and hence it is a dimensionless quantity.

Stress is the force between two objects that acts parallel to their interface. For example, the horizontal force on a floor caused by dragging an object across the floor is a *stress*. Stress is measured as force per unit area, mathematically given as:

$$\text{stress} = \frac{\text{force}}{\text{area}} \qquad (9.6)$$

The SI units of stress are newtons per meter squared (N m^{-2}) or pascals (Pa), which are the same as the unit of pressure.

If the stress on an object exceeds the *elastic limit*, then the object does not return to its original length and the object fractures. This is called the *breaking point*. The ultimate strength of a material is the maximum stress that it can withstand before breaking. The ratio of the maximum load to the original cross-sectional area is called *tensile strength*.

9.2 Sound waves

Sound waves have very modest energy. To appreciate how small this energy is, let us consider a one-watt light bulb, usually used in night lights. Assume there is a global

---

[2] Tensile force is the measure of the ability of material to resist a force that tends to pull it apart.

population of one trillion (that is a million million) people. If you were to split the light from that dim one-watt bulb equally between all those people, each would receive a radiant power of 1×10^{-12} watts or 1 watt per 1 000 000 000 000 people. Next, let us say that each of those people spreads this power over an area of one square meter. This is a very, very tiny energy and at this tiny energy, the eardrum vibrates over a distance smaller than the diameter of a hydrogen atom (120 picometer or 1.2×10^{-10} m)!

What is truly amazing is that for a person with normal hearing, even this tiny energy can be detected! Even more amazing is that a person could detect a sound wave a thousand times more powerful, or even a million times more powerful, or even a billion times more powerful than this, all before it even begins to get painful!

9.2.1 The amplitude of sound waves

Sound waves are produced by the vibration of an object, say a vibrating string or a drum. These waves then travel through a medium, most commonly air, by creating a pressure disturbance. This pressure disturbance consists of a pattern of compressions and rarefactions that travel from one molecule to the next primarily by direct collisions between air molecules. The amount of energy transferred to the medium depends on the amplitude of the initial vibrations of the source.

Figure 9.5 shows the motion of the medium as a sound wave passes through it. The molecules move as a result of the pressure compression and then go back to their original position; thus, the medium reverts back to its original state after the wave has passed through it.

Suppose the external force in a string is applied by plucking. Plucking displaces the string from its equilibrium position, which puts energy into the system. This energy is transferred to the surrounding air and is carried away by the molecules of the medium (air). The loudness of the sound wave depends on the initial applied force: when the initial applied force is large, it results in a larger displacement of the string and hence more energy is imparted to the surrounding medium. The relationship between amplitude and the energy transmitted is shown in figure 9.6. In summary, we can say:

1. A small amplitude implies that the transmitted energy is low.
2. A large amplitude implies that the transmitted energy is high.

9.3 The intensity of sound waves (*I*)

The amount of energy that passes through any given area of a medium per unit time is called the ***intensity of a wave***. As the amplitude of the vibrations increases, the rate at which energy is transported through the medium increases.

Figure 9.5. Motion of a medium as a sound wave passes through it.

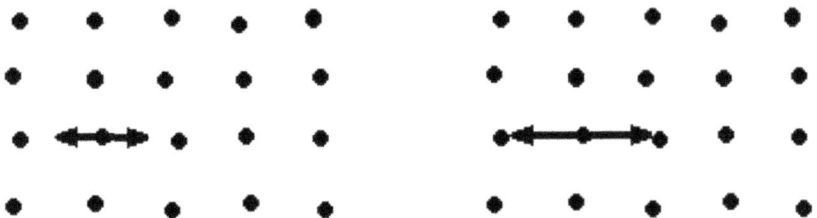

Figure 9.6. The relation between amplitude and energy transmitted: (right) small amplitude; (left) large amplitude.

For a wave that travels in three dimensions, the intensity is a measure of the average power per unit area carried by the wave past a surface perpendicular to the wave's direction of propagation. In other words, intensity can also be called the energy per time per area, mathematically expressed as

$$\text{intensity} = \frac{\text{energy}}{\text{time} \times \text{area}}. \tag{9.7}$$

We also know from equation (9.4) and the work–energy relation that

$$\text{power} = \frac{\text{work}}{\text{time}} = \frac{\text{energy}}{\text{time}}. \tag{9.8}$$

Combining the above with equations (9.7) and (9.8), we can rewrite intensity simply as

$$\text{intensity} = \frac{\text{power}}{\text{area}}. \tag{9.9}$$

The SI units for the intensity of a sound wave are W m^{-2}. Please note that the intensity of a sound is an objective quantity and can be measured by instruments.

Example 9.4. A spherical sound source radiates at 25 W. What is the intensity of the sound wave 5 m from the source?
Solution:
Use equation (9.9); we know that for a spherical source area, $A = 4\pi r^2$, therefore we can find the intensity as follows:

$$\text{intensity} = \frac{\text{power}}{\text{area}} = \frac{25 \text{ W}}{4\pi(5)^2 \text{ m}^2} = 0.079 \text{ W m}^{-2}.$$

9.3.1 Intensity versus distance from a wave source

The intensity of sound waves, like that of most waves, decreases as the distance from the source increases. There are two main reasons for this decrease:
1. As a sound wave propagates, some of its energy is absorbed by the medium. The amount of energy absorbed depends on the medium. However, air absorbs relatively little sound energy, so we can generally hear sounds generated far away from us.

2. Another reason that the intensity of sound decreases with distance is that the wave spreads out and therefore the energy is spread over a larger and larger area, as shown in figure 9.7.

When we drop a stone into water, a wave is created on its surface. This wave propagates outward in increasing concentric circles whose circumferences are ($C = 2\pi r$). The original energy in this system came from the energy of the rock hitting the surface of water, and according to the law of conservation of energy, the energy contained in the wave must remain the same as the wave propagates. This means that as the wave spreads out, this same energy must also spread out over a larger and larger area. When the circle's radius is doubled, the energy is spread over a wave front twice as long, while the intensity of the wave is halved. We can say that the intensity of the wave is inversely proportional to the radius of the circle, as shown in figure 9.7.

On the other hand, 3D waves, such as those of sound and light, travel and expand in all directions, and their waves expand as spheres rather than circles. The area of a sphere is given mathematically by ($4\pi r^2$), where r is the radius of the sphere. This means that when you double the radius, the area quadruples, which implies that the original energy of the wave is spread over four times the area as a result of doubling the radius. However, conservation of energy again says that the total energy over the ever-expanding sphere must remain constant.

For an isotropic source,[3] the average power (energy per unit time) emitted remains constant. Assuming no energy is absorbed by the medium and there are no obstacles to reflect or absorb the sound, the intensity (I) at a distance r from an isotropic source is mathematically given by

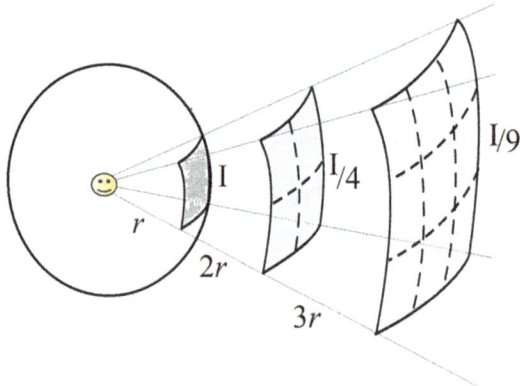

Figure 9.7. The expansion of 3D sound waves. As the radius becomes larger, the intensity of the wave decreases.

---

[3] An isotropic source is a point source that radiates uniformly in all directions. Hence, an isotropic sound source radiates sound uniformly in all directions.

$$\text{intensity} = \frac{\text{power}}{\text{surface area}} = \frac{W}{4\pi r^2} = \frac{\text{watt}}{\text{m}^2}. \qquad (9.10)$$

This is called the '**inverse square law**' for a spherical area surrounding a source. As the name implies, the inverse square law states that the intensity of a sound wave is inversely proportional to the square of the distance from the source. This means that as the distance from the source is doubled, the intensity is quartered, and energy passing through one square meter of spherical surface per second must drop off according to $1/r^2$:

$$\text{intensity} = I(r) \propto \frac{1}{r^2}. \qquad (9.11)$$

All waves that propagate in 3D obey the *inverse square law*, with the possible exception of laser light. Laser beams are very special, since they are made of *coherent* light that is confined to a very narrow beam which does not spread like those produced by other light sources. The intensity of laser light does not decrease according to the inverse square law, which allows a very powerful laser beam to be visible on the Earth after it has been reflected back from a reflector on the surface of the Moon.

9.4 Decibels

Human ears are very sensitive to intensity changes. The faintest sound a normal human ear can detect has an intensity of 10^{-12} W m^{-2}. This intensity corresponds to a pressure wave that displaces the molecules in the compression region by 0.3 billionth of atmospheric pressure! This faintest sound that the human ear can detect is known as the *threshold of hearing*. The most intense sound that the ear can safely detect without suffering physical damage is more than one billion times more intense than the threshold of hearing.

The range of intensities that the human ear can detect is very large, hence physicists generally use a scale based on multiples of ten, called a *logarithmic scale*,[4] to measure intensity. The scale used to measure sound intensity is called the '**decibel scale**' (abbreviated dB) and is named after Alexander Graham Bell. The decibel is a dimensionless unit since it is a ratio of two quantities under consideration.

---

[4] A logarithmic scale is a scale of measurement that uses the logarithm of a physical quantity instead of the quantity itself. When the data covers a large range of values, the logarithm reduces it to a more manageable range. Our sense of hearing operates logarithmically, which makes the use of logarithmic scales particularly suitable for these quantities. On most logarithmic scales, small values (or ratios) of quantities correspond to small values of the logarithmic measure and large values of quantities correspond to large values of the logarithmic measure.

9.4.1 The sound intensity level (dB SIL)

Sound intensity (I) is a physical measure of sound wave amplitude; the sound intensity level (SIL) is its psychological measure. This means that we can experimentally measure the sound intensity, whereas the SIL describes the way in which humans react to said sound intensity. Sound intensity is measured in watts per meter squared and SIL is measured in dB. The equation that relates sound intensity (I) to the SIL is:

$$\text{sound intensity level, (dB SIL)} = 10 \log\left(\frac{I_2}{I_1}\right), \quad (9.12)$$

where SIL is the number of decibels, I_2 is the sound intensity being compared, and I_1 is the reference sound intensity. Remember, intensity (I) is measured in W m^{-2}. The SIL in equation (9.12) is the decibel difference between these two intensities. To see the advantage of using logarithms, let us compare the intensities, as follows:

1. When I_2 is twice as large as I_1, then the SIL is:

$$\text{SIL}_1 = 10 \log\left(\frac{I_2}{I_1}\right) = 10 \log \frac{2 \times I_1}{I_1} = 10 \log(2) = 3 \text{ dB SIL}. \quad (9.13)$$

2. When I_2 is ten times louder than I_1, this corresponds to:

$$\text{SIL}_2 = 10 \log\left(\frac{I_2}{I_1}\right) = 10 \log \frac{10 \times I_1}{I_1} = 10 \log(10) = 10 \text{ dB SIL}.$$

3. When I_2 is a hundred times louder than I_1, this corresponds to:

$$\text{SIL}_3 = 10 \log\left(\frac{I_3}{I_1}\right) = 20 \log \frac{100 \times I_1}{I_1} = 10 \log(100) = 20 \text{ dB SIL}.$$

4. When I_2 is a thousand times louder than I_1, this corresponds to:

$$\text{SIL}_4 = 10 \log\left(\frac{I_4}{I_1}\right) = 10 \log \frac{1000 \times I_1}{I_1} = 10 \log(1000) = 30 \text{ dB SIL}.$$

5. When I_2 is ten thousand times louder than I_1, this corresponds to:

$$\text{SIL}_5 = 10 \log\left(\frac{I_5}{I_1}\right) = 10 \log \frac{10\,000 \times I_1}{I_1} = 10 \log(10\,000) = 40 \text{ dB SIL}$$

and so on...

This can be summarized as follows: *for each power of ten increase or decrease in the intensity of sound with respect to a reference intensity, there is a (± 10 dB) change in the SIL.*

Example 9.5. If we ignore other effects, such as reflections from the ground or other materials, the presence of wind, etc. by how much does the sound intensity change if the distance doubles from r to $2r$?

Solution:
We know from equation (9.10) that the intensity falls off from the sound source according to $1/r^2$. We also know that I_2, the higher sound intensity, is found at distance r, whereas I_1, the lower sound intensity, is found at $2r$. Using equation (9.12), it follows that:

$$I(2r) - I(r) = 10 \log\left(\frac{I(2r)}{I_0}\right) - 10 \log\left(\frac{I(r)}{I_0}\right)$$

$$10 \log\left(\frac{I(2r)}{I(r)}\right)$$

$$= 10 \log\left(\frac{W/(2r)^2}{W/r^2}\right).$$

Since the power is related to the source and is the same in both cases, we are left with:[5]

$$I(2r) - I(r) = 10 \log\left(\frac{1/(2r)^2}{1/r^2}\right)$$

$$= 10 \log\left(\frac{1/4r^2}{1/r^2}\right)$$

$$= 10 \left(\frac{\log 1 - \log(4r^2)}{\log 1 - \log(r^2)}\right)$$

$$= 10 \left(\frac{0 - \log(4r^2)}{0 - \log(r^2)}\right)$$

$$= 10 \left(\frac{\log(4) + \log r^2}{\log(r^2)}\right)$$

$$= 10 \log 4$$
$$= 6.02 \text{ dB SIL}.$$

This means that there is a 6 dB drop in SIL every time the distance is doubled.

9.4.2 Sound intensity at the threshold of hearing I_0

If we consider sounds at the threshold of hearing, say at 1×10^{-12} W m^{-2}, then equation (9.12) may be written as

$$\text{sound intensity level (dB SIL}_0\text{)} = 10 \log\left(\frac{I}{I_0}\right) = 10 \log \frac{1 \times 10^{-12}}{1 \times 10^{-12}} = 10 \log(1) = 0 \text{ dB SIL}.$$

Please note that 0 dB SIL does not mean that there is no sound, it only means that we cannot hear it. In addition, negative dB values of sound intensity are just a fraction of the intensities that humans can hear. Hence, we can find the intensity of all sounds with respect to the threshold of hearing using the formula:

$$\text{sound intensity level (dB SIL)} = 10 \log\left(\frac{I}{I_0}\right), \quad (9.14)$$

---

[5] Use logarithmic properties 1 and 2 given in appendix C to solve this example: (1) the **product property:** $\log(AB) = \log A + \log B$, (2) the **quotient property:** $\log(\frac{A}{B}) = \log A - \log B$,

where $I_0 = 1 \times 10^{-12}$ W m^{-2} is the lowest sound intensity detectable by humans.

Example 9.6. Convert a measured sound intensity of 10^{-9} W m^{-2} to a sound intensity level (dB SIL).
  Solution:
  Using equation (9.14); we see that

$$\text{sound intensity level (dB SIL)} = 10 \log \frac{I}{I_0} = 10 \log \left(\frac{10^{-9}}{10^{-12}} \right) = 10(3 \log 10) = 30 \text{ dB SIL}.$$

Hence, the SIL that corresponds to a sound intensity of 10^{-9} W m^{-2} is 30 dB SIL with respect to the threshold of hearing.

9.4.3 The sound intensity level as a function of the number of sources

The total intensity of two sources that each have an intensity of (I) is simply the sum of the individual source intensities. For example, if two violins are played equally loudly, the sound intensity at a certain point in space due to both violins is $2 \times I$. We know from equation (9.13) that there is a 3 dB increase in the SIL (dB SIL) when we double the number of instruments. When we increase the number of instruments to ten violins, because of our logarithmic hearing, they sound twice as loud as when only one violin was playing. In general, if there are multiple sound sources, we can calculate the total SIL at a point using the formula:

$$\text{total level} = \text{level of a single source} + (10 \times \log(\text{number of sources})). \quad (9.15)$$

This equation says that the total level produced is numerically equal to the sum of the SILs of the individual sources and ten times the log of the number of sources.

Example 9.7. If one violin produces 70 dB SIL and we add a second violin that also produces 70 dB SIL, what is the total SIL?
  Solution:
  We can use equation (9.15) to find the total SIL for two violins:

$$\text{total level} = \text{level of a single source} + (10 \times \log(\text{number of sources}))$$
$$= 70 \text{ [dB SIL]} + (10 \log(2)) \text{ [dB SIL]}$$
$$= 70 \text{ [dB SIL]} + 3 \text{[dB SIL]}$$
$$= 73 \text{ [dB SIL]}.$$

Hence, the new level is 73 dB SIL.

Example 9.8. If 60 violins, each of which produces the same SIL, combine to produce 120 dB SIL, what is the individual SIL of each violin?
  Solution:
  We can use equation (9.15) to find the SIL of an individual source as follows:

total level = level of a single source + (10 log(number of sources))
120 dB SIL = L + 10 log(60)
120 dB SIL = L + 17.78 dB SIL
$\Rightarrow L = 102.22$ dB SIL.

Hence, the SIL of each violin is about 102 dB SIL.

Review questions 9.1.
1. What is the difference between pressure and force?
2. What is the pressure of the atmosphere on our bodies when we are standing at the beach?
3. What is the difference between power and energy?
4. What are the units of sound intensity? Mathematically, should sound intensity have units?

Exercises 9.1.
1. During a competition, contestants are required to take blocks of wood up some stairs (h = 10 meters) in the least amount of time. Player A takes five blocks of wood (each of mass 10 kg) in 30 s, player B takes five blocks in 40 s, whereas player C completes the task in 20 s.
   (a) Which player does the most work?
   (b) Which player delivers the most power?
2. A spherical sound source radiates at 100 W. What is the intensity of the sound wave:
   (a) 1 m from the source?
   (b) 2 m from the source?
   (c) 5 m from the source?
3. A bell radiates 1.5 W of sound power while playing a particular note. If the bell has an area of 0.2 m^2, find the average intensity and the SIL at the bell.
4. Ten trumpets, each of which produces the same SIL, combine to produce a 100 dB SIL. What is the individual SIL of each trumpet?

9.5 The speed of sound versus particle velocity

A sound source is defined as an object that emits sound energy into the surrounding environment. This energy radiates outwards from the sound source in the form of alternating compressions and rarefactions over an increasing area, as we saw in figure 9.7. These compressions and rarefactions are temporary changes in atmospheric pressure that are pushed through the medium. The magnitude of this change in the local atmospheric pressure caused by the vibration of the sound source is called the *sound pressure* or *acoustic pressure*. The SI unit of sound pressure is the pascal (Pa), and the sound pressure is mathematically written as

$$\text{sound pressure } (p) = \frac{\text{force (N)}}{\text{area (m}^2)}, \qquad (9.16)$$

where the force is provided by the sound source and the area is the surface area over which the force acts.

As the sound source transfers its energy to the surrounding medium, the molecules in the medium are displaced from their equilibrium positions. This affects the velocities of the particles of the medium and accelerates them. This is called the *molecule velocity* or the *particle velocity* and is directly related to the kinetic energy of the moving sound wave.

It is important to note that the particle velocity is different from the speed of sound. The speed of sound tells us how fast the wave is traveling through the medium and is a scalar, whereas the particle velocity is the velocity of the individual vibrating particles of the medium that are excited by the sound energy and is a vector quantity. The particles of the medium move, transfer this energy to the adjacent particles, and return to their equilibrium state.

The speed of sound depends on the properties of the medium, whereas the particle velocities do not. The particle velocities depend on the magnitude of the sound energy. Sounds with different energies that correspond to different sound pressures travel at the same speed throughout the medium; however, particle velocities change as the energy spreads over a larger and larger area.

9.6 Sound power (W)

Sound power is the acoustical energy emitted by a sound source per unit time. It is an absolute value and is not affected by the environment. Sounds audible to humans have sound powers with a lower threshold value of 10^{-12} W and a higher threshold value of 100 W. This means the ratio between the highest and the lowest is threshold value is 100 $W/10^{-12}$ $W = 10^{14}$!

Unlike sound pressure, sound power (W) is neither room dependent nor distance dependent. The sound power is solely a characteristic of the sound source and is the total power produced by the source projected in all directions.

Example 9.9. A fire engine generates a sound wave that has intensity of 0.09 W m^{-2} when it is 12 m away. What is the sound intensity 20 m away from the fire engine? Assume the siren is an isotropic source.

Solution:

The fire engine generates a sound wave with an intensity of 0.09 W m^{-2} at 12 m. We can use equation (9.10) to find the power (W) of the engine:

$$W = I_{(12\,m)}(4\pi r^2) = (0.09 \text{ W m}^{-2})(4\pi(12)^2 \text{ m}^2) = 162.86 \text{ W}.$$

We then use equation (9.10) again since the power of the engine does not change with distance:

$$I_{(20\,m)} = \frac{W}{4\pi r^2} = \frac{162.7 \text{ W}}{4\pi(20)^2 \text{ m}^2} = 0.032 \text{ W m}^{-2}.$$

The sound intensity 20 m away from the fire engine is 0.032 W m^{-2}.

9.7 Sound pressure level (dB SPL)

There is a direct relation between the threshold intensity (I_0) of human hearing and the smallest pressure oscillation (p_0) that can be detected by the ear. Intensity is proportional to the square of the pressure amplitude. In this form, the intensity of a wave is given by

$$\text{intensity } (I) = \frac{p^2}{\rho v}, \tag{9.17}$$

where p is the pressure of the wave, ρ is the density of air, and v represents the speed of sound. For room temperatures (around 20 °C), the product (ρv) is approximately 410 to 420 kg m^{-1} s^{-1}, but for most calculations we tend to set it to 400 kg m^{-1} s^{-1}. Hence, in terms of the pressure, equation (9.12) becomes:

$$\text{sound intensity level (dB SIL)} = 10 \log \left(\frac{I}{I_0}\right)$$

$$= 10 \log \left(\frac{p^2/\rho v}{p_0^2/\rho v}\right)$$

$$= 10 \log \left(\frac{p^2}{p_0^2}\right)$$

$$= 10 \log \left(\frac{p}{p_0}\right)^2$$

Use property of log, $\log(x^n) = n \log x$

$$= 20 \log \left(\frac{p}{p_0}\right)$$

$$\equiv \text{Sound pressure level (dB SPL)}.$$

Hence, the SPL is defined as

$$\text{sound pressure level (dB SPL)} = 20 \log \left(\frac{p}{p_0}\right). \tag{9.18}$$

The threshold intensity $I_0 = 10^{-12}$ W m^{-2} corresponds to a threshold pressure of $p_0 = 2 \times 10^{-5}$ Pa or 20 μPa. Since the atmospheric pressure (p_{atm}) = (1.01 $\times$ 10^5 Pa), we can calculate that the ear is able to detect pressures equal to

$$\frac{p_0}{p_{\text{atm}}} = \frac{2 \times 10^{-5}}{1.01 \times 10^5} = 2 \times 10^{-10}. \tag{9.19}$$

This means that the detectable pressure is 2×10^{-10} times smaller than atmospheric pressure (p_{atm})!

Example 9.10. What sound pressure level corresponds to a sound pressure of 0.01 N m^{-2}?

Solution:
Using equation (9.18),

$$\text{sound pressure level (dB SPL)} = 20 \log \frac{p}{p_0} = 20 \log \left(\frac{0.01}{2 \times 10^{-5}}\right) = 54 \text{ dB}.$$

9.7.1 Sound power (*W*) versus sound pressure (*p*)

'Sound power' (*W*) and 'sound pressure' (*p*) are the two most commonly confused characteristics of sound. The main source of confusion comes from the fact that they both have the same unit of measure, the decibel (dB), and the term 'sound level' is commonly misused. However, they are completely different from each other. Sound power (*W*) is the sound energy emitted by a vibrating object, whereas sound pressure describes what is heard at a specific location. The radiated sound power (*W*) is the cause, and the sound pressure (*p*) is the effect. A mechanical engineer is most interested in the sound power (the cause), whereas a sound engineer is particularly interested in the effect, i.e. sound pressure, particularly if it is a noise the sound engineer is interested in minimizing!

Most of us are more familiar with light than sound. So, I will use light to help us distinguish between the two confusing terms. In terms of light, sound power (*W*) can be thought of as the wattage rating of a light bulb, and sound pressure (*p*) corresponds to the brightness in a particular part of the room. Both the sound power and the wattage measure a particular and fixed amount of energy. Please note that the brightness of light is measured using a light meter, whereas sound pressure is measured using a sound meter. Both sound meters and light meters give us numerical values for magnitudes.

Another example is the difference between the power of an air conditioner (AC) and the cooling it produces. If you stand close to an AC in a large room, you can feel the cool air it provides. However, if you are far away, you probably do not feel anything. The cooling effect that the AC produces depends on your distance from the it, even though the AC itself is putting out the same cooling power. Sound power corresponds to the power of the AC, while sound pressure corresponds to the temperature that it produces.

9.7.2 Measuring sound pressure levels

A sound level meter is used to measure sound or noise. Sound level meters, like the one shown in figure 9.8, generally measure sound pressure levels. They are

Figure 9.8. A sound level meter.

made up of a microphone (which picks up the sound by measuring the changes in air pressure produced by the sound source), a suitable amplifier, electric circuits to convert sound into an electrical signal, and a meter calibrated to give a reading in decibels (dB). Sound meters are calibrated such that they read 0 dB for the threshold of hearing and 120 dB for the threshold of pain (extremely loud sounds).

The sensitivity of the ear varies with frequency. To account for this, sound-level meters are normally made such that they can be switched between a scale that reads sound intensities uniformly for most frequencies, called an *unweighted scale*, or a scale that introduces a frequency-dependent weighting factor, which gives an almost human response.

A low-frequency sound at a certain power does not seem as loud as a higher-frequency sound of the same power. To account for this difference, a weighted scale called the 'A-weighted' scale was developed. Sound power levels adjusted by this specific weighting scale are called *A-frequency weighted*. The A-frequency-weighting scale is used to describe the effect of complex noises on people. Thus, the A-scale is recognized for measurements that help in the prevention of the harmful effects of excessive noise in our environments.

The A-weighted scale is the most commonly used standard, but other scales exist as well, including the B-, C-, D-, and Z-frequency-weightings, some of which are shown in figure 9.9.

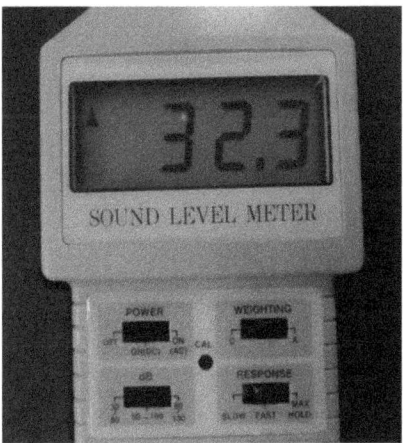

Figure 9.9. A sound level meter with weighted scales.

9.8 Summary

- **Sound pressure:** $(p) = \frac{\text{force}}{\text{area}}$.

  Sound pressure (amplitude) falls so that it is inversely proportional to the distance $1/r$ from the sound source; this is called the $1/r$ law or the inverse distance law.

- **Sound intensity:** $(I) = \frac{\text{sound power (W)}}{\text{area}}$

  Sound intensity (energy) falls so that it is inversely proportional to the square of the distance $1/r^2$ from the sound source; this is called the inverse square law.

- **Sound pressure level (dB SPL):** sound pressure (p) can be converted to sound pressure level (dB SPL) $= L_p = 20 \log\left(\frac{p_2}{p_0}\right)$.

  Sound pressure level decreases by 6 dB when the distance from the source is doubled, falling to half (50%) of the initial sound pressure value.

- **SIL (dB SIL):** sound intensity (I) can be converted to SIL (dB SIL): $= 10 \log\left(\frac{I_2}{I_0}\right)$

  The SIL decreases by 6 dB when the distance from the source is doubled, falling to a quarter (25%) of the initial sound intensity value.

Review questions 9.2.
1. Sound pressure and sound intensity are not the same. Explain this in your own words.
2. Neither the sound power nor the sound power level decreases when the distance is doubled. Why is this so?
3. What are the units of sound intensity and the SIL?
4. Mathematically, should the SIL have units?

Exercises 9.2.
1. If the sound intensity of a screaming baby were 0.01 W m^{-2} at 1 m away, what would it be at 6 m away?
2. An orchestra with 75 performers has an acoustic power of about 70 W. Determine the SIL at 10 m.
3. What is the corresponding SIL when the sound intensity is equal to 10^{-10} W m^{-2}?
4. If sound pressure is equal to 1000 μPa, what is the corresponding sound pressure level?

9.9 The sound power level (dB SWL)

The sound power level (dB SWL or L_w) is the acoustic energy emitted by a source which produces a sound pressure level at some distance. The dB SWL is a logarithmic measure of the power of a sound relative to a reference value:

$$\text{sound power level (dB SWL}_0) = 10 \log\left(\frac{W}{W_0}\right), \qquad (9.20)$$

where W[6] is the sound power measured in watts and W_0 is the reference sound power (a commonly used value is 1×10^{-12} watt).

The dB SWL does not depend on distance, position, or environment and it is a theoretical value, so it is not directly measurable. A noise source has the same sound power irrespective of where it is placed. This provides us with a way to directly compare two sound sources. If the powers of the two sources are W_2 and W_1, respectively, we can directly compare them using the equation

$$\text{sound power level (dB SWL)} = 10 \log\left(\frac{W_2}{W_1}\right). \qquad (9.21)$$

The SWL is very useful in quantifying how noisy a source is. Therefore, it can be used to predict the noise impact of a source before it is installed or used, without having to measure it. If you are asked to provide noise data for something, your acoustician will always appreciate data which is given as a dB SWL.

9.9.1 The sound pressure level (SPL) versus the sound power level (SWL)

The dB SWL is theoretical and cannot be measured using a sound level meter. A sound power is measured in watts (W) and is the logarithmic ratio of the sound power to a reference sound power (W_0), whereas the SPL depends on distance, the position of the source, and things in the environment such as objects or surfaces in the room, reflections from the ground or walls, the cubic capacity of the room, etc. The SPL is a value that we can physically measure using a sound level meter. Most noise level parameters are based on values expressed as SPL levels. The relationship between the dB SWL and the SPL of the source depends on the nature of the sound field and is given by

$$(\text{dB SPL}) = (\text{dB SWL}) - \log\left(\frac{Q_D}{4\pi r^2}\right), \qquad (9.22)$$

where Q_D (called the '**directivity factor**') is defined as the ratio of the sound intensity at a distance (r) in front of the source to the sound intensity averaged over all directions.

---

[6] Recall that I have changed the generally employed symbol of power from P to W the avoid confusion with the symbol for pressure (p).

9.9.1.1 Sound fields

1. **Free field**: for a point source or any source that radiates equally in all directions, the intensity of the sound decays according to $1/r^2$, where r is the distance from the source. An environment in which there are no reflections is called a *free field* and is shown in figure 9.10. A spherical source that radiates equally in all directions has a directivity factor of $Q_D = 1$. Substituting this value into equation (9.22) gives the equation for a free field, equation (9.23), as follows:

$$(\text{dB SPL}) = (\text{dB SWL}) - \log\left(\frac{1}{4\pi r^2}\right). \tag{9.23}$$

A method used to estimate the dB SWL in a free field at a distance (r) from the source is to measure the SPL at distance (r) and then solve equation (9.23) for the dB SWL. In a free field, the dB SWL decreases by 6 dB each time the distance from the source is doubled. The sound power level *one meter away from a source* in a free field is 11 dB less than the sound power level of the source. From this information, we can obtain the SIL (in dB) and the SPL (in dB) as follows:

$$(\text{dB SIL}) = (\text{dB SPL}) = (\text{dB SWL}) - 11 \text{ dB}. \tag{9.24}$$

2. **Hemispherical field**: in real life, a hemispherical field is more common than a free field. In a hemispherical field, the sound source rests on a hard, sound-reflecting surface and radiates hemispherical waves into the air, as seen in figure 9.11. For example, a speaker mounted on a wall or a radio placed on the floor forms a hemispherical field. The directivity factor of a hemispherical field is $Q_D = 2$. When the source is in a hemispherical space, equation (9.22) reduces to:

$$\text{dB SPL} = \text{dB SWL} - \log\left(\frac{2}{4\pi r^2}\right). \tag{9.25}$$

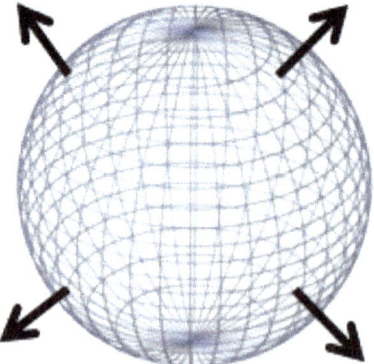

Figure 9.10. In a free field, a source radiates equally in all directions.

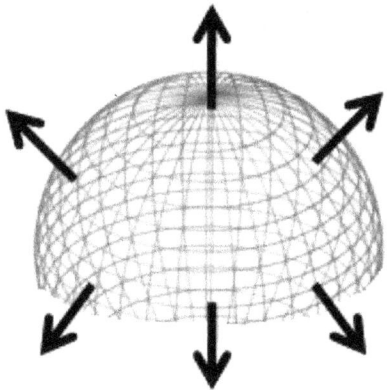

Figure 9.11. A source in a hemispherical field.

In a hemispherical field, the SIL decreases by 6 dB each time the distance from the source is doubled. The SIL (or sound pressure level) at a distance of **one meter from a source** in a free field is 8 dB less than the dB SWL of the source.

$$(\text{dB SIL}) = (\text{dB SPL}) = (\text{dB SWL}) - 8 \text{ dB} \quad (9.26)$$

Example 9.11. A power generator in your basement rests on a concrete floor and is very loud and is annoying your family. You get a sound meter and measure the sound pressure level one meter from the generator: it is 110 dB. Find the dB SWL and the sound power (W) of the generator.

Solution:
Since the generator is resting on the floor, it creates a hemispherical field. For hemispherical fields, we know that the SIL or SPL at a distance of one meter from a source in a free field is 8 dB less than the SWL of the source. Hence, we can write

$$(\text{dB SWL}) = (\text{dB SPL}) + 8 = 110 + 8 = 118 \text{ dB}.$$

Hence, the sound power level is 118 dB.

To find the sound power (W), use equation (9.20):

$$\text{sound power level (dB SWL)} = 10 \log \frac{W}{W_0}$$

$$\frac{(\text{dB SWL})}{10} = \log \frac{W}{W_0}$$

$$\frac{118}{10} = \log \frac{W}{10^{-12}}$$

$$\log^{-1}\left(\frac{118}{10}\right) = \frac{W}{10^{-12}}$$

$$6.3 \times 10^{11} = \frac{W}{10^{-12}}$$

$$6.3 \times 10^{11}(10^{-12}) = W.$$

Hence, the sound power is calculated to be 0.63 W.

9.10 Loudness and loudness level

We know that the sound pressure of a sound is an objective quantity that can be measured by a sound meter. Sound loudness, on the other hand, is entirely a psychological construct, which means that loudness cannot be measured with a meter. However, we still use decibels (dB) as a unit of loudness. Just like pitch, the loudness of a sound is a subjective quantity that only exists in the minds of people.

9.10.1 Loudness

When you increase the volume of your stereo, you are technically increasing the amplitude of the vibrations of molecules, which the in turn brain interprets as an increase in the loudness of the sound. The loudness of a sound is a subjective quantity most closely related to the SIL or sound pressure level. Though not always true, in general we say that sounds with a greater SIL or SPL sound louder to us.

This may seem not very important, but it is essential for our understanding of sound. Several odd anomalies exist in the mental representation of sound amplitude, for example:

1. Loudness is not additive in the way that amplitudes are; rather, loudness is logarithmic, like pitch.
2. Loudness varies with the frequency of the incoming sound wave.
3. Loudness also varies with sound quality and ear sensitivity. This means the same sound may seem to have different loudness depending on the individual hearing it.

Another interesting fact is that the human ear tends to amplify sounds at frequencies between 1000 Hz to 5000 Hz. This means that sounds in this frequency range *seem* louder to us. Hence, two sounds with the same intensity but different frequencies can be perceived to have different loudness. Despite this distinction between intensity and loudness, it is probably safe to assume that more intense sounds are also the ones we perceive to be the louder.

9.10.2 Loudness as a function of distance

Loudness, which is a subjective quantity, falls off as the distance from the source increases. The perception of loudness depends on the sound intensity, the frequency, and the duration of sound. On the other hand, sound intensity and sound pressure are objective and can be measured by instruments. Sound intensity is proportional to the sound energy and to the mean square of the fluctuations in pressure. We recall from section 9.3.1 that the sound intensity for waves in 3D falls off according to $1/r^2$, which is called the *inverse square law*.

Loudness, as a type of human perception, cannot be truly made quantitative, but a good rule of thumb is that for a sound to be perceived as twice as loud at a certain distance, the intensity of the sound must be increased tenfold. This is known as the *tenfold rule* and it states that loudness (L) and sound intensity (I) are related by

$$L \approx I^{0.301} \propto \left(\frac{1}{r^2}\right)^{0.301} = \frac{1}{r^{0.602}}. \tag{9.27}$$

Example 9.12. How far do you have to move away from an earsplitting noise at 10 m for it to be tolerable?
Solution:
From equation (9.27), we can say that loudness at 10 m of an earsplitting noise is given by

$$L_{10} \approx \frac{1}{10^{0.602}},$$

and a tolerable loudness of sound would be about four times less loud at, say, distance r, which is given by

$$L_r = \frac{L_{10}}{4} \approx \frac{1}{r^{0.602}}.$$

Comparing the above equations,

$$\frac{L_{10}}{L_r} = \frac{L_{10}}{\frac{L_{10}}{4}} = \frac{4 \times L_{10}}{L_{10}} = \frac{r^{0.602}}{10^{0.602}}$$

$$4 = \frac{r^{0.602}}{10^{0.602}}$$
$$15.99 = r^{0.602}$$
$$15.99^{1.66} = r.$$

Solving for r, we see that we must be at least 100 m away from the source of noise for it to be tolerable.

9.10.3 Other units for loudness and loudness level

Loudness is sometimes also expressed using a unit called a *sone* and the loudness level is expressed in a unit called a *phon*.

Sone: the sone, proposed in 1936 CE by Stanley Smith Stevens, is a unit sometimes used for loudness; however, it is not an SI unit. The sone scale was created to provide a linear scale for loudness, and loudness (in sones) expresses a subjective rating of loudness. Doubling the perceived loudness doubles the sone value and by convention, the loudness of a 1000 Hz sine wave at a sound level of 40 dB equals 1 sone.

Phon: the phon, a dimensionless unit, was proposed as a unit of loudness level (L_N) for pure tones by S S Stevens. The loudness level (in phons) expresses the sound pressure level of a tone that sounds equally loud when compared to a 1000 Hz tone. The purpose of the phon scale is to compensate for the effect of frequency on perceived loudness of tones. By definition, the loudness level in phons mathematically equals the sound pressure in decibels. 1 phon numerically equals 1 dB SPL at a frequency of 1000 Hz, and orchestral music is generally considered to have values from 40 to 100 phons. The general relationship between the phon and the sone is given in table 9.1.

Table 9.1. Relationship between the phon and the sone.

sone	2^0	2^1	2^2	2^3	2^4	2^5	2^6	2^7	2^8	2^9	2^{10}
	1	2	4	8	16	32	64	128	256	512	1024
phon	40	50	60	70	80	90	100	110	120	130	140

Review questions 9.3.
1. What is a free field? How does the sound level decrease in a free field when the distance from the source is increased?
2. What is loudness of a sound? What factors does it depend on, and what are the SI units of loudness?
3. What is a loudness level? What are its units?

Exercises 9.3.
1. If a sound pressure is equal to 10 000 μPa, what is the corresponding SPL?
2. If a sound power is equal to 100 W, what is the corresponding db SWL?
3. What loudness levels in phons correspond to loudnesses of 1 sone, 4 sones, and 32 sones?
4. The ambient noise on a construction site is 88 dB; however, the noise power level measured at a particular point on the site when a jackhammer is operating is 118 dB. How many times louder is this noise than the ambient noise?

Further reading
- Resnick R, Halliday D and Krane K 2002 *Physics* vol 1 5th edn (New York: Wiley)
- Berg R E and Stork D G 1994 *The Physics of Sound* 2nd edn (Englewood Cliffs, NJ: Prentice-Hall)
- Rossing T D, Moore R F and Wheeler P A 2002 *The Science of Sound* 3rd edn (Upper Saddle River, NJ: Pearson Education)

- Emanuel D C and Letowski T 2009 *Hearing Science* (Philadelphia, PA: Wolters Kluwer)
- Hall D E 1991 *Musical Acoustics* 2nd edn (Pacific Grove, CA: Brooks/Cole)
- Fletcher N H and Rossing T D 1998 *The Physics of Musical Instruments* (Berlin: Springer)

IOP Publishing

The Physics of Sound and Music, Volume 1
A complete course text (Textbook)
Samya Bano Zain

Chapter 10

The human factor

All different kinds of waves, for example, light, sound waves, etc. have similar properties, such as frequency (f), wavelength (λ), and speed of propagation (v). These properties are related by the general expression for the speed of a wave:

$$v = \lambda f. \tag{10.1}$$

In this chapter, we attempt to discuss the human auditory system in some depth and explain how it is extraordinarily complex but amazingly remarkable.

10.1 The ranges of human hearing and sight

In this section, let us briefly compare a human's ability to see with their ability to hear.

10.1.1 The range of visible light

All electromagnetic radiation is called light, but we can only see a small portion of it, called the *visible portion* of light. The rest of the electromagnetic spectrum has wavelengths that are either too large or too small for our biological limitations. A typical human eye responds to wavelengths from about 390 to 750 nm, which corresponds to a frequency band of 400–790 THz (THz = terahertz = 10^{12} Hz). Red light has the lowest frequency (4×10^{14} Hz) and the longest wavelength, while violet light has the highest frequency (8×10^{14} Hz) and the shortest wavelength.

The maximum sensitivity of the human eye generally corresponds to light of a wavelength of around $\lambda = 555$ nm (5.5×10^{-7} m) and a frequency of 540 THz, which is in the yellow–green region of the visible spectrum. This is because our Sun emits most of its electromagnetic energy at around this wavelength, so humans have evolved such that light at this wavelength appears 'brightest' to us.

An electromagnetic wave with a frequency of less than 4×10^{14} Hz is called infrared (IR) radiation; it is invisible to the human eye. Similarly, an electromagnetic wave with a frequency higher than 8×10^{14} Hz (in the range of ≈300 to 400 nm) is called ultraviolet (UV) radiation and it is also invisible to the human eye. However, many animals can see light in the ultraviolet region; for example, bees can see light in the ultraviolet region. So, plants that depend on insect pollination have evolved such that they are visible in the UV range for bees rather than being colorful in the visible region for humans.

10.1.2 The range of hearing

The human auditory system is extremely complex in structure, and most hearing takes place in the ear; however, recent studies have shown that hearing additionally depends on data processing in the central nervous system. For humans, hearing is limited to frequencies between about 20 and 20 000 Hz (20 kHz). Frequencies capable of being heard by humans are called *audio* or *sonic*. Frequencies higher than audio frequencies are called *ultrasonic*, while frequencies below the audio range are referred to as *infrasonic*. Dogs can hear ultrasounds, which is the principle of 'silent' dog whistles, while snakes can sense infrasounds through their bellies. The intensity ratio between sound that causes pain for our ears and the weakest sound that can be heard is more than 10^{12} (1 000 000 000 000)!

Example 10.1. The frequency range that can normally be detected by the human ear in air is between 20 Hz and 20 000 Hz. What wavelengths do these frequencies correspond to at room temperature?
Solution:
Use equation (10.1):
$$v = \lambda f.$$
Given that the speed of sound at room temperature is 344 m s^{-1}, we can find the wavelength as follows:
For the lower limit, 20 Hz:
$$\lambda_{20} = \frac{v}{f} = \frac{344 \text{ m s}^{-1}}{20 \text{ Hz}} = 17.2 \text{ m}.$$
For the upper limit, 20 000 Hz:
$$\lambda_{20\,000} = \frac{v}{f} = \frac{344 \text{ m s}^{-1}}{20\,000 \text{ Hz}} = 0.0172 \text{ m} = 1.72 \text{ cm}.$$

10.1.3 Infrasound and ultrasound

Some animals primarily use sound to communicate with each other. Each animal species has a range of normal hearing for both loudness and pitch which is beyond that range that humans can hear. Dogs can hear sound waves up to 40 000 Hz, and cats can hear up to 70 000 Hz. In this section, we will look at the difference between ultrasound and infrasound.

1. **Ultrasound:** ultrasounds have frequencies greater than 20 000 Hz, which is the upper limit of the human hearing range. For this reason, humans cannot hear ultrasounds but dogs can. Hence, dog whistles emit frequencies in the range of 18 to 22 kHz, which is beyond the audible range of most humans. Bats use ultrasounds to navigate and hunt in nearly complete darkness. Ultrasounds have many other applications, including the testing of products during construction to detect issues that might be invisible to the naked eye and medical imaging. Ultrasounds at frequencies equal to or greater than 2 MHz are extensively used to safely and noninvasively image internal organs, such as the human heart or the human fetus in the womb. Imaging using ultrasonic sounds is called *sonography*, and is used because it is perfectly safe for human use. Figure 10.1 shows a sonogram of Sami at about 16 weeks.
2. **Infrasound:** infrasound is sound that has frequency of less than 20 Hz, which is the lower limit of human hearing. It is also called low-frequency sound, and the study of low-frequency sound waves is called *infrasonics*. The infrasonic range lies from 0.001 Hz to 20 Hz, and the advantage of such low frequencies is that they have long wavelengths, which gives them the ability to cover long distances and bypass obstacles easily. Animals such as whales, elephants, hippos, etc. are known to use infrasound to communicate over large distances. Elephants produce infrasound waves that travel through solid ground and are sensed by other herds miles away using their feet. Elephant calls range in frequency from 15 to 35 Hz and can be as loud as 117 dB, which gives a possible range over which they can be 'heard' through air of around six miles. These calls are used to coordinate the movement of herds and allow male elephants to find mates. Similarly, whales are known to 'talk' over hundreds of miles using infrasonic sound signals.

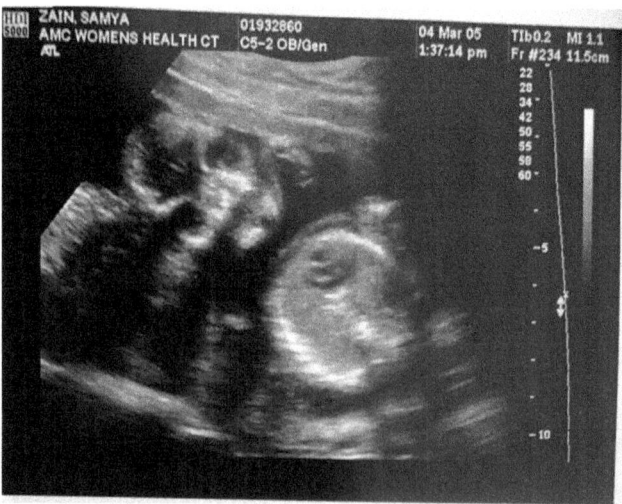

Figure 10.1. A sonogram of Sami at about 16 weeks.

10.2 Speech production in humans

The primary source of sound in humans is the lung. During speech, air that originates in the lungs is forced upwards through the trachea (the windpipe) and the oral (mouth) and nasal cavities (nose), as shown in figure 10.2.

Speech production may be divided into four separate but interconnected processes:

(a) The initiation of an air stream in the lungs.
(b) Vocalization of said air stream in the larynx through the operation of the vocal folds. The primary function of the vocal cords is to regulate and control the airflow by rapidly opening and closing. This motion causes a buzzing that produces sounds. The fundamental frequency of vibration in humans is around 110 Hz for adult men, 200 Hz for adult women, and around 300 Hz for children.
(c) The direction of the air stream by the velum into either the oral cavity (mouth) or the nasal cavity (nose).
(d) The verbalization of the air stream in the oral cavity, which is mainly done by the tongue. The oral cavity (mouth) is one of the most important parts of the vocal tract. Humans control acoustics by controlling the movement of the tongue, lips, cheeks, and teeth.

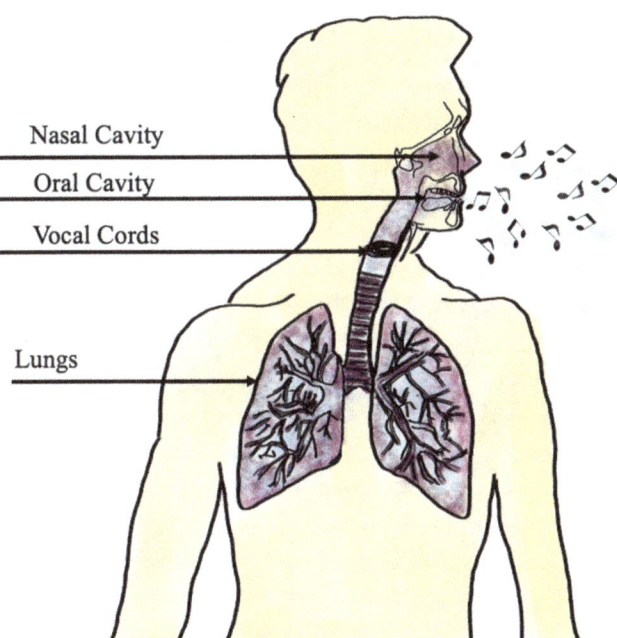

Figure 10.2. A schematic of the human vocal tract.

10.3 Auditory systems

The human auditory system is remarkable in its ability to respond to a wide range of incoming stimuli. When we talk about the auditory system, we have to subdivide it into two main parts, as shown in figure 10.3 that are distinct but interconnected:

(a) Signals processed in the ear (the peripheral auditory system).
(b) Signals processed in the brain (the auditory nervous system).

10.3.1 Signals processed in the ear (the peripheral auditory system)

Sound enters the outer ear and travels down the auditory canal to the eardrum. To understand the signal processing performed by the ear, an overview of the structure of the human ear is necessary, as shown in figure 10.4. The human ear can be subdivided into three basic parts: the outer ear, the middle ear, and the inner ear.

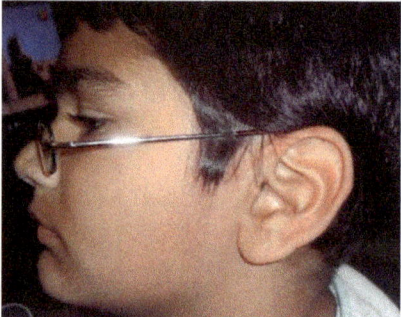

Figure 10.3. Sami's auditory system—his brain & his ear!

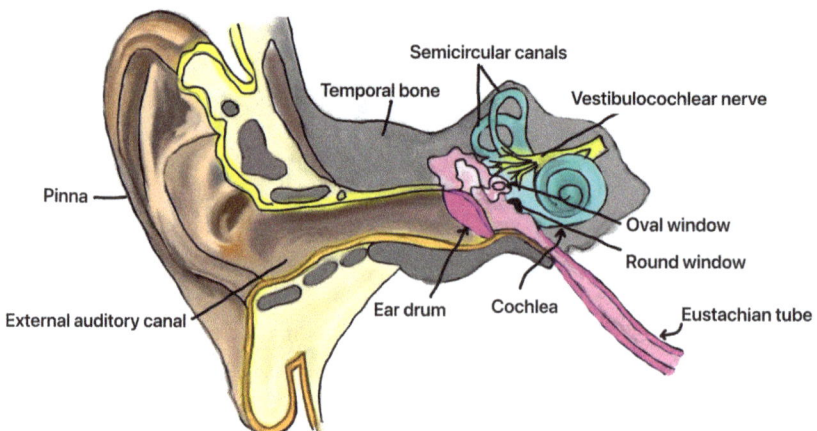

Figure 10.4. The parts of a human ear.

The outer and middle ear receive acoustic energy (sound), collect it, and then transmit it to the inner ear. The inner ear converts the acoustic energy into neural impulses that are further sent to the brain, where they are interpreted as sound.

All parts of the ear work together to convert sound from its original form, namely sound waves in the air, to bioelectrical energy that is transmitted to the brain to be interpreted. In this section, I will very briefly discuss each part of the ear; please refer to other textbooks for more details.

(a) **The outer ear:** the outer ear consists of the external pinna and the auditory canal.
  i. **The pinna**: the pinna acts like a funnel and collects sound waves, concentrating them at the opening of the auditory canal. The pinna helps with the localization of the source direction; due to its shape, it is better at collecting sounds from the front of the face than the back of the head.
  ii. **Auditory canal**: resonances in the auditory canal boost the sensitivity of the ear in the frequency range between 2 and 5 kHz, which is a crucial range for speech recognition.

(b) **The middle ear:**

   The middle ear consists of an eardrum, also called the *tympanic membrane,* which is made up of stretched circular and radial fibers that oscillate with the incoming sound waves and convert them to mechanical waves. These mechanical waves are then transferred to three small bones in the middle ear called the hammer, the anvil, and the stirrup; these bones act together as a lever. The eardrum is connected to the hammer, which is connected to the stirrup through the anvil. The three small bones are collectively called the *auditory ossicles*. The force ($\vec{F}$) exerted by the eardrum on the hammer is amplified by 1.5 to 2.0 times the original force by the time it reaches the oval window of the cochlea in the inner ear. However, since the area of the cochlea is one twentieth of the area of the eardrum, the pressure ($P = \vec{F}/A$) increases to about 4000% of the original pressure, which means that the sound waves become 40 times more powerful than the original wave! This is why the middle ear is also called an *amplifying system*.

   The air pressure inside the middle ear should be the same as it is outside, to allow the eardrum to vibrate freely and painlessly. This is accomplished naturally by venting the middle ear to the outside world through a small tube, called the Eustachian tube or the auditory tube, which runs from the middle ear to the back of the throat. The Eustachian tube opens to equalize the middle ear pressure each time you swallow.

   The auditory ossicles additionally help to protect the ear from damage. If a damagingly loud sound enters the ear, a muscle pulls the stirrup away from the oval window, and the loud sound is not transmitted any further. At the same time, another muscle increases the tension in the eardrum. Together, these two changes temporarily make the ear less sensitive to sounds, and the eardrum is protected from harm that could be caused by the incoming loud sound. This system of protection is called the *acoustic reflex*. Please note that this reflex action is very temporary, and the ear rather quickly returns to its original sensitivity. The only downside is that this muscular response takes a few milliseconds, so it provides little protection against very sudden loud sounds.

(c) **The inner ear:**
   The inner ear is very complex; it consists of the semicircular canals and the cochlea. The semicircular canals help us with maintaining vertical balance, while the cochlea converts the pressure impulses from the middle ear into neural impulses that are sent to the brain to be decoded. The cochlea is filled with two liquids:

   i. Perilymph, which is present in the semicircular canals and is similar to spinal fluid, and
   ii. Endolymph, which is contained in the cochlear duct and is similar to the fluid within cells.

The two liquids are kept separate by two membranes, Reissner's membrane and the basilar membrane.

   i. Reissner's membrane is a very thin membrane that is only two cells thick.
   ii. Even though Reissner's membrane is very thin, the basilar membrane is even thinner. In addition, the basilar membrane is held under a large tension near the oval and round windows. The thickness of the membrane gradually increases from one end to the other, and its tension decreases with increasing thickness. High-frequency incoming waves cause the membrane to vibrate at its maximum amplitude near its thin, high-tension end, whereas low-frequency waves cause the membrane to vibrate at its maximum amplitude near its thicker, lower-tension end, as shown in figure 10.5.

   On the top of the basilar membrane, there is a very delicate membrane called the *organ of Corti*, which consists of several rows of tiny hair cells attached to auditory nerve fibers. These tiny hairs bend when the basilar membrane vibrates in response to an incoming sound. This bending in turn excites the neurons (see the schematic shown in figure 10.6) in the auditory nerve, and the signal is transmitted to the brain.

The location of the maximum-amplitude vibrations is one way that the ear determines frequency. For low-frequency sounds, mostly up to 1 kHz, the ear sends periodic nerve signals to the brain at the frequency of the sound wave. For more complex sounds, which may consist of the superposition of many since waves at different frequencies, the ear naturally performs a spectral analysis and decomposes the complex sound into

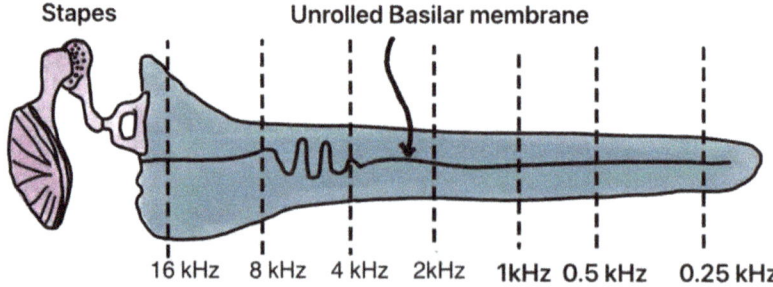

Figure 10.5. The response of the basilar membrane to an incoming sound wave at a frequency of around 3.5 kHz.

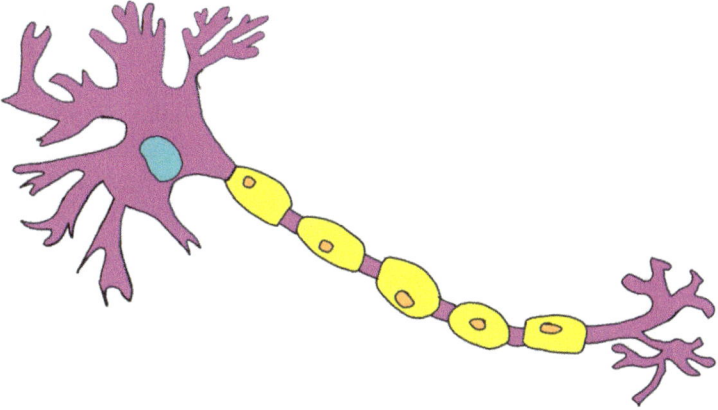

Figure 10.6. A neuron. Courtesy: Eman Zain.

a number of constituent frequencies. The mathematical field of spectral analysis involves the decomposition of a continuous *spectrum* into a sum of discrete (essentially individual) components. This process is called *spectral decomposition*.

10.3.2 Signals processed in the brain (the auditory nervous system)

Nature has given us two ears for signal reception and one brain to decipher it. When a sound transmitted through air molecules bombards the eardrum, it causes the eardrum to vibrate in response. The vibration of the eardrum depends on (i) the impact force on the eardrum (volume per amplitude of sound) and (ii) the frequency of the sound waves. But there is nothing in the molecules that tells the eardrum where they came from or which ones are associated with which sources. The molecules that were set in motion by a baby crying do not carry an identifying tag that says 'baby,' and they may arrive at the eardrum at the same time and in the same region of the eardrum as a sound from your TV. So what does the eardrum do?

Fortunately, we process auditory data in the central nervous system. Our brain identifies the objects in the world using the movement of the eardrum as a guide. The field of psychoacoustics, discussed further in chapter 11, is responsible for the study of this process. The central building block of the nervous system is called the *neuron*, as shown in figure 10.6, which transmits and processes neural impulses. Every human on Earth has about ten billion neurons, to which human intelligence is attributed.[1] Neurons have two major components:

---

[1] It is difficult to appreciate the complexity of the human brain because the sheer numbers of neurons. The average human brain consists of 100 billion neurons. To give some context, suppose each neuron is one dollar and you stand on the street corner trying to give away dollars to people as fast as you can; you give away one dollar per second for 24 hours a day. It would take you more than 2000 years to give away all of your dollars! In addition, neurons do not work individually. The real power and complexity of the human brain comes from the network, i.e. the connections that neurons make and form. Each neuron is connected to approximately a thousand to ten thousand others. Just four neurons can be connected in 63 ways for a total of 64 possibilities. As the number of neurons increases, the number of possible combinations grows exponentially. For just six new neurons, there are 32 768 possible ways in which they can be connected!

1. Receptors called *dendrites*, which receive information from other neurons.
2. *Axons*, which transmit information to other neurons.

Our brains use this interconnection of neurons to understand the world around us, and our brains learn continuously by deciphering patterns found in nature over repeated interactions. Two of the very important processing functions in the nervous system are:
1. **Autocorrelation:** autocorrelation is the comparison of an incoming pulse with a previous pulse to pick out repetitive features.
2. **Cross-correlation:** cross-correlation describes a comparison between signals on two different nerve fibers, for example, a comparison between the signals coming from the cochleas of the right and left ears. This could be responsible for the phenomena of localization and binaural hearing, discussed in chapter 11.

Any given auditory nerve fiber carries two basic types of information:
1. **The place theory of hearing:** the *place* at which a neuron is fired corresponds to the location of excitation along the basilar membrane. As previously discussed, the location of the excitation along the basilar membrane corresponds to the incoming sound, which the brain interprets as the pitch/frequency of the sound. This is called *place theory*, and it gives us information about incoming frequencies.
2. **The duration/temporal theory of hearing,** also called frequency theory or timing theory, states that the human perception of sound depends on the timing sequence (the time intervals between nerve spikes) at which neurons respond to incoming sounds in the cochlea. For incoming sound, the time distribution of impulses contains information about the periodicity and the vibration pattern of the sound waves. This is called the *periodicity pitch*.

Sound signals received by the right ear are nearly always transmitted to the left side of the cerebral cortex of the brain and vice versa. However, both sides of the cortex are connected, and the information is processed throughout the brain. Scientists studying the brain have evidence that suggests that for 97% of the human population, the left-hand side is the dominant side, and it is used for speech processing. The right side of the brain is the dormant/recessive side, and it is specialized for nearly all non-speech-related processing, for example, music, which requires holistic or synthetic processing. The left side of the brain is more analytical, since speech processing requires analytic signal processing of incoming sounds.

On the other hand, the recognition of melodies requires both sides of the brain. It has been experimentally found that melody is recognized better by non-musically inclined people if sound is heard in the left ear, whereas for musically inclined people, melody recognition is better if sound is heard in the right ear. This suggests that musicians learn to process melodies in the dominant, analytical side of the brain. Sometimes, trauma patients are able to sing songs learned before the onset of trauma, even if they are not able to speak the same words.

10.4 Critical bands

The term critical band was first used by Harvey Fletcher in the 1940s, and the concept was later developed by Georg von Bekesy in the 1960s. To understand how the auditory system responds to different incoming stimuli, one must understand the concept of critical bands. Critical bands are a phenomenon created by the cochlea within the inner ear. The cochlea contains the basilar membrane and has a logarithmic spiral shape as shown in figure 10.7 (left). If the cochlea were unrolled and its frequency response mapped onto it, the result would be as shown in figure 10.7 (right). Because of the basic natural structure of the basilar membrane, different incoming frequencies resonate at different points along it. However, the basilar membrane is a continuous element, so it is generally incapable of resolving and analyzing inputs when the frequency difference is smaller than a certain limit. This limit is referred to as the *critical band*.

The critical bands in humans can be defined in two different ways: (i) in terms of the structure of the human ear and (ii) in terms of the incoming sound, as follows:

1. **Critical bands in terms of the structure of the ear:** 'critical band' refers to the specific area on the basilar membrane that goes into vibration in resonance with an incoming simple tone. Its length is determined by the elastic properties of the basilar membrane. Experiments carried out over the years have concluded that there are about 24 critical bands along the Basilar membrane. Each critical band is about 1 mm long and is composed of about 1300 neurons.
2. **Critical bands in terms of incoming sound:** when two pure tones are so close in frequency that there is considerable overlap in their amplitude envelopes on the basilar membrane, they are said to lie within the same *critical band*.

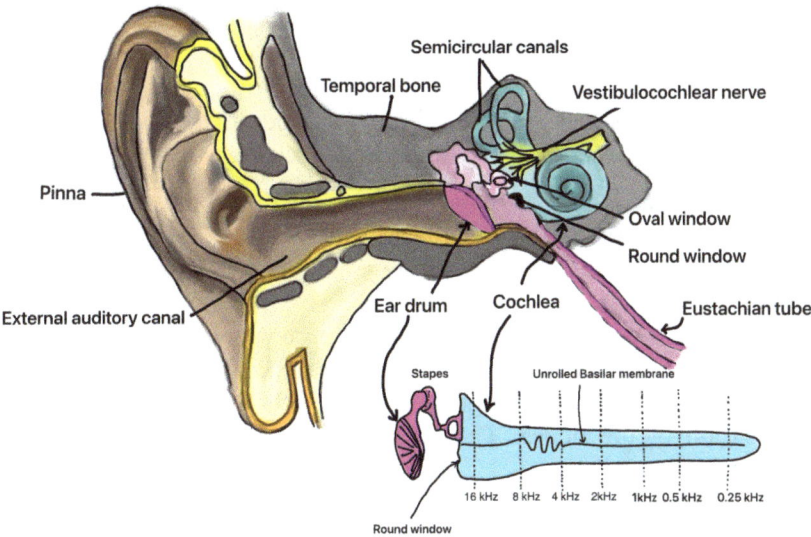

Figure 10.7. Critical bands: (left) the cochlea within a human ear; (right) a frequency map of the basilar membrane.

10.4.1 Visual and auditory illusions

The brain faces three main issues when it tries to identify external stimuli. First, the information arriving at the receptor may be indistinguishable from other information arriving at the same time. Second, the information may be ambiguous; that is, different objects can give rise to similar patterns at the eardrum. Third, the incoming information might be incomplete. Based on these difficulties, the brain has to make educated guesses based on prior information and it has to do it quickly and mostly subconsciously. Helmholtz called this process *unconscious inference*. It basically says that what we 'see' or 'hear' is the end of a long chain of effects that gives us an image of the physical world that may or may not exist in reality.

The processing in which only the neuron circuit considers the incoming information is called *bottom-up processing*, whereas when the frontal lobe, which is the more sophisticated part of the brain, processes the incoming information, it is called *top-down processing*. Bottom-up processing is simple and uncomplicated, whereas top-down processing allows our brains to predict what is coming up, say in music, based on prior knowledge or information. The bottom-up and top-down processes work together at the same time to analyze incoming stimuli. From these, our brain makes a number of inferences which can be based on incomplete or ambiguous information, and sometimes these inferences turn out to be wrong. That is what visual and auditory illusions are, false perceptions given to outside stimuli such as real sounds.

Many auditory illusions are caused by hearing sound patterns that are highly probable, even though they may or may not actually be present in the stimulus. Many composers use the spatial components of music to alter the overall sound experienced by the listener. One common method of doing this is to use combination tones, which are illusions that are not physically present as sound waves; rather, they are created by the listener's own interpretation of sounds they hear.

10.5 Bone conduction

During speaking and singing, two different methods of hearing are utilized to transmit sound to the brain. These two methods include:

1. **Air conduction**: as discussed above, air conduction is the process by which an acoustic signal travels through the structures of the outer ear and middle ear and arrives at the cochlea before being transmitted to the brain for interpretation.
2. **Bone conduction**: bone conduction is the process by which acoustical signals vibrate the bones of the skull to simulate the cochlea. The vibration of the skull bones can result from either or both acoustical and mechanical simulation of the bones. The primary component of bone conduction is the inner ear.

    Hearing by bone conduction plays a very important role in speaking and is the reason why a person's voice sounds different to them when it is recorded and played back. The recorded sound of your voice sounds very unnatural to you because only the airborne sound is received by the microphone, whereas you are used to hearing both components in your own voice. Bone conduction tends to amplify the lower frequencies, so most people perceive their voice as being at a lower frequency or pitch than the pitch at which others hear it.

The sounds of humming or clicking one's teeth are heard almost entirely by bone conduction. If you stop your ears with your fingers, thus interfering with the air path, humming may actually sound louder to you. Bone conduction also plays a very important role in the perception of direct vibrations and high-intensity sounds. Intense sounds may cause vibrations of the skull and create an auditory sensation in the cochlea. If a sound is intense enough, can reach the cochlea despite the most effective earmuffs. This is the reason that most earplugs or earmuffs on their own, i.e. without active noise cancelation, are not effective against loud noises. The hearing threshold and our ability to localize sound sources are reduced underwater even though the underwater speed of sound is faster than in air. This is because underwater hearing takes place entirely by bone conduction.

Example 10.2. Before we move on, let us quickly talk about some interesting facts related to sound.
1. Why do your ears pop when you go to higher altitudes?
   Solution:
   Slow atmospheric air pressure changes can produce pressure differences between the atmosphere and the inner ear. This pressure difference creates a force on the eardrum. When the force on the eardrum is sufficiently large, it is painful for our ears. This effect is often felt at the beginning and end of flights, i.e. during altitude changes. When you go up to high elevations, the change in pressure causes your ears to pop.
2. How do we keep our balance?
   Solution:
   Near the top of the cochlea are three loops called the semicircular canals. These canals are full of liquid. When you move your head, the liquid moves. It pushes against hair-like nerve endings, which send messages to your brain. From these messages, your brain can tell how your body is moving.
   When you feel dizzy after spinning around, it is because the liquid inside the semicircular canals is still in motion after you have stopped moving. This makes the hairs of the sensory cells bend in all different directions, so the cells send incorrect signals and confuse your brain.
3. What is an earache?
   Solution:
   Too much fluid pressure on your eardrum causes an earache. Earaches are often the result of an infection, allergies, or a virus. Babies can get earaches because of milk backing up in the Eustachian tubes, which causes bacteria to grow and may cause hearing problems later in life. This is why most doctors do not recommend babies be given milk as a method to help them fall asleep.

Review questions 10.1.
1. What is the frequency ratio between the highest and lowest audible frequencies and frequencies that we can see?
2. What are the main parts of the ear?
3. How does the basilar membrane respond to incoming sound? What part of the basilar membrane responds to low-frequency vibrations?
4. How does our brain localize sounds of low and high frequencies?

Exercises 10.1.
1. Can you think of two or possibly three effects in the middle ear that can decrease the amount of vibration that reaches the inner ear?
2. Assume the speed of sound is about 344 m s^{-1} at room temperature. Find the wavelengths that correspond to following frequencies in meters and in cm:
   (a) 440 Hz
   (b) 1000 Hz
   (c) 10 000 Hz
3. The effective area of an average adult eardrum is approximately 0.56 cm^2. Calculate the force exerted on the eardrum if it is subjected to a sound pressure variation of about 10^{-2} N m^{-2}.

Further reading
- Berg R E and Stork D G 1994 *The Physics of Sound* 2nd edn (Englewood Cliffs, NJ: Prentice-Hall)
- Rossing T D, Moore R F and Wheeler P A 2002 *The Science of Sound* 3rd edn (Upper Saddle River, NJ: Pearson Education)
- Emanuel D C and Letowski T 2009 *Hearing Science* (Philadelphia, PA: Wolters Kluwer)
- Hall D E 1991 *Musical Acoustics* 2nd edn (Pacific Grove, CA: Brooks/Cole)

IOP Publishing

The Physics of Sound and Music, Volume 1
A complete course text (Textbook)
Samya Bano Zain

Chapter 11

Psychoacoustics

The scientific study of sound perception in humans is called psychoacoustics. More specifically, psychoacoustics, a term coined by Gustav Fechner in 1860 CE, is the branch of science that studies the psychological and physiological responses associated with sounds that include noise, speech, and music. Basically, psychoacoustics is the study of how sound is perceived and how it affects our bodies and minds.

Sound starts as a mechanical wave produced by a vibrating object. This mechanical sound wave travels through the air and reaches the ear as a mechanical sound wave. In the ear, it is transformed into neural stimuli before it travels to the brain as a nerve impulse. Once the brain receives this nerve impulse, it is interpreted as sound, and we respond to it either positively or negatively, which affects our moods. This means that both the human ear and the human brain are key stakeholders in our auditory experiences. In this chapter, we discuss hearing in humans, which includes localization, echolocation, masking, selectivity, binaural hearing, etc. We also discuss the effects of noise on humans.

11.1 Hearing in humans

Psychoacoustics is divided into two main areas: perception and cognition. The two systems are interrelated and influence each other to a large extent. Perception deals with the human auditory system, for example, the range of human hearing, equal loudness contours, masking, localization, etc. and may be understood relatively easily, whereas cognition, which is a lot more complicated, focuses on what happens in the brain. Cognition, for example, deals with questions like 'Why does an octave have to be in a 2:1 frequency ratio?' Answering this is not easy and would require a thicker book with a ton of mathematics, so I will not address it here. But I suggest that any budding musician should try to understand these concepts, even at a fundamental level, to achieve better music production and experience.

11.1.1 Spatial hearing or localization

On Earth, we are always surrounded by sounds, whether we are indoors or outdoors. The sound resulting from chirping birds is generally considered to be a pleasant sound, whereas most people agree that busy highway traffic is noisy. To a listener, the apparent loudness of a sound depends on the sound source and the distance between the receiver and the source. The loudness of a sound also depends on other factors, such as the direction of the wind or the number of obstacles present between the source and the receiver.

When multiple sounds reach our ears, they either interfere constructively or destructively with each other. One issue is that there are many, many sounds reaching our ears at any given time, so how does our ear tell where the sound comes from? To answer this question, let us look at the tools we have at our disposal to locate the origin of sound:

1. The first is the pinna of the outer ear (discussed in section 10.3.1). As we saw, the pinna acts like a funnel; it collects the mechanical sound waves from the air and concentrates them at the opening of the auditory canal. Because of the shape of the pinna, humans naturally prefer sounds coming from the front of the face rather than the back of the head. Hence the shape of the pinna helps with the front–back localization of the incoming sounds.
2. For high-frequency sounds, i.e. frequencies greater than 4 kHz, the fact that our two ears are located on different sides of the head results in a difference in intensity. The head casts a *sound shadow*, as shown in figure 11.1, so sound coming from the left has a larger intensity at the left ear than at the right ear.
3. For low-frequency sounds, the ears use both the difference in the arrival time and the phase difference between the waves to find the location of the sound.

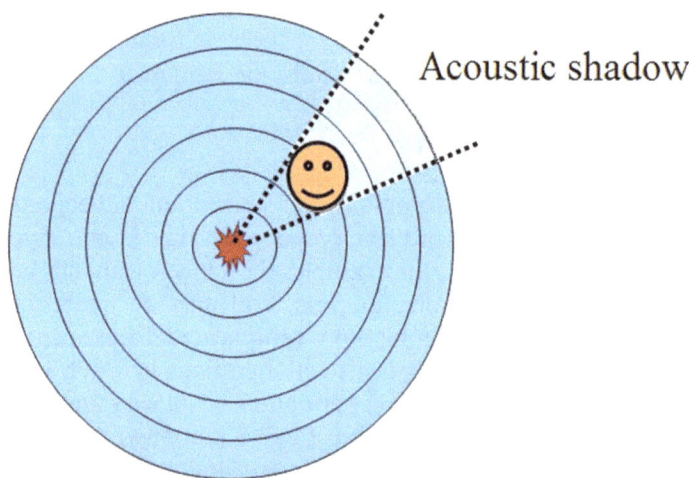

Figure 11.1. An acoustic shadow.

Utilizing these facts in relatively quiet environments, healthy humans with normal hearing can localize the direction of sound sources to an accuracy of just a few degrees. Comparative measurements in mammals have demonstrated that humans have amongst the best localization performance. Of all species measured, only dolphins and elephants perform as well or better, which is an amazing skill for humans to possess.

11.1.2 Auditory masking

Two tones that exist at the same time influence each other. If one tone is more intense than the other, it changes the perception of the first tone. This phenomenon, in which the perception of one sound is obscured by the presence of another sound, is called *auditory masking*. One of the effects of auditory masking is to reduce the loudness of sound. In general: (i) low-frequency tones mask high-frequency tones, (ii) the energy needed to mask a tone increases with delay and the duration of the tone, and (iii) no masking occurs beyond 100 to 200 milliseconds.

Auditory masking is known as *frequency or spectral masking* when it occurs in the frequency domain, and it is called *temporal masking* when it occurs in the time domain. Masking plays a very important role in the perception of combinations of tones. It allows humans to reject unwanted sounds, for example, the sounds of the HVAC in the background of a classroom or a theater. In most circumstances, the lower tone is called the *masked tone* (signal) and the louder tone is called the *masking tone*. Usually, but not always, a masked tone is variable in frequency, whereas the masking tone is fixed in frequency.

The *masking level* is the sound intensity level to which the masked tone must be raised for it to be heard in the presence of the masking tone. For two identical frequencies, the *masking tone* is very effective and the masking level is very high. On the other hand, when the difference in frequencies between the masking and the masked tones (signal) becomes large, the masking level is reduced, such that even low-intensity instances of the masked tones may easily be heard. We can generalize this to say that the masking level drops as the frequency difference between the masked tone and the masking tone increases.

Masking is related to the critical band in our ears. When two successive signals fall within the same critical band, the masking effect is apparent. In music, when melodic lines are closely spaced, we hear a melody with rich harmonic texture, and as the lines become more spaced, each can be heard more clearly as an individual melody line.

There are four main types of masking, which are briefly discussed here:
1. **Forward masking:** masking prevents us from being able to compute the perceived loudness of a complex tone as the sum of its components. Two successive signals that lie within the same critical bands may show masking effects. For example, noise can mask a tone if the noise is sufficiently loud and the delay between the noise and the tone is short. *Forward masking* occurs when the masker precedes the signal. Forward masking is also called post-masking.

2. **Backward masking:** backward masking, also called pre-masking, refers to temporal masking by quiet sounds (maskers) that occur moments before a louder sound (signal). In backward masking, the masker follows the signal and both sounds occur within 20 milliseconds of each other. Please note that backward masking may not be purely involuntary, since trained subjects may show no backward masking.
3. **Partial masking:** sounds are almost never heard in isolation. The presence of other sounds not only raises the threshold for hearing a given sound but generally also reduces the loudness of sound. This is called *partial masking*.
4. **Central masking:** the masking of a tone in one ear can be caused by noise in the other ear, and this is called *central masking*.

11.1.3 Auditory adaptation and auditory fatigue

Both auditory adaptation and auditory fatigue are considered to be mostly reversible to a certain extent. Adaptation is a feature of the auditory neurons, in which neurons reduce their responses to recurring (maybe loud) sounds that reach the ears. Like most other sensations, sensitivity to loudness decreases with prolonged stimulation. Such a decrease is called *adaptation*. Under most listening conditions, loudness adaptation experimentally appears to be very small. A steady 1000 Hz tone at 50 dB causes little adaptation, whereas the loudness of a tone that alternates between 40 dB and 60 dB appears to decrease in loudness over the first two or three minutes of exposure. Please note that adaptive processes are not always harmful; they can be very useful, in fact, they are essential for auditory rehabilitation. For example, they allow patients that need hearing aids or cochlear implants to adapt to them after a little while.

Exposure to loud sounds affects our ability to hear other sounds at a later time. This is called *auditory fatigue* and may happen when someone is exposed to noise levels that are much higher than those of normal speech levels. Auditory fatigue is most commonly experienced by people who work in noisy environments. Sound levels of more than 110 dB can produce permanent damage very quickly. People who experience auditory fatigue usually report tiredness and/or hearing a whistling or buzzing in their ears, which is a sign of a temporary sensory hearing loss. This temporary hearing loss usually lasts about a day or two immediately after exposure to extreme noise. However, if the exposure is prolonged, this temporary hearing loss may lead to tinnitus,[1] which might become permanent if the exposure continues.

11.1.4 Selectivity

A remarkable quality of the auditory system is *selectivity*. From a noisy room crowded with people, it is possible to pick out the dialog of a single speaker. During sleep, we tune ourselves out from the sounds of heavy traffic but awaken to the

---

[1] Tinnitus is characterized by a noise in a patient's ears which does not exist externally. Most patients describe the sound as a ringing or buzzing, but other patients have reported sounds that include wheezing, hissing, clicking, squealing, and ticking.

sound of an alarm clock. A mother can sleep through a hurricane but will awaken at the slightest sound of her baby crying.

We know that for humans, lower frequencies are heard as low pitches and higher frequencies are heard as high pitches. However, our sensitivity varies greatly over the audible range. For example, a 50 Hz sound must reach 43 dB before it is perceived to be as loud as a 4000 Hz sound at 2 dB. In this case, we require the 50 Hz sound to have 13 000 times the actual intensity of the 4000 Hz sound in order to have the same perceived intensity!

11.1.5 Binaural hearing

Binaural hearing refers to the ability of humans to integrate the information the human brain receives from two ears. Binaural literally translates to 'having or relating to two ears.' Binaural hearing helps humans and animals to localize sound sources even when the environment is noisy and complex.

For sounds whose frequencies are lower than about 1.5 kHz, the location of the source is determined by the difference in the arrival times of sound waves at both ears, as seen in figure 11.2. The maximum difference in the arrival time (Δt) of the sound is given by the formula:

$$\Delta t = \frac{L_2 - L_1}{v}, \tag{11.1}$$

where L_1 is the distance to the ear closer to the source, L_2 is the distance to the ear farther away from the source, and v is the speed of sound at the particular temperature (331 m s^{-1} at 0 °C and 344 m s^{-1} at 21 °C). For frequencies higher than 1.5 kHz, the location of the source is determined by the spectral amplitude difference between the ears, since the head masks high-frequency sound from the further ear.

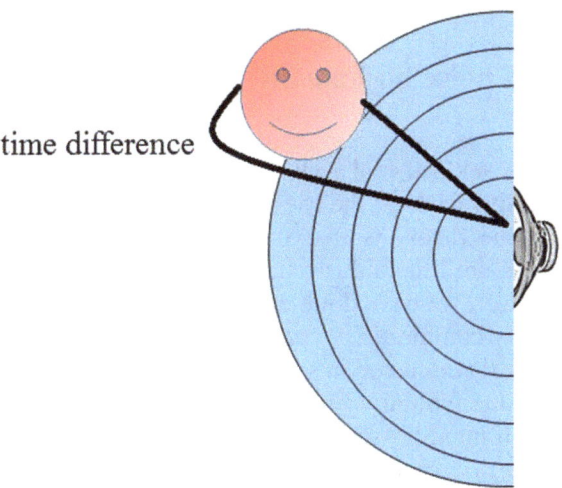

Figure 11.2. Binaural hearing.

As a rule, binaural hearing is superior to monaural (through one ear) hearing because binaural hearing increases signal loudness, localization stimulus, and speech intelligibility and improves sound quality. In normal rooms, it is essential to have two ears to understand speech and to counteract the echoes that occur naturally in closed spaces.

11.1.6 Echolocation

Echolocation is an ability developed by certain animals that helps them to detect objects in their surroundings by emitting sounds and then perceiving the echoes bouncing from objects in the environment. To perform echolocation, animals have developed a biotransmitter and two receivers that are placed slightly apart. Animals using echolocation emit sounds and then wait to hear echos in the two receivers. The returning echoes arrive at the two receivers at different times and different loudness levels depending on the position of the object. This time delay and difference in loudness allow the animal to 'see' their environment by providing information about the distance and direction of the objects in it.

Bats emit ultrasound for echolocation and use it to navigate and hunt, most often in total darkness. Blind freshwater dolphins live in highly muddy parts of the Indus river, Pakistan. The water is so muddy that they have lost the use of their eyes, but they compensate for this loss with their highly developed sonar system. These dolphins emit sounds that are reflected back when they collide with an object in their path. These reflected sounds are received by tiny openings (2 to 3 mm in diameter) at the back of their eyes by which the blind dolphins 'see.' In fact, the whole skull, including the lower jaw, helps this dolphin to capture reflected sounds which are then transmitted to the brain for processing. When a blind dolphin wants to have an overall image of something, it issues low-frequency signals that can propagate over several hundred meters but are not very precise. To see a detailed image, it uses high-frequency signals that do not go far but provide an impressive amount of information, from the shape of an object to its texture!

11.2 The effect of noise on humans

11.2.1 Ear damage

Our ear is a highly developed and well-designed organ; however, unlike the eye, which has two protective devices, the iris and the eyelid, the ear does not have any natural protection. One theory is that during their evolution, humans were not exposed to loud noises until the industrial age, so we never needed 'natural built-in ear protectors.' However, we have double protection for our eyes because exposure to bright light was very common. One protective mechanism that the ear has is the *acoustic reflex*, previously discussed in section 10.3.1. The acoustic reflex, also called the *auditory reflex*, is an involuntary muscle contraction that happens in the middle ear in response to high-intensity sound.

Overall, the ear is not generally damaged by loud noises except in unusual circumstances such as a loud, abrupt sound resulting from an explosion that may rupture the eardrum. The rupture of the eardrum may cause temporary hearing loss and/or a possible middle-ear infection. The membrane generally heals over time; however, if the scaring is large, it may lower the sensitivity to high-frequency sounds.

After extensive studies of noise exposure in humans and animals, most of the scientific community agrees that if the ear is permanently damaged, it is most likely to occur in the organ of Corti. Excessive noise exposure first destroys the delicate hair cells on the organ of Corti and then eventually the entire organ is damaged. One theory is that overexposure to loud noise causes the hairs to work at a higher rate. If this continues for a long time, it results in the death of the cells themselves. Unfortunately, these kinds of cells do not regenerate. If they are lost once, they are lost for life.

11.2.2 Measuring hearing loss

The main way to measure hearing loss is by means of an 'audiometer.' An audiometer measures the hearing threshold at various test frequencies increasing from lower to higher intensities because hearing sensitivity varies with the input sound frequency. A pure-tone audiometer can test the hearing by applying several selected pure tones at various frequencies to each ear individually. The results from the audiometer are then compared to normal hearing at each frequency. There are many harmful effects of high-intensity sounds or noise on humans, some of which are listed here:

1. **Temporary hearing loss:** temporary hearing loss is also called noise-induced temporary threshold shift (NITTS or TTS for short). If the noise is loud but its duration is small, the ear has a remarkable ability to desensitize itself to it. Once the overload is removed, the threshold returns to its normal level remarkably quickly.
2. **Permanent hearing loss:** permanent hearing loss is called noise-induced permanent threshold shift (NIPTS). If the duration of the loud noise is long or it is repeated often, the temporary hearing loss becomes permanent and a nonreversible shift in threshold occurs. As an example, consider a person who works in an environment where they are exposed to loud noise. Initially, hearing loss takes place mainly around 2000–8000 Hz, which does not interfere with normal speech (which mainly consists of sounds in the 'speech band,' 300–3000 Hz). However, as the exposure continues, hearing loss becomes more apparent. One interesting side note is that NIPTS first occurs at frequencies greater than the speech band and hence preventative action can be taken before dangerous levels are reached and cause damage.
3. **Sleep interruption:** sleep interruption is a very common but highly underrated side-effect of a noisy environment. Almost everyone has experienced sleeplessness due to outside noises, such as construction and/or traffic in big cities, or early birds chirping outside the window in the morning in the country. Stimuli of 50 dB are usually found to be uncomfortable, and stimuli of 70 dB awaken sleepers as well as causing problems for people trying to get to sleep.

    Alarm clocks are constructed to make loud sounds with sudden tone changes because the brain interprets such tonal change as a dangerous thing,

causing us to wake up. Modern alarm clocks have radios that can be set to start playing at specified times and are known as clock radios. There are advantages to clock radios over traditional alarms Even though traditional alarms are loud, if we hear them often enough our brain *adapts* to them and no longer responds to them. Clock radios play whatever station the radio is tuned to, with programming that is different every day, so that the brain does not have a chance to *adapt* to it.
4. **Speech interruption:** humans communicate with each other by speaking to each other. Speech communication depends on many factors, such as environmental sound levels, the distance between the speaker and the listener, clarity, and room acoustics. If the environment is loud and noisy, loud sounds are just a nuisance at first, but prolonged exposure results in significant hearing loss.
5. **Other physiological effects:** a number of experiments have been performed in order to predict and study the effects of noise on subjects. As with all human experimentation, it is hard to definitively conclude the effects of noise; however, some general conclusions may be drawn. First, steady sounds above 90 dB do not affect the performance of subjects; however, the same sounds are disruptive if they are irregular. Second, high-frequency noise is more disruptive then low-frequency noise, which makes sense considering that humans are more sensitive to higher-frequency sounds. Third, sudden noises have a tendency to startle, a phenomenon utilized by horror movie-makers to produce shock value and terror in their movies. Fourth, noise is less likely to reduce the quantity of work done by subjects but more likely to reduce the accuracy of the work performed.

11.3 Noise control

Some of the noisiest environments are factories, in which thousands of people work. In environments such as these, ambient noise reduction can be a major asset. The simplest way to talk about noise control is to divide it into three basic elements:
1. Source, i.e. the cause of noise, for example, noisy machinery, heavy traffic.
2. Path, i.e. the space between the source and the receiver. This may be a direct air path or a complex path through a building, which includes but is not limited to noise transfer through windows, etc.
3. Receiver, the affected individuals who have to live in the noisy environment.

The first step is to stop noise at the source, which involves quieting noisy components of machines, damping vibration panels, and installing soundproof barriers. The next line of defense could be to reduce sound as much as possible along the path to the receiver, say by installing soundproof walls, etc. Last but not least, we need to protect the receiver. This may include wearing ear protectors while working in a noisy factory or riding a bike in heavy traffic. Another approach that has proven useful is limiting the time of exposure to the noise.

11.3.1 Sound absorption

Since energy and matter can be neither created nor destroyed, the absorption of sound involves the conversion of sound energy into a different form of energy. The most common way is to convert sound into heat energy. To absorb sound traveling in air, we use porous insulating materials made up of many small fibers. The air-carrying sound waves move back and forth inside the material, the resulting friction generates heat, and sound is absorbed. Since sound waves do not have too much energy, the temperature rise in the material is small and not measurable.

11.3.2 Noise regulations

In the US, the Noise Control Act of 1972 was the first major piece of legislation passed for noise control. This act directed the USEPA (United States Environmental protection Agency) to work on research and development and the distribution of information to make the population aware of the problem of noise pollution and its harmful effects.

Nearly all states in the US have noise-control ordinances, which may or may not be similar among states; for example, in Pennsylvania, noise disturbance is defined as any noise which:
1. Endangers the safety or health of humans or animals, or
2. Annoys or disturbs a reasonable person of normal sensibilities, or
3. Jeopardizes property values or harms the environment, or
4. Is in excess of the established allowable noise level.

Review Questions 11.1.
1. What is the threshold of audibility for humans?
2. What is the threshold of pain?
3. How do people determine the source of a sound?
4. Explain echolocation in your own words.
5. How does our brain localize sounds of low and high frequencies?

Exercises 11.1.
1. What is the dB ratio between the threshold of pain and the threshold of audibility? Hint: use 3 dB as the threshold of audibility.
2. Find the distance between your ears. (Hint: the formula for circumference $C = 2\pi r$.)
3. Calculate the maximum difference in arrival time for you if a sound comes directly from your right side. (Hint: sound travels at $344 \, \text{m s}^{-1}$ at room temperature.)

Further reading

- Berg R E and Stork D G 1994 *The Physics of Sound* 2nd edn (Englewood Cliffs, NJ: Prentice-Hall)
- Rossing T D, Moore R F and Wheeler P A 2002 *The Science of Sound* 3rd edn (Upper Saddle River, NJ: Pearson Education)
- Emanuel D C and Letowski T 2009 *Hearing Science* (Philadelphia, PA: Wolters Kluwer)
- Daniel J L 2006 *This Is Your Brain on Music: The Science of a Human Obsession* (New York: Dutton Penguin)
- Hall D E 1991 *Musical Acoustics* 2nd edn (Pacific Grove, CA: Brooks/Cole)

IOP Publishing

The Physics of Sound and Music, Volume 1
A complete course text (Textbook)
Samya Bano Zain

Chapter 12

The acoustics of rooms

In this chapter, we address the acoustics of rooms, or room acoustics, which is a subfield of acoustics that deals with sound behavior in closed spaces. We briefly discuss acoustic spaces, the acoustic environment in which sound is heard by a listener, and how good architectural designs bring out the best sounds. Improving room acoustics benefits everything one does in any space. It increases focus and helps with concentration, which leads to increased productivity in your work. With good acoustics, everything sounds better, which lifts everyone's mood and improves mental health.

12.1 Sound propagation

We know from chapter 9, section 9.3 that in 3D, sound waves from an isotropic source travel away from the source equally in all directions. Assuming no energy is absorbed by the medium and there are no obstacles to reflect or absorb the sound, the intensity (I) at distance (r) from an isotropic source obeys the *inverse square law* for a spherical area surrounding a source. This means that as the distance from the source doubles, the intensity is quartered and the energy passing through one square meter of surface on a sphere per second drops off according to $1/r^2$. In terms of sound intensity level, as we proved in example 9.5, each time the distance is doubled, the sound intensity level decreases by 6 dB.

This chapter focuses on the acoustical properties of different spaces. The acoustical properties of spaces fundamentally depend on the ultimate purpose of the space. Requirements for rooms at home are very different from the requirements of theaters and/or music recording studios. A good acoustical design does not happen by accident and must be planned carefully. Special care must be taken when considering the final purpose and the use of the space.

12.1.1 Background noise

Background noise, also called *ambient noise*, is the sound pressure level naturally present at any location, outside or inside. Background noise is used as a reference to study any produced sound in a space. For example, to measure the loudness of a guitar, we first need to measure the background noise, such as the sound produced by the heating, ventilating, and air-conditioning (HVAC) unit, and take that into account.

Background noise is measured using a sound level meter (previously discussed in section 9.7.2) and its SI unit is the pascal (Pa). Noise levels are generally measured using the A-weighting scale for the frequency weighting filter and are expressed in dB(A). The A-weighting is used because it is very efficient in measuring low-level sounds (around 40 phon; section 9.10.). The average normal background noise level varies from situation to situation; for instance, in recording studios, background noise is kept to around 25 dB, in classrooms it is about 30 dB, in offices and homes it is about 45 dB, and in noisy spaces such as restaurants it can be more than 50 dB.

12.1.2 Indoor sound propagation

The speed of sound is approximately 344 m s^{-1} at normal temperature (20 °C). So, this means it takes between 0.02 and 0.2 s for a sound to reach the receiver directly from the source, depending on the distance between the receiver and the source. This sound is called *direct sound*. However, indoor sound waves travel only a short distance before they are reflected from the walls, ceilings, and other objects in the room. Walls and other things around a room are obstacles to the path of sound and do two things: (i) they reflect the sound and (ii) they absorb the sounds present in the room. Once the source stops emitting, the sound pressure level decreases at a constant rate until it reaches the inaudibility level in about a millisecond. Since a millisecond is such a short interval of time, this phenomenon is not apparent during normal speech.

Once reflected, called the *first reflection*, as seen in figure 12.1, sound returns back to a receiver with a certain time delay, which is called an *echo*. Please note that a distinct echo is only detected when a minimum time of 50 to 100 milliseconds has elapsed after the previous sound; hence, not all echos present in a room are detectable by humans. The first groups of reflections that reach the receiver are called *early sound*.

12.1.2.1 Reverberant sound
A sound source supplies sound energy that travels through the air and reaches all parts of a room. Some of this energy is absorbed by obstacles (such as furniture, people in the room, etc.), while most of it is reflected. After the first reflection, the following reflections become smaller and closer over time until they merge together. At this point, they are called *reverberations* or *reverbs* for short. When the rate at which the source emits sound energy equals the rate at which sound energy is being absorbed by the environment, we call it *reaching the reverberation level*. Reverberation is the natural result of reflections that arrive at a spot in a

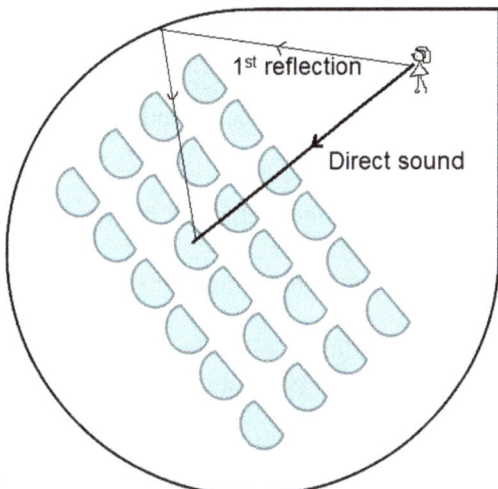

Figure 12.1. Types of sound in a room traveling from a source to a receiver: direct sound and the first reflection.

sequence that takes less than about 50 milliseconds; they depend directly on the frequency of sound. Please note that reverberations continue even after the source stops emitting. However, they decrease in amplitude over time until they reach zero and stop.

Reverberant sound reinforces direct sound and adds to the overall loudness of a sound. Reverberation is a characteristic of all rooms that most people are familiar with. Our brains estimate the size of an enclosed space based on the reverberation and echoes present in the signal that reaches us. Even though few of us understand the mathematical equations necessary to describe how one room differs from another, all of us can, with our eyes closed, tell whether we are standing in a small bathroom, a medium-sized room, a concert hall, or a large theater with high ceilings. This also extends to recorded voices; humans are able to decipher the size of the room where a recording was made using reverberations. This gives recording engineers an advantage when recording songs and/or movies.

In a well-designed large auditorium or concert hall, direct sound should always be louder than the background noise at all locations, even towards the back of the room. So, large auditoriums should have good reverberation but not too much, because too much reverberant sound results in a loss of clarity in the space.

12.1.2.2 The reverberation time
We define the *reverberation time* (RT) as the time required for a sound to *fade away* in an enclosed area after the source of the sound has stopped producing it. The optimum RT for any space depends on the performance requirements in the space.

Spaces primarily used for speeches need a shorter RT so that speech can be understood more clearly. However, if we make the RT too short, we may negatively affect the loudness and tonal balance of the sound. Reverberation effects are often used in music studios to add depth to musical sounds produced in a room. The basic factors that affect the RT include:
1. The size and shape of the room
2. The materials used in the construction of the room, wallpaper, pictures on the wall, etc.
3. Objects in the room, the furniture, number of people, etc.

For a full-size orchestra, the direct sound has experimentally been found to be ideal when one is seated about 20 m (60 feet) away from the stage. This is roughly about the center of many concert halls. In general, we see that larger rooms have longer RTs. The overall behavior of reverberant sound is rather complicated, but the *RT* at mid-frequency ($\approx$ 500 to 2000 Hz) gives a good estimation of the *liveliness* of an auditorium or concert hall. A feeling of liveliness or reverberation is especially important at low frequencies to support bass notes.

In summary, reverberation naturally occurs whenever a sound is present in a space with reflective surfaces for sound; in other words, reverbs occur pretty much wherever reflection exists!

12.1.2.3 Reverberation time 60 (RT60)
A metric called reverberation time 60 (RT60) is sometimes used when we want to make objective measurements of reverberation. RT60 is defined as the time it takes for the sound pressure level to reduce by 60 dB after a sound source stops. It is important to know why 60 dB is used. The main reason is that the loudest sound level in orchestral music is typically 100 dB, while 40 dB is a reasonable background noise level for listening to music. So RT60 measures the time it takes for the loudest noise in a concert hall to fade to the background level.

12.2 The precedence effect

When a sound emanating from the source reaches the listener directly, its level only depends on the physical distance between the source and the receiver. As we saw in chapter 9, this sound intensity level decreases by 6 dB each time the distance between the source and receiver doubles. We noted in section 11.1.1 that humans have the remarkable ability to localize sound by determining the direction of a sound source even in the presence of other sounds. Localization is generally unimpeded by the presence of other sounds. For low-frequency sound, i.e. less than 1000 Hz, localization is mainly achieved by observing the times of arrival at our two ears. For high-frequency sounds, i.e. greater than 1000 Hz, localization mainly uses the difference in sound level observed at our two ears. This is due to the sound shadow cast by our head.

However, if sounds reach our ears from many directions, our system gets confused, especially if the listener is far away from the source. Remarkably, our

auditory processors still continue to somehow figure out the direction of the source from the early sound reaching our ears. This remarkable ability of our auditory system is called the *precedence effect*. Sound is perceived by our auditory system to originate from the direction from which the first sound arrives provided that:
1. The successive sounds arrive within 35 ms.
2. The successive sounds have similar spectra and time envelopes to those of the early sound.
3. The successive sounds are not too much louder than the first sound.

Recent studies have shown that early reflections from the walls are different from reflections from the ceiling. In rooms that have high ceilings, such as concert halls, the sounds reflected from the walls usually reach the listener much sooner than the reflections from the ceiling.

12.3 Room acoustics

It may not seem like a deal-breaker if a room has bad acoustics, for example, if it echoes, magnifies background noise, or has really long reverberations. Bad acoustics are tolerable, but with a bit of consideration the same room could have great acoustics, which represent added value for any room. In this section, we look at the acoustics of rooms.

12.3.1 The acoustics of an ideal room

An ideal room is defined as any space in which sound energy is uniformly distributed throughout. In a completely empty ideal room, all surfaces absorb the same amount of energy and the RT is found to be directly proportional to the volume of the room and inversely proportional to the absorbing surface area of the room. Hence, for an ideal room of volume (V) with walls that do not absorb much sound and a window with surface area (A) through which sound may escape, the RT is given by

$$\text{reverberation time (RT)} = k\left(\frac{V}{A}\right), \tag{12.1}$$

where k is a proportionality constant that depends on the dimensions of the room. In a room which absorbs all sound incident on it, k is calculated to be 0.161 s m^{-1}. The decay of reverberant sound in an ideal room with uniform energy distribution is an exponential decay curve, as shown in figure 12.2.

12.3.2 The acoustics of a real room

A real room has many surfaces that absorb sound and each has its own absorption rate. A good rule of thumb is that if a surface absorbs water, it also absorbs sound. So surfaces such as glass and steel are good reflectors of sound, while carpets and curtains are good absorbers of sound. The absorption coefficient, also called the noise reduction coefficient (NRC), of each surface is a measure of how much of the incoming sound the surface absorbs. If a surface has an NRC of 0.45, it absorbs 45%

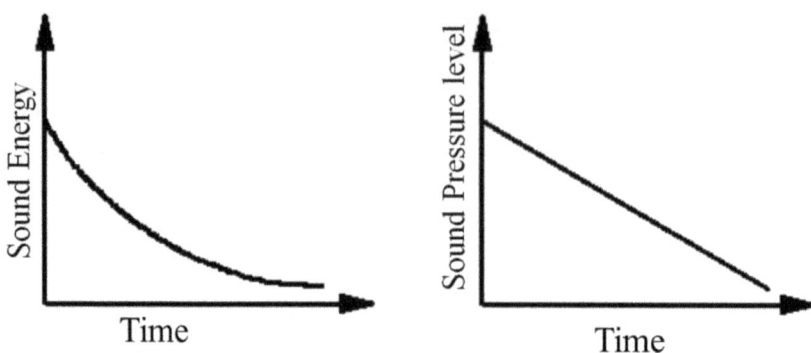

Figure 12.2. The decay of reverberant sound for an ideal room with uniform energy.

of the sound incident on it and reflects 55%. An NRC rating of 1.0 means that the surface absorbs 100% of the sound waves.

If all the different surfaces have surface areas denoted by $A_1, A_2, \ldots$ and their absorption coefficients are denoted by $a_1, a_2, \ldots$, then the total absorption of the room is given by

$$A_{abs} = a_1(A_1) + a_2(A_2) + \cdots. \tag{12.2}$$

The SI units for absorption of the room (A_{abs}) are m^2; however, absorption is also sometimes expressed in units called '**sabins**' or '**metric sabins**,' named after Wallace Sabine, who did pioneering work in room acoustics. In general, one sabin numerically equals the absorption of one square foot of an open window and one metric sabin numerically equals the absorption of one square meter of an open window.

Example 12.1. Calculate the RT for a 4000 Hz frequency in an empty ideal room with absorbing walls. Its dimensions are 20 m × 15 m and it has a ceiling height of 5 m.

Solution:

To find the RT, we can use equation (12.1). We know that $k = 0.161$ s m^{-1} at room temperature:

$$\text{reverberation time} = k\left(\frac{V}{A_{abs}}\right) = 0.161\left(\frac{V}{A_{abs}}\right).$$

We have to find the absorption surface area (A) and volume (V) of the room:

$$A_{abs} = 2(20 \times 5) + 2(15 \times 5) = 350 \text{ m}^2$$

$$V = 20 \times 15 \times 5 = 1500 \text{ m}^3.$$

Substituting these values into the above equation gives

$$\text{reverberation time} = 0.161(\text{s m}^{-1})\left(\frac{1500 \text{ m}^3}{350 \text{ m}^2}\right) = 0.69 \text{ s}.$$

Hence, the RT in a completely empty room is 0.69 s.

Example 12.2. Calculate the RT for the room given in example 12.1 if a carpet is installed in the room. Assume that the absorption coefficient of the carpet is 0.6.

Solution:

In a carpeted room, we have to calculate a new value for the total absorption (A_{abs}) using equation (12.2):

$$A_{abs} = a_1 A_1 + a_2 A_2 + \cdots = a_{carpet} A_{floor} = (0.6)(20 \times 15) = 180 \text{ m}^2.$$

Then

$$A_{abs} = A_{walls} + a_{carpet} = 350 \text{ m}^2 + 180 \text{ m}^2 = 530 \text{ m}^2.$$

We then substitute this value into equation (12.1):

$$\text{Reverberation time} = k\left(\frac{V}{A_{abs}}\right) = 0.161\left(\frac{1500}{530}\right) = 0.456 \text{ s}.$$

Hence, the RT in a carpeted room is equal to 0.456 s.

12.3.3 The absorption of sound by air

Sound absorption by air is negligible for frequencies less than 1000 Hz and hence can be ignored. However, if the room is very large (for example, an auditorium) and high frequencies are present, air absorption cannot be neglected. The absorption of sounds by air depends on many factors, including temperature and relative humidity, and for a large room it is given by

$$\text{reverberation time (RT)} = k\left(\frac{V}{A_{abs} + 4mV}\right), \quad (12.3)$$

where m is the air absorption constant, also called the attenuation constant (in SI units of m^{-1}), which depends on the frequency of the sound. Some values of the absorption constant are provided in table 12.1 but for more details please refer to other textbooks.

Increasing the absorption decreases both the reverberation level and the RT. The optimum RT is a compromise between the following three things:

1. Clarity, which needs a short RT.

Table 12.1. Absorption constant in air.

	Unit	2000 Hz	4000 Hz	8000 Hz
Absorption constant in air				
At 20 °C, 30% humidity	m^{-1}	0.012	0.038	0.136
At 20 °C, 50% humidity	m^{-1}	0.010	0.024	0.086

2. Loudness, which requires a high reverberation level.
3. Liveliness, which requires a long RT.

Example 12.3. Calculate the RT for the room described in example 12.1 but now consider air absorption to be a factor.
Solution
To find the RT, we can use equation (12.3). We know that $k = 0.161$ s m^{-1} at room temperature and from table 12.1, the air absorption constant (m) for a frequency of 4000 Hz = 0.038 m^{-1} at 20 °C and 30% humidity.

$$\text{Reverberation time (RT)} = 0.161 \left(\frac{V}{A_{\text{abs}} + 4mV} \right).$$

We have found the following values for the room in examples 12.1 and 12.2:

$$V = 20 \times 15 \times 5 = 1500 \text{ m}^3$$

$$A_{\text{abs}} = 530 \text{ m}^2.$$

Substituting these values into the above equation gives:

$$\text{reverberation time} = 0.161 \left(\frac{V}{A_{\text{abs}} + 4mV} \right) = 0.161 \left(\frac{1500}{530 + 4(0.038)(1500)} \right) = 0.319 \text{ s}.$$

Hence, the RT for this room is 0.319 s when air absorption is considered.

12.3.4 The optimum RT60 for a room

As discussed in section 12.1.2.3, RT60 is defined as *the time required for the sound level to decrease by 60 dB*. In any room, the optimum RT60 depends on the use of the room. When RT60 is more than 2 s, the room is considered too echoey; on the other hand, if the RT60 is less than 0.3 s, the room is called acoustically dead. So, for every room we have to find the 'sweet spot,' which depends on the main use of the room.

For example, an RT60 of less than 1 s is good for classrooms, where it is important for speech to be sharp and clear; however, a classroom with RT60 <1 s is not appropriate for singing or making music. In fact, for music to sound warm, full, and rich, the RT60 value should be closer to approximately 3.5 s, but at this value it is difficult to understand speakers at a distance. On the other hand, very long RTs (RT60 values of 8 s and above) are used in large cathedrals because this value is particularly suited to playing organ music or singing without the aid of musical instruments. Hence, the optimum RT depends on the size and the design of the room. For example, auditoriums used primarily for plays and speeches should have relatively shorter RTs, whereas auditoriums designed primarily for musical performances should have longer RTs.

12.4 Problems in acoustical design

There are some basic things that should be avoided when designing a room. These include:

1. **Focusing:** curved surfaces in a room, such as domes or curved walls, produce focal points at which any sound in the room is much louder than the sound in the surrounding area. This leads to a *whispering chamber effect*, which means that if you are the one playing at the focal point in a room, audiences hear you louder than all the other players on stage, and you in turn hear even the smallest noise made by the audience. These are not good situations for either of the parties involved, i.e. whether you are the performer or the audience.
2. **Echoes:** echoes must be avoided, especially those created by sounds reflected from behind the performer. The reflected sounds reach the audience later than the direct sound and both are superimposed on each other, causing all sorts of problems for the listening audience.
3. **Shadows:** acoustical shadows are quiet regions in a room; they are produced when there are extended structures, for example, long overhanging balconies, walls, or columns in a room. In these situations, sound waves undergo diffraction around the structure, which leads to the distortion of sound. This distortion depends on the frequency of the sound, which means that certain sounds are more distorted than others. This is very disturbing for the listeners in the room, and utmost care must be taken to avoid acoustical shadows when designing a space.
4. **Resonances:** resonances become a huge factor in smaller spaces, such as showers. In rooms where the dimensions of the room are only a few wavelengths long, there is a node at each of the walls and the sound resonates like a tube closed at both ends, as discussed in section 5.3, which may or may not be desirable at all frequencies. A way to avoid some room resonances is to avoid parallel walls, hence some small music practice rooms are constructed with oblique opposing walls.
5. **Background noises:** no amount of architectural detail and careful design is of any advantage if there is external noise that interferes with the sound produced in the room. Some permanent noise sources cannot be avoided, such as HVAC systems and busy highways. HVAC systems are often sources of white noise, which is rich in lower frequencies. These low frequencies may mask out the higher-frequency sound of music, which is clearly undesirable.

    One way to control the noise level is to match the size of the room to the size and type of performance. This is usually achieved by either adding separating walls that may be removed or added or by providing sufficient sound absorption.

12.5 The criteria for good acoustics

The acoustic requirements for various venues may be different, but some common requirements for good acoustics must be met for all spaces:

1. **Adequate loudness:** all listeners must be able to hear the performance, and the room should be not too large or too sound absorbent.
2. **Uniformity:** the sound should reach all listeners equally, independent of their physical location in the room. *Dead spots* should be avoided and sounds from different instruments in an orchestra should blend seamlessly and uniformly.
3. **Clarity:** for sound clarity, the masking of sounds by reverberant sounds must be avoided. This means adequate absorbing surfaces are needed in the room.
4. **Reverberance and liveliness:** reverberance is the sound that persists in a room after a tone is suddenly stopped, whereas liveliness is related to the RTs for frequencies of more than 350 Hz. A feeling of liveliness is vital at low frequencies to give support to bass notes in a performance. Please keep in mind that clarity and liveliness of sound sometimes conflict with each other, so care must be taken to strike a balance between them.
5. **Echo-free space:** reflected sound should arrive in time to reinforce direct sound and should not sound like a separate echo. Lateral reflections arriving within 25 to 80 ms of a direct sound appear to add the feeling of spaciousness, and overhead reflections in the same time period add to the early sound.
6. **Background noise:** background (white) noise, such as that produced by HVAC systems, should be minimized as much as possible so that it does not interfere with the performance.
7. **Wide sweet spot:** when using more than one speaker, the *sweet spot* is the focal point for sound between the two speakers. The listening sweet spot should be as wide as possible in a good acoustically designed room.

12.6 Designing spaces

When designing a space, one must take into consideration the primary use of the space because the sound requirements for each space are quite different, and they depend on what the space is used for. Some of the criteria for different rooms are briefly discussed in this section:

1. **A small room at home:** in a small room, the walls are close to each other and many reflections arrive from different sides within a few milliseconds of each other. In this case, it is not important to differentiate between early sound and reverberant sound. According to Sabine's formula, the RT is proportional to room volume and inversely proportional to the absorption (A_a). For a room with volume (V), we can increase the RT by making absorption as small as possible. For a rectangular room of length (L), width (W), and height (H), the frequencies of the various resonances set up in the room are given by

$$f_{lmn} = \frac{v}{2}\sqrt{\left(\frac{l}{L}\right)^2 + \left(\frac{m}{W}\right)^2 + \left(\frac{n}{H}\right)^2}, \qquad (12.4)$$

where v is the speed of sound and l, m, and n, are integers (0, 1, 2, ...) that determine the various modes. The three modes in which the most acoustic

energy is stored (because the sound travels farther between reflections) are detailed below:
  (a) *Oblique modes:* oblique modes are modes for which none of the integers l, m, and n are zero ($l, m, n \neq 0$). Standing waves in oblique modes reflect off all three pairs of boundaries in a cubic room with six sides.
  (b) *Tangential modes:* tangential modes are modes for which one of the integers equals zero, that is, either $l = 0$, or $m = 0$, or $n = 0$. Standing waves in tangential modes reflect off two boundaries, like balls on a pool table.
  (c) *Axial modes:* axial modes are modes for which all integers equal zero, that is, (l, m, and $n = 0$). Standing waves in axial modes propagate back and forth between two pairs of room boundaries and reflect from the third pair of boundaries.

Example 12.4. Find the lowest four resonances of a room with dimensions $(L \times W \times H) = (6 \times 5 \times 4)$ m. (Hint: the lowest resonance is at $f100$; the next harmonics are $f010$, $f001$, and $f110$. Assume that the speed of sound is 344 m s^{-1} at room temperature.)
Solution:
We can use equation (12.4) to find the required frequencies:

$$f_{lmn} = \frac{v}{2}\sqrt{\left(\frac{l}{L}\right)^2 + \left(\frac{m}{W}\right)^2 + \left(\frac{n}{H}\right)^2}.$$

The lowest resonance ($f100$) occurs at the frequency

$$f_{100} = \frac{344}{2}\sqrt{\left(\frac{1}{6}\right)^2 + \left(\frac{0}{5}\right)^2 + \left(\frac{0}{4}\right)^2} = \frac{344}{2}\sqrt{\left(\frac{1}{6}\right)^2} = 28.6 \text{ Hz}.$$

The next resonance ($f010$) occurs at the frequency

$$f_{010} = \frac{344}{2}\sqrt{\left(\frac{0}{6}\right)^2 + \left(\frac{1}{5}\right)^2 + \left(\frac{0}{4}\right)^2} = \frac{344}{2}\sqrt{\left(\frac{1}{5}\right)^2} = 34.4 \text{ Hz}.$$

The second resonance ($f001$) occurs at the frequency

$$f_{001} = \frac{344}{2}\sqrt{\left(\frac{0}{6}\right)^2 + \left(\frac{0}{5}\right)^2 + \left(\frac{1}{4}\right)^2} = \frac{344}{2}\sqrt{\left(\frac{1}{4}\right)^2} = 43 \text{ Hz}.$$

The third resonance ($f110$) occurs at the frequency

$$f_{110} = \frac{344}{2}\sqrt{\left(\frac{1}{6}\right)^2 + \left(\frac{1}{5}\right)^2 + \left(\frac{0}{4}\right)^2} = \frac{344}{2}\sqrt{\left(\frac{1}{6}\right)^2 + \left(\frac{1}{5}\right)^2} = 45 \text{ Hz}.$$

2. **Classrooms:** the requirements for a classroom are simple: students must clearly understand the instructor (who generally stands at the front of the classroom). They also need to understand conversations between students. In a classroom, three main things should be controlled to achieve good acoustics:
   (a) reverberation, which increases when the room size increases and reduces when more sound-absorbent surfaces are introduced;
   (b) noise from HVAC units; and
   (c) external noise. If the background noise is at 36 dB, a teacher using a normal voice produces a sound level of about 46 dB for a student who is 30 feet away.
3. **Libraries:** libraries are usually very quiet spaces. In quiet spaces, the RT should be half a second or less in order to avoid speech interference.
4. **Home theaters:** for a lot of people, installing a home theater means purchasing a 52-inch flat-screen TV and a surround-sound system and putting them randomly in a room. However, there are three additional concerns when we talk about home theaters: (i) the optimal dimensions of the room itself, (ii) the placement of the speakers, and (iii) how to limit the production of reflections.

    One main consideration is the size and the shape of the room. In general, you want to choose a room that is not too large; in fact, it should be an *acoustically small room*. Generally, rectangular rooms with dimensions in the ratio of 5:3:2 for length, width, and height are known to work the best. For example, a room with dimensions of $L = 15'$, $W = 9'$, and $H = 6'$ is acoustically good.

    The next important thing is the placement of the speakers. The standard surround-sound system uses a subwoofer and five loudspeakers, namely the right, center, left, left-surround, and right-surround speakers. However, this is impractical in a home, so in a home we usually have a central loudspeaker that primarily reproduces dialog and speech. The left and right speakers are about 60° apart, and the central loudspeaker is placed in the middle at 0°, as shown in figure 12.3.

    In addition, we must control early reflections in the room. Early reflections interact with the direct sound, and this results in a *hilly* response in the sound spectrum. Basically, this response produces constructive and destructive interferences in the sound spectrum that are unwanted. So, we ideally want to reduce early reflections in the room, which is achieved by signal processing rather than room acoustics. The other factors that influence sound quality in a room include:

    (a) The construction materials used for the walls, floor, and ceiling.
    (b) The types of rugs or furniture in the room.
    (c) The placement of the rugs or furniture in the room.
5. **Movie theaters:** there is an optimal place to experience the sound system in most theaters. Movie theaters use a central speaker behind the screen,

Figure 12.3. A home theater.

primarily to centrally localize the spoken dialog. When in a movie theater, to have the best audio experience:

(a) Try to find a row two thirds of the distance away from the front of the screen. The reason for this is simple: when checking the audio levels, most technicians check the audio levels from the center seat about two thirds back from the screen. Hence, this is probably the spot that will give you the best sound.

(b) Try to find a seat one or two seats from the exact center of the row. There are generally two speakers on either side. The sounds produced by the right and left speakers are equalized for the center, so you want to sit slightly off-center to enhance the stereo effect.

6. **Concert halls:** among the world's greatest musical halls are those constructed in a rectangular shape, sometimes referred to as a *shoebox* design. These are generally considered to have excellent acoustics. One of the reasons for this is that they all have strong, early reflections from the sidewalls that reach all of the audience equally, leading to a sense of spaciousness. Some examples are the Symphony Hall in Boston and the Concertgebouw in Amsterdam. Generally, it is thought that the best acoustics are provided in halls that have seating capacities of less than 2000 seats. Many modern concert halls have *tuning features*. These features include moveable reflecting and absorbing surfaces that may be changed, removed, or added to customize the hall for different playing conditions.

7. **Churches:** churches share the same requirements for good acoustics as those for concert halls. In old cathedrals in Europe, the pulpit was placed in the

center of the congregation and had a large canopy to reflect the preacher's voice. This, combined with the fact that most cathedrals of the time had long RTs, made for bad acoustics, which may be why preachers of the day had loud, booming voices just to be heard over the terrible acoustics! In modern times, background noise is the biggest concern and must be reduced.
8. **Recording studios:** recording studios are very carefully designed to keep the recorded sound as close to the original as possible. This process involves:

    (a) Very meticulous calculations of the correct physical dimensions of the rooms to get the optimum room response to sound.
    (b) The use of absorbent and diffusional materials on the surfaces in the rooms, for example, on the walls and other furniture.
    (c) Proper soundproofing, to provide audio isolation from unwanted outside sounds.

    Keeping these considerations in mind, recording studios these days normally comprise three rooms:

    (a) The studio where the sound is created, also called the *live room*.
    (b) The control room, where the sound from the studio is recorded and manipulated.
    (c) The machine room, where noisier equipment that may interfere with the recording process is kept.

12.7 Loudspeakers

There are two very important considerations for loudspeakers: placement and directivity. In this section, we talk briefly about both.

12.7.1 Loudspeaker placement

Loudspeaker placement is very important when designing an acoustically acceptable room. A single large source is generally made up of a cluster of loudspeakers with a directivity factor (Q_D), as discussed in section 9.9.1.1, chosen to give the best coverage for the audience. For most auditoriums, a single source system located in the center of the room, near the front, and over the speaker's head may be enough. However, if the room is long, two loudspeakers may be required, one in the front and one at the back. This kind of placement should include an appropriate electronic time delay for the speakers at the back. Otherwise, the direct sound arrives at the back after the sound from the back speaker, causing a distracting echo.

Loudspeakers should not be placed along the sidewalls of an auditorium. If they are placed on the sidewalls, they produce cross fire, which causes the audience to hear several loudspeakers at the same time. Two poorly phased loudspeakers that receive signals from the same source but are placed far away from each other could possibly create issues for the creation of a good central sound image. The standard surround-sound setup uses a subwoofer and five loudspeakers, namely the right, center, left, left-surround, and right-surround speakers, as shown in figure 12.3.

12.7.2 Loudspeaker directivity

Commonly used loudspeaker arrays include:
1. **Cone radiators:** an example of a cone radiator is shown in figure 12.4 (left). The directivity factor (Q_D) for cone radiators depends on the input sound frequency. When the wavelength of sound is much greater than the cone size, sound radiation is uniform and $Q_D = 2$. In this case, the sound level in the forward direction is about 10 to 15 dB greater than it is in the backward hemisphere, which is almost the same. For higher frequencies (low wavelengths), sound is mainly radiated in the forward direction, and depending on the speaker size and sound wavelength, the directivity factor value can be greater than 20 ($Q_D > 20$).
2. **Line or column radiators:** line or column radiators are simply cone radiators placed one on top of another. An example of a line radiator is shown in figure 12.4 (middle). This method of placing cone radiators on top of another is equivalent to increasing the cone speaker radiation in the vertical direction. The sound beam does not spread out in the vertical direction but spreads more in the horizontal direction, as we saw for line radiators in chapter 2, figure 2.9. Line or column radiators are mostly used because they increase Q_D while maintaining the broad distribution of a single speaker in the horizontal direction.
3. **Horn radiators:** horn radiators, also called conical or hyperbolic radiators, are generally designed to have greater directivity and efficiency than cone radiators in converting electrical energy into acoustical energy. However, horn radiators have a relatively poor low-frequency response, are large in size, and are hence harder to transport. An example of a horn radiator is shown in figure 12.4 (right).

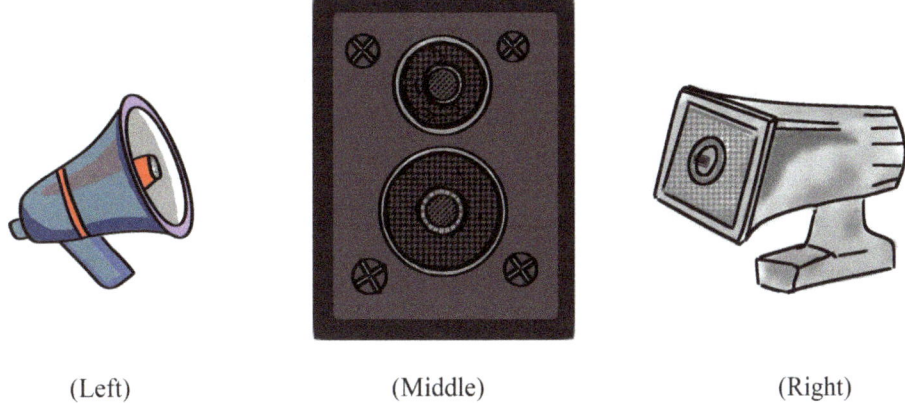

(Left)   (Middle)   (Right)

Figure 12.4. Types of radiator: (left) a cone radiator; (middle) a line or column radiator; (right) a horn radiator.

12.8 Outdoor sound systems

The considerations for an outdoor system are very different from the considerations employed for indoor ones. In outdoor venues where we do not have a reverberant sound field for reinforcement, sound power acquires great importance. For good sound transmission outdoors, multiple intentionally deployed arrays of loudspeakers and large strategically placed power amplifiers are needed.

Consider New York's Central Park, designed to hold concerts for 100 000 people. Its loudspeakers are built into 40-foot portable towers. Each tower has eight woofers in four bass horn cabinets, six mid-range radial horns, and 12 tweeters along with 80 W amplifiers to supply electrical power. The system is designed to provide peak sound levels of 105 dB at 45 m and 99 dB at 90 m from the stage and to cover an area of about 80 000 m^2. The system has a built-in mechanical reverberation unit that provides *liveness* for the outdoor environment.

Review Questions 12.1.
1. What should be taken into consideration in the design of a concert hall?
2. What are the differences in design conventions between small home theater settings and large commercial theater settings?

Exercises 12.1.
1. Calculate the RT for a frequency of 2000 Hz in an empty room with dimensions of 33 × 33 feet and a ceiling height of 12 feet.
   Hint: 1 meter = 3.3 feet.
2. Calculate the RT for the room described in Q1, which now has 100 empty seats and 100 occupied seats. (Hint: the absorption coefficients at f = 2000 Hz are 0.39 m^2 for unoccupied seats and 0.56 m^2 for occupied seats.)
3. Find the lowest two resonances of a room that has dimensions of $(L \times W \times H) = (12 \times 20 \times 9)$ ft versus those of a room whose dimensions are $(L \times W \times H) = (20 \times 12 \times 9)$ ft. Are they the same? Hint: the lowest resonances are at $f100$ and $f010$.

Further reading

- Berg R E and Stork D G 1994 *The Physics of Sound* 2nd edn (Englewood Cliffs, NJ: Prentice-Hall)
- Rossing T D, Moore R F and Wheeler P A 2002 *The Science of Sound* 3rd edn (Upper Saddle River, NJ: Pearson Education)
- Emanuel D C and Letowski T 2009 *Hearing Science* (Philadelphia, PA: Wolters Kluwer)
- Daniel J L 2006 *This Is Your Brain on Music: The Science of a Human Obsession* (New York: Dutton Penguin)

Part V

Of sound and music

IOP Publishing

The Physics of Sound and Music, Volume 1
A complete course text (Textbook)
Samya Bano Zain

Chapter 13

Musical tones, pitch, timbre, and vibrato

13.1 Musical tones and pitch

A sound wave that is described by a single frequency is called a *pure tone*. Most sounds are not pure tones; rather, they are a combination of many pure tones. Standing waves in a string expressed as frequencies may be written as a *harmonic sequence*, $f_1, f_2, f_3, \ldots$ where f_1 is called the fundamental frequency and the other frequencies are called harmonics. Harmonics are mathematically related to the fundamental frequency as follows: $f_2 = 2f_1, f_3 = 3f_1$, and so on.

Many sounds are a combination of frequencies that are harmonically related. One way to characterize such combinations is a property called *pitch*. Pitch, like loudness, is a subjective construct that totally depends on the listener; in other words, it depends on how humans perceive sound. However, frequency is an objective, scientific concept, which can be measured by instruments. As we saw in chapter 10, sound processing in the human brain is complicated and depends on the individual brain, hence pitch is very difficult to understand.

Under most circumstances, we can establish the following things about pitch with a degree of certainty:

1. The pitch of a pure tone that has a periodic sinusoidal waveform is equal to the frequency of the pure tone.
2. The pitch of a combination of harmonically related tones is associated with the fundamental frequency. Harmonically related tones can be tones produced by, say, a plucked guitar string.
3. The pitches of complex sounds, such as those produced by an orchestra, corresponds to the repetition rate of the produced periodic sound, that is, how often the pattern repeats. A human brain perceives complex tones as a collection of tones at different pitches. The pitches of complex tones can sometimes be hard to distinguish, which means that a listener can perceive two or more different incoming pitches; this perception totally depends on the listener.

In a nutshell, we have to relate the frequency of vibration produced by a sound-producing object to the pitch that we hear in our brains.

13.1.1 Pitch

Historically, the study of pitch and the perception of pitch has been the main field of study in psychoacoustics (chapter 11), and it has been instrumental in forming and testing theories of sound representation, processing, and perception in the human auditory system. The American Standards Institute (1960) defines pitch as:

> *'That attribute of the auditory sensory sensation in terms of which sounds may be ordered on a scale extending from low to high.'*

Pitch allows sounds to be ordered on a scale primarily based on the frequency of the object producing the sound. Pitch is determined by how quickly the sound wave makes the air around it vibrate and has almost nothing to do with the intensity of the wave. *High pitch* means very rapid oscillations of the air molecules and *low pitch* means slower vibrations of air molecules. High pitch corresponds to high frequency and low pitch corresponds to low frequency. It is worth noting that sound waves themselves do not have pitch; although their oscillations can be measured to obtain a frequency, it takes a human brain to decipher the internal quality of pitch. The basic unit of pitch in most musical scales is the *octave*. Notes judged to be an octave apart have frequencies nearly (but not always) in the ratio 2:1. Pitch can be broadly divided into two basic categories:
1. **Definite pitch:** definite pitch is a sound or a musical note for which it is possible or relatively easy to differentiate the pitch. Sounds with definite pitch have harmonic frequency spectra or near-harmonic spectra.
2. **Indefinite pitch:** indefinite pitch is a sound or a musical note for which it is impossible or extremely difficult to separate out a particular pitch from the multiple pitches produced. Sounds with indefinite pitch do not have harmonic spectra or have altered harmonic spectra, a characteristic known as *inharmonicity*. Examples include percussion instruments that do not have a particular pitch.

Because pitch is a subjective sensation, two people hearing the same sound may assign it a different position on a pitch scale. In fact, listeners may assign a different pitch to a sound depending on whether they hear it with the left ear or the right ear. This is called *binaural diplacusis*.

13.1.2 The history of pitch development

The pitches used in music have remained essentially the same since the time of the ancient Greeks, with the possible exception of the development or rather the refinement of the equal-tempered scale (discussed in section 14.4) around the 1600s. One of the earliest methods used to produce a particular frequency was the *siren*, similar to the one shown in figure 13.1. A siren is made of a circular disk mounted on an axle through the center with equally spaced holes at the circumference. The disk in a siren spins freely, and a nozzle attached to a source of compressed air is arranged such that it blows air perpendicular to the holes.

When the disk spins, the puffs of air escaping through the holes produce a tone. When the disk revolves slowly, our ear is able to distinguish successive puffs of air, but increasing the spin speed to ten revolutions per second merges the produced sound into a uniform note. As the speed of the disk is increased, the frequency goes up, hence the pitch also increases, and the sound becomes shrill. If we know the number of holes and the rotational frequency of the disk, the frequency (and hence the pitch) of the produced tone can be found mathematically.

The modern way of producing a tone with a given frequency is to connect a speaker to an audio oscillator and to produce a known frequency. We find an unknown frequency by changing the known frequency until it coincides with the unknown frequency. Note that as the two frequencies come closer to each other, we observe the beating phenomenon, as discussed in section 5.5.6. The beats become slower and disappear completely when the two frequencies are exactly equal.

13.1.3 Pitch versus frequency

It might seem as though we should simply say pitch is the same as frequency. We really should not say that pitch and frequency are the same; a frequency is the vibration of air molecules in the physical world, whereas pitch is a mental response to the frequency of vibration. This means that pitch is in our heads and corresponds to a human response to frequencies that exist in the world. Sound waves which are molecules of air vibrating at different frequencies do not themselves have pitch. Their motion is measured by experiments and instruments, but it takes a human brain to map them to the internal quality we call pitch. Think of it in terms of taste. Suppose you want to eat an ice cream; ice cream only has taste when it comes in contact with our tongue and is interpreted by our brains. It does not have any taste when it is sitting in the refrigerator or once it is in your stomach. Pitch is the same; pitch is how our brains interpret the sound waves coming to our ears.

13.1.4 Pitch and loudness

Pitch is not independent of loudness. As the loudness of a pure tone increases, the pitch may either increase or decrease, depending on its original frequency. For example, for pitches with frequencies above 4000 Hz, increasing the loudness increases the pitch. However, for frequencies between 1000 and 3000 Hz, pitch is relatively independent of loudness, whereas for frequencies between 20 and 1000 Hz, pitch actually decreases when loudness is increased. One might think that this effect would cause problems in musical melodies. What we have to keep in mind is that these effects are only valid for pure tones and do not hold for complex tones, such as

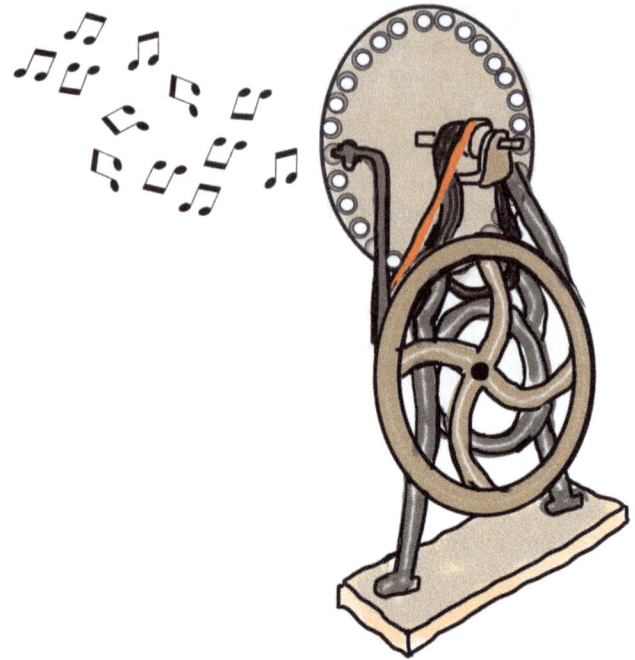

Figure 13.1. A depiction of a siren.

those of music and musical instruments. Most musical instruments produce pitches that are remarkably stable despite differences in loudness.

13.1.5 Pitch perception

For humans to perceive the pitches of tones, stimuli must last for a certain minimum time. This minimum duration varies with frequency. For a pitch corresponding to a frequency of 50 Hz, the tone must last for 60 milliseconds, whereas for a frequency of 1000 Hz, the tone must last for at least ten milliseconds to be perceived by the human ear. Stimuli lasting less for than one second are usually perceived as a click rather than a tone. Some people think that clicks have no pitch, which is not true, as some clicks sound higher in pitch than others.

13.1.6 Concert pitch

Concert pitch is the pitch to which musical instruments are tuned for a performance. Concert pitch has varied widely over history, and it may also vary from ensemble to ensemble. These days, the A above middle C, usually set at 440 Hz and written as 'a4,' 'A = 440 Hz,' or 'A440' is used as the concert pitch. Historically, the A has been tuned to a variety of higher and/or lower pitches. For example, in 1619 CE, the frequency used for a4 was 567.3 Hz in the North German church pitch and in 1636 CE, the two pitches used for a4 in Paris were 563.1 Hz (Mersenne's chamber pitch) and 503.7 Hz (Mersenne's *ton de chapelle*). The two pitches used as a4 around the 1880s in the US were 451.7 Hz (Chickering's standard fork) and 458 (Steinway's pitch). In 1939 CE, at the International Pitch Conference, the frequency of 440 Hz was adopted as a standard pitch for a4.

13.1.7 Relative pitch

Relative pitch is a skill which can be learned through intense ear training and practice. Many musicians have quite good relative pitch. With enough practice, it is possible for humans to listen to a single known pitch once and reliably identify pitch by comparing heard notes to the stored memory of the tonic pitch. Relative pitch depends on a musician's ability to identify the intervals between given tones. It allows singers to correctly sing a melody by pitching each note in the melody according to its distance from the previous note.

13.1.8 Absolute pitch

Absolute pitch (AP), or perfect pitch, is the ability of certain people to name or reproduce a tone without needing a reference to an external standard. Generally, people with absolute pitch have the ability to do some or all of the following things:
1. They are able to identify and name individual pitches played on various instruments.
2. They can identify all the tones of a given chord.
3. They can name the pitches of common everyday sounds, for example, alarms, telephone bells, etc.
4. They are able to sing a pitch perfectly without needing an external reference.
5. They can name the key of a given piece of tonal music just by listening to it.

13.1.9 Virtual pitch

In 1970 CE, Professor Ernst Terhardt from the Technical University of Munich was the first person to use the term *virtual pitch*. Virtual pitch is the experimentally established phenomenon in which one's brain interprets tones in music that don't actually exist.

13.2 Pitch perception theories

There are four types of pitch perception theories, which are presented here in a highly simplified form:
1. **Telephone theory:** telephone theory proposes that the ear is merely a microphone that just converts sound waves into electrical signals. This theory says that the ear does no analysis or signal processing; rather, it is merely a tool for passage and all processing is done in the brain. This theory was proposed by Rutherford in 1886 CE, and it has a few major flaws. First, it says that all the important functions reside in the brain, but recent experiments have proved that the signal that the brain receives is very different from the signal that initially enters the ear cavity.

2. **Place theory:** place theory is so named because it associates different pitches with differnet places on the basilar membrane. The place theory idea was first proposed by Helmholtz in 1863 CE and was experimentally tested by von Bekesy about 70 years later. According to place theory, different frequencies stimulate different areas of the basilar membrane, causing resonance to occur, which causes specific neurons to fire. The brain then gets the pitch information from the particular nerve endings that are simulated.

    The biggest objection to place theory is that it does not explain how humans are able to assign a unique pitch to incoming complex tones. If we only consider place theory, we should be able to hear many pitches in a complex incoming tone, one corresponding to each region on the basilar membrane. It is not good enough to just claim that our brains just choose the lowest pitch (the fundamental of the harmonic series) and ignore all the others. When we present a complex tone to the ear from which the fundamental is completely absent, the perceived pitch is still that of the fundamental. However, in this case, the basilar membrane is not vibrating in this region at all! The same phenomenon is also seen when the first several harmonics are missing or are masked by other noise. The ear still responds to the *missing harmonic*. This *missing harmonic* allows us to *hear* bass notes from a pocket transistor radio whose speakers produce practically nothing below 200 Hz.

3. **Periodicity theory:** periodicity theory states that the messages transferred from the cochlea to the brain contain more than just information about harmonic components. Periodicity theory accounts for pitch judgment of the tones heard in music that consist of sets of Fourier components shifted in frequency such that they are evenly spaced but no longer form integral multiples of the fundamental. Simple periodicity theory still is unable to explain all the experimental data. For example, it predicts that in order to determine pitch, higher harmonics should be more important. However, experiments have proved that lower harmonics actually dominate in pitch perception.

4. **Pattern recognition theory:** both place theory and periodicity theory have had their successes, but so far, pattern recognition theory has gained most favor. It was first discovered and published around 1973 CE by a number of scientists working independently. It, like the telephone theory, emphasizes that the brain is our central processing unit and the ear's primary purpose is to transmit information to the brain. Most scientists agree that the brain searches for order among the incoming frequencies. If there is a harmonic series present in the incoming sound, the brain assigns a pitch corresponding to the fundamental of the harmonic series. One of the strongest pieces of evidence for central pattern recognition is that the information from both ears is combined when it arrives at the brain. The signals arriving at the ears are called *dichotic signals*.

13.3 Vibrato

A sine wave of a given frequency (say 512 Hz) can be *frequency modulated* by another sine wave at a frequency of 1 Hz. The result of this frequency modulation is the production of third sine wave, centered at the frequency of the original wave (512 Hz), which has a tone that varies gently up and down in frequency depending on the amplitude of the modulating wave. This effect is called *vibrato*, and it manifests itself as a pulsating sound effect produced by small but rapid changes in the frequency of a note. Hence, vibrato is the periodic change in the frequency of a musical tone caused by the addition of a modulating wave. Vibrato adds a distinctive flavor to the tone of a melody and has been used over the centuries to enhance musical performances. The American National Standards Institute defines vibrato as follows:

> *'vibrato is a family of tonal effects in music that depend on periodic variations of one or more characteristics in the sound wave.'*

Vibrato is mainly characterized by two parameters: the first is depth, which is the amount of variation in frequency, and the second is speed, which describes how quickly the frequency is changed. True vibrato is most often achieved either manually (called 'hand vibrato' or 'finger vibrato') or mechanically. Manual vibrato is a technique in which the fretting hand bends the string smoothly and gently up and down to produce slight alterations in pitch. When a trombone (chapter 17) player wiggles the slide in and out, it is almost a *true pitch vibrato*; similarly, vibrato on a stringed instrument (chapter 15), say a guitar, is close to a *pure pitch vibrato*. However, to produce mechanical vibrato, some electric guitars are equipped with mechanically operated vibrato systems, most often in the form of a rocking bridge assembly operated by a hand lever. Please note that singers use a little more depth of frequency in vibrato than instrumentalists. Vibrato helps to cover up small errors in frequency, especially when multiple singers or instrumentalists are singing or playing together. Vibrato also varies with various performances. Experimentally, it has been found that the average rate for both singers and instrumentalists is around 7 Hz. At rates higher than 12 Hz, the sound becomes a rather unpleasant combination of more than one note.

Vibrato is also sometimes used interchangeably with periodic changes in amplitude (volume). However, periodic changes in amplitude in a melody should be correctly referred to as *tremolo*, not vibrato. Tremolo is a 'trembling' or 'shuddering' sound effect produced by slight and rapid changes in the amplitude/volume of a note. A singer's voice is usually a mixture of true vibrato and tremolo.

13.3.1 Frequency modulation versus amplitude modulation

The term *frequency modulation* (FM) refers to modulation in the frequency of an original wave, and the term *amplitude modulation* (AM) is used to refer to modulation in the amplitude of a wave. Let us discuss them briefly here in terms of radio waves:

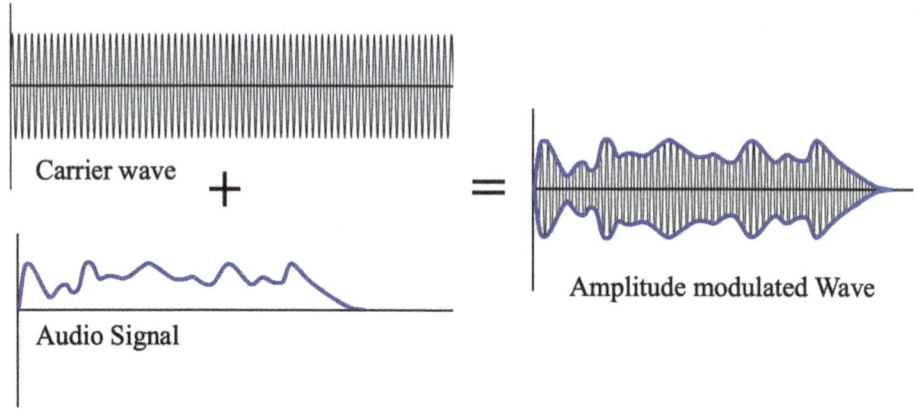

Figure 13.2. Amplitude modulation, as used for AM radio: (a) a carrier wave at the station's basic frequency; (b) an audio signal at much lower audible frequencies. (c) The amplitude of the carrier is modulated by the audio signal without changing its basic frequency.

1. **Amplitude modulation (AM):** AM radio waves are used to carry commercial radio signals in the frequency range from 540 kHz to 1600 kHz, a range of about 1 MHz. In the AM method, information is carried by the radio waves in the range given above. Each radio station is assigned a particular frequency at which the station transmits radio signals. This is called a *carrier wave*. The carrier wave is varied or modulated in amplitude by the addition of the audio signal, which results in a final radio wave that has a constant frequency but which varies in amplitude as a function of the information it carries, as seen in figure 13.2(c).

   A radio in your car is a receiver which you can tune to the resonant frequency of the radio station you want to listen to. When you tune your car radio to the resonant frequency transmitted by the radio station, it picks up the radio signal, which it then transforms to sound that you can enjoy while driving. The car radio rejects all other frequencies incident on its antenna until you tune it to another frequency.

2. **Frequency modulation (FM):** FM radio waves are used for radio transmission in the frequency range between 88 MHz and 108 MHz or about 100 million cycles per second and are another method of carrying information on carrier waves. In this case, the carrier wave is modulated in frequency by the audio signal, producing a wave that has a constant amplitude but varies in frequency, as seen in figure 13.3(c).

13.3.2 AM versus FM radio signals

Radio signals are electromagnetic waves that are either amplitude modulated (AM) or frequency modulated (FM). Let us talk about some of their advantages and disadvantages here. One advantage of AM signals is that the wavelengths of AM waves are large, so AM signals go over and around large obstacles, such as buildings

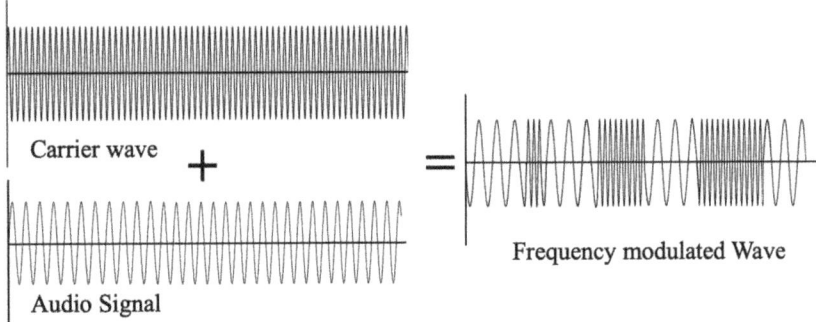

Figure 13.3. Frequency modulation, as used for FM radio: (a) a carrier wave at the basic frequency of the station; (b) an audio signal at much lower audible frequencies. (c) The frequency of the carrier is modulated by the audio signal without changing its amplitude.

and/or mountains. However, to receive the best FM signal, a direct line of sight between the broadcast antenna and the receiver is essential. To achieve this, FM signals are usually transmitted from tall structures. On the other hand, FM broadcast antennas are structurally much smaller than those used for AM signals. So there are advantages and disadvantages to both AM and FM signals.

AM radio signals are more susceptible to additional noise from other radio sources than FM signals. This is because the amplitudes of waves add together and AM receivers interpret noise as added information in the carrier wave. On the other hand, FM receivers can easily reject amplitudes other than those of the carrier wave and interpret only variations in frequency. Hence, it is easier to reject noise from FM signals. Please note that AM without FM is possible, however it is virtually impossible to have FM without AM.

13.3.3 TV signals

TV signals are also electromagnetic waves, but in this case the electromagnetic waves must carry video as well as audio data. Remember that the TV video signal is AM, while TV audio is FM. So, in this case, each channel requires a larger range of frequencies than simple radio transmission. TV channels are called 'very high frequency' (VHF) channels when they utilize frequencies in the range of 54 MHz to 88 MHz and 174 MHz to 222 MHz; they are called 'ultra-high frequency' (UHF) channels when they utilize frequencies in the range of 470 MHz to 1000 MHz. Please note that these frequencies correspond to old-fashioned roof antennas only; satellite dishes and cable TV actually use significantly higher frequencies.

Example 13.1. a. Does the frequency 1450 kHz correspond to an AM or FM radio channel?
  Solution:
  The given frequency is in the kHz range, hence it is an AM radio signal channel.
  b. Does the frequency 100.1 MHz correspond to an AM or FM radio channel?

Solution:
The given frequency is in the MHz range, hence it is an FM radio signal channel.
c. Calculate the wavelengths of a 1450 kHz radio signal and a 100.1 MHz radio signal. (Hint: the speed of light c = 3 × 10⁸ m s⁻¹.)
Solution:
The wavelength of a 1450 kHz (1450 × 10³ Hz) radio signal is given by:

$$v = \lambda f$$

$$\lambda = \frac{v}{f} = \frac{3 \times 10^8}{1450 \times 10^3} = 206 \text{ m.}$$

The wavelength of a 100.1 MHz (100.1 × 10⁶ Hz) radio signal is given by:

$$v = \lambda f$$

$$\lambda = \frac{v}{f} = \frac{3 \times 10^8}{100.1 \times 10^6} \approx 3 \text{ m.}$$

d. Compare these wavelengths to the signals above; what does this tell you?
The wavelength of a 1450 kHz radio signal is 206 m, whereas the wavelength of a 100.1 MHz radio signal is about 3 m. The wavelengths in this example represent AM and FM wavelengths, which tell us how each is broadcasted and how it travels. The most efficient length for a linear antenna is $\lambda/2$. Thus, a very large antenna is needed to efficiently broadcast typical AM radio, as its carrier wavelengths are on the order of hundreds of meters.

13.4 The just-noticeable difference (JND)

We have talked about the lowest threshold of hearing (chapter 10), but we have not talked about the smallest frequency change that we can hear. Another question that we should ask is: is this smallest frequency change somehow limited by the critical band?

To answer these questions, let us start the discussion by considering two sine waves. Suppose one sine wave (f = 440 Hz) is *frequency modulated* by another sine wave that has a frequency of 1 Hz. If the amplitudes of both sine waves are the same, we get a resultant wave that varies slowly up and down in frequency, an effect called vibrato (section 13.3). If the amplitude of the pitch variation is now decreased, there comes a point at which the change in frequency becomes undetectable. Similarly, if we start from no frequency difference between the modulating frequency and the main frequency (f = 440 Hz) and slowly increase the difference, there comes a time at which the pitch change eventually becomes noticeable. The particular frequency at which the difference becomes audible is called the *frequency just-noticeable difference* or *frequency JND*. We can measure the JND for nearly all frequencies in the audible range by varying the original frequency with respect to the modulating frequency. Experimentally, it has been found that frequency JND is between 0.5% to 0.6% over most of the audible range. However, this variation is larger for very high and very low frequencies.

The next question was about the basilar membrane. The basilar membrane in our ear vibrates in response to the incoming frequency of the sound. For most frequencies, the JND is much smaller than the critical band, which means that even though a large area of the basilar membrane vibrates in response to the frequency, humans can distinguish between changes in frequency that generally correspond to one-tenth of the width of the basilar membrane. This neural process of assigning a single frequency to the wide band of excitation along the basilar membrane is called *sharpening*. This effect is more pronounced in complex waves, such as square waves. As a result of this sharpening, the human ear is able to distinguish between two frequencies that are remarkably close to each other. The JND is also called the limit of discrimination, difference threshold, or difference limen (DL).

13.4.1 Measuring the frequency JND

One way to experimentally determine the frequency JND is to compare a fixed reference sound with several test sounds. The JND is be the smallest difference between the test sound and the reference sounds, i.e. the difference that is just audible. Multiple test subjects should also be used because the human perception of sound varies with age. Using a relative scale, the JND can be mathematically expressed as

$$JND = \left| \frac{\Delta f}{f_0} \right|, \quad (13.1)$$

where Δf is the absolute value of the difference threshold and f is the reference frequency. For two pure tones, the JND is about 7% for low frequencies and about 15% for high frequencies. Most studies place it around 3% in the 100 Hz range but only at 0.5% in the 2000 Hz range. Musicians can tell the difference between two notes played one half step apart and can perceive them to be two separate notes.

Example 13.2. Find the JND if the reference frequency is 3000 Hz and the absolute change in signal frequency is 30 Hz.
Solution:
Using equation (13.1),

$$JND = \left| \frac{\Delta f}{f_0} \right| = \left| \frac{30}{3000} \right| = 0.01,$$

and the JND is calculated to be $0.01 \times 100\% = 1$ Hz.
This means that the smallest noticeable difference between the two tones is 1 Hz.

13.4.2 Intensity and the JND

The ability to detect changes in intensity (the *just-noticeable difference in loudness*) is proportional to the original intensity of the sound. If you and a friend are in a very quiet exam room, you can hear their whisper because it is significant with respect to the background noise. However, at a football game, you may not be able to hear the same friend sitting next to you shouting. This is because the added sound of your friend is now insignificant with respect to the existing background sound level. Simply put, as the background sound gets louder, we need a bigger change in intensity in order to detect it. The JND can be considered to be a sort of a confusion measure. One can approximately say that for a particular measurement method, two stimuli that are one JND apart will be confused 50% of the time.

Ernst Heinrich Weber[1] experimented to find out how much a particular stimulus (I_R) needs to change before the difference (ΔI_R) becomes noticeable to humans, a quantity that he called the JND. Through experimentation, he found that the JND is proportional to the initial stimulus itself, which is expressed mathematically as

$$\Delta I_R = \text{JND} \propto (I_R + I_0) = k\,(I_R + I_0), \tag{13.2}$$

where k is a constant of proportionality, called the Weber constant, I_0 is the threshold of perception of sound intensity (which depends on the initial stimulus), and I_R is the initial stimulus. The Weber constant (k) is different for each sense and is called the *Weber fraction*. The Weber fraction depends on the characteristics of the stimulus itself and is given mathematically by

$$k = \frac{\Delta I_R}{I_R}. \tag{13.3}$$

Please note that the Weber constant (k) is sometimes expressed as a percentage, so please take care when solving numerical questions. A Weber constant of 0.028 (or 2.8%) implies that $\frac{\Delta I_R}{I_R} = 0.028$, i.e. $\Delta I_R = 0.028I$, where I_R is the initial stimulus. Suppose you present two sounds to a test subject, each of intensity 70 dB. A k of 2.8% means that a listener detects a difference between the two stimuli at about 71.96 dB, which means that the JND in this case is (71.96 dB minus 70 dB) = 1.96 dB.

Example 13.3. Suppose Sami is presented with two sounds, each of which has an intensity of 80 dB. The intensity of one of the sound sources is then increased very slowly. Calculate the JND if the k value is 3%.
Solution:
Given a k value of 3%, Sami will note a difference between the two stimuli at

$$0.03 = \frac{\Delta I}{80} = \frac{I - 80}{80}$$

---

[1] Ernst Weber was a 19th century German experimental psychologist who pioneered work on studies of human sensations.

$$(0.03)(80) + 80 = I$$

$$\Rightarrow I = 82.4 \text{ dB},$$

which implies the JND in this case is (82.4 dB minus 80 dB) = 2.4 dB.

13.4.3 Cents and mel scales

A *cent*, which is a ratio between two close frequencies, is a logarithmic unit of measure used for musical intervals. Cents are fundamentally used to measure extremely small but finite intervals. In fact, a one cent interval is too small to be heard between successive notes. For reference, in an equally tempered semitone, the interval between two adjacent piano keys spans 100 cents! The twelve-tone equal temperament divides an octave into 12 semitones, each of which has 100 cents. Cents are also used to compare the sizes of comparable intervals in different tuning systems. It is difficult to know exactly how many cents are distinguishable to humans, since this varies from person to person, but one experiment has proved that humans can distinguish a difference in pitch of about 5 to 6 cents.

$$c = 1200 \log_2\left(\frac{f_2}{f_1}\right) \tag{13.4}$$

The *mel* scale, named by Stevens, Volkman, and Newman in 1937 CE is a perceptual scale of pitches judged by listeners to be equal in distance from one another. The name *mel* comes from the word *melody* to indicate that the scale is based on pitch comparisons. One formula used to convert frequency (f) in hertz (Hz) into *mel* is:

$$m = 2595 \log\left(1 + \frac{f}{700}\right). \tag{13.5}$$

Example 13.4. Convert *frequencies of* 100 Hz and 1000 Hz into mel.
Solution:
Using equation (13.5),

$$m = 2595 \log\left(1 + \frac{f}{700}\right) = 2595 \log\left(1 + \frac{100}{700}\right) = 150 \text{ mel}$$

and

$$m = 2595 \log\left(1 + \frac{f}{700}\right) = 2595 \log\left(1 + \frac{1000}{700}\right) = 1000 \text{ mel}.$$

13.5 Timbre or tone quality

All instruments, such as the flute, piano, or a violin can play the same note with an identical fundamental frequency. When each instrument is played, we hear the same pitch from each. So why is it that each instrument sounds very different to us? Basically, this difference in sound arises because of a property called the *timbre*[2] of each instrument. The timbre of the sound is the principal feature that distinguishes the growl of a lion from the purr of a cat. Humans can recognize hundreds of different voices and can tell when people are happy, angry, or sad or when someone is coming down with a cold based on the timbre of their voice.

When a musician plays a musical tone or sings a song, only a particular set of frequencies is heard. Each note produced by the instrument is a mixture of different pitches, called *harmonics*. These harmonics are so well blended that we do not hear them as separate notes and they give the note its *color*. In order to understand the color of a note, it helps to think about paint colors that we see and how we feel about a room once we paint it red, green, or blue. We define color based on its characteristics, for example, dull, light, smooth, bright, deep, clear, bold, and/or rich. Similarly, the color of a sound is defined according to its characteristics and how it makes us feel: reedy, brassy, piercing, mellow, hollow, focused, transparent, breathy,[3] or full. Color helps us to tell the difference between notes played on different musical instruments. A note played on the violin sounds different from the same note played on a piano.

Timbre is basically a consequence of overtones. Each instrument has its own overtone profile, which is like a fingerprint. It is a complicated pattern that can be used to identify each particular instrument. The contribution to the total sound arising from the overtones varies from instrument to instrument, from note to note on the same instrument, and even for the same note, say, if the player produces that note differently by blowing a bit harder or a bit softer. In some instruments, such as the bassoon and the bagpipes, the overtones contribute more significantly to the power spectrum, giving the timbre a more complex sound. In other cases, such as the flute or violin, the power due to the overtones is less prominent and the timbre has a very pure sound. To the trained ear, there are even differences for the same instrument. For example, all trumpets do not sound alike, and what distinguishes one particular trumpet from another is that the overtone profile of each trumpet differs slightly from that of another. But of course, they do not differ as much as they differ from the profiles of other instruments such as the violin or piano.

In summary, timbre in acoustics is the characteristic quality of a sound. It is independent of the pitch and loudness of the sound. Timbre depends on the relative strengths of the components of different frequencies, which are determined by resonance. In music, timbre is the characteristic quality of sound produced by a particular instrument or voice. In phonetics, timbre is the distinctive tone quality

---

[2] TAM-brr pronounced as *amber*. The root of the word timbre is from a French word that means *note of a bell* and from an Old French word for 'drum.'
[3] pronounced BRETH-ee.

that differentiates one vowel or consonant from another. The American Standards Association (1960) defines timbre as:

'that attribute of sensation in terms of which a listener can judge that two sounds having the same loudness and pitch are dissimilar.'

Timbre depends on the number, relative intensity, and distribution of the partials in a tone source. We normally say that timbre depends on the following characteristics: waveform, the spectrum of the stimulus, sound pressure, frequency, and the temporal characteristics of the stimulus.

13.5.1 The effect of timbre on loudness

The effect of loudness on timbre is small, and for a long time scientists and musicians thought that it was not important. We know that our ear responds to very loud sounds in such a way as to avoid damage to the ear. This means that beyond a certain loudness, any increase in the amplitude of a sound does not produce a corresponding increase in loudness; this is called *nonlinear* behavior. So, in a sound spectrum, if a particular sound is played loudly, this part of the sound spectrum drives the ear into nonlinear behavior, while other notes with relatively small sound energies are not audible in the nonlinear range. The net effect is that the relative strengths of the overtones change depending on whether the same sound is played softly or loudly, that is, the timbre changes. These effects are small but nonzero. This is why a note played loudly on a violin has a different sound spectrum than the same note played on a violin softly; that is, the tonal structure of the note changes. Therefore, it is essential to separate the physical effect from the subjective effect arising from the nonlinear behavior of the ear. One way to do this is to record the instrument and then play the note at a low amplitude/volume and then play it again at very high amplitude/volume. We observe that as the loudness changes, the timbre starts to change from the original sound. It is found that the best sound is actually achieved when the recorded sound is played at a loudness level close to the loudness level of the original sound.

Review Questions 13.1.
1. What is pitch? In what way is this different from the frequency of a sound?
2. What does the term 'absolute pitch' mean?
3. What pitch do we hear if the following frequencies are played together: 200 Hz, 400 Hz, and 800 Hz?
4. What is the difference between absolute and relative pitch?
5. What is the difference between vibrato and timbre?
6. What are the various pitch perception theories? Can you list some pros and cons of each?
7. Does playing a tone backwards change its timbre?

Exercises 13.1.
1. What are some advantages and disadvantages of absolute pitch for a musician?
2. Convert $f = 440$ Hz and $f = 2000$ Hz into mel.
3. Find the JND if the reference frequency is 3000 Hz and the absolute change in signal frequency is 10 Hz.
4. Suppose you are presented with two sounds, each of which has a frequency of 440 Hz. We then slowly increase the frequency of one of the sound sources. If you first notice the difference in pitch at 442 Hz, calculate the k value.

Further reading

- Roy D P, Etienne G and Thomas C W 2010 *The Perception of Family and Register in Musical Tones* (New York: Springer)
- Mari R J, Richard R F and Arthur N P 2010 *Music Perception* (Berlin: Springer)
- Hartmann W M 1997 *Signals, Sound, and Sensation* (Berlin: Springer)
- Wever E G 1949 *Theory of Hearing* (New York: Wiley)
- Levitin D J 2016 *This is your Brain on Music: The Science of Human Obsession* (Dutton)
- Emanuel D C and Letowski T 2009 *Hearing Science* (Philadelphia, PA: Wolters Kluwer)

IOP Publishing

The Physics of Sound and Music, Volume 1
A complete course text (Textbook)
Samya Bano Zain

Chapter 14

A musician's graph paper and musical scales

In chapters 9 and 10, we saw that the frequency response of the ear is not linear but rather logarithmic. This means that our ears do not respond linearly to the frequencies of external stimuli but rather they respond logarithmically. We also noted that humans perceive the same ratios between two frequencies to be the same musical interval. It is worth noting here that a similar type of nonlinearity exists in our ears' response when the incoming signals have varying amplitudes. To quantitatively handle these nonlinear behaviors, we have to turn to the mathematical formalism called the logarithm.

14.1 Logarithms

Logarithms have very useful properties that help us when we deal mathematically with the frequency and amplitude responses of our ears. A logarithmic scale is also called a *compressed scale*, because it allows us to plot widely varying numbers on a more manageable scale. Some advantages of using a logarithmic scale are mentioned below:
  1. The frequency response of the human ear is expressed on a logarithmic scale because our sense of hearing perceives equal ratios of frequencies as equal differences in pitch.
  2. As we saw in chapter 9, the decibel scale used to express the sound pressure level (SPL), sound power level (SWL), and sound intensity level (SIL) is based on logs.
  3. The musical scale is logarithmic; in other words, each step has a ratio of particular frequencies. Hence, the keyboards of pianos and other musical instruments are made to be logarithmic.

Mathematically, the logarithm to the base ten of any number x is the power to which ten must be raised in order to equal x. This means:
$10 = 10^1$. The logarithm of 10 (to the base 10) is 1 $\Rightarrow \log_{10} 10 = 1$.

$100 = 10^2$. The logarithm of 100 (to the base 10) is 2 $\Rightarrow \log_{10} 100 = 2$.
$1000 = 10^3$. The logarithm of 1000 (to the base 10) is 3 $\Rightarrow \log_{10} 1000 = 3$.
$10\,000 = 10^4$. The logarithm of 10 000 (to the base 10) is 4 $\Rightarrow \log_{10} 10\,000 = 4$.

14.2 The musical stave or staff

Graph paper that has one logarithmic axis and one linear axis is called *semi-logarithmic*. We use semi-logarithmic graphs to represent quantities that are functions of frequency. When frequency is represented on the logarithmic scale, octaves (that have a frequency ratio of 2:1) become equidistant anywhere on the scale. For example, the distance from 20 to 200 Hz is the same as that from 200 to 2000 Hz, which is the same distance as that from 2000 to 20 000 Hz on a logarithmic graph. A piano keyboard is made using an approximately logarithmic scale, which means that any particular octave requires equal reach (i.e. the same distance) anywhere on the keyboard, as shown in figure 14.1.

14.2.1 The musical staff

Music is written on a staff or stave, such as the one shown at the top of figure 14.2. The term 'staff' is used in US English, whereas the term 'stave' is mainly used in British English; however, both forms of English use the term 'staves' as the plural form. The staff can be thought of as a musical graph on which music notes, musical symbols, etc. are placed to indicate the specific pitch of a note. A musical staff plots frequency on the vertical *y*-axis (the logarithmic axis), whereas time is placed on the horizontal (linear) *x*-axis. Interestingly, the units on the *x*-axis are not seconds but rather another unit called a *measure* in the USA or a *bar* in England. A measure is a segment of time defined by a given number of beats, each of which is assigned a particular note value. Dividing music into measures actually provides regular

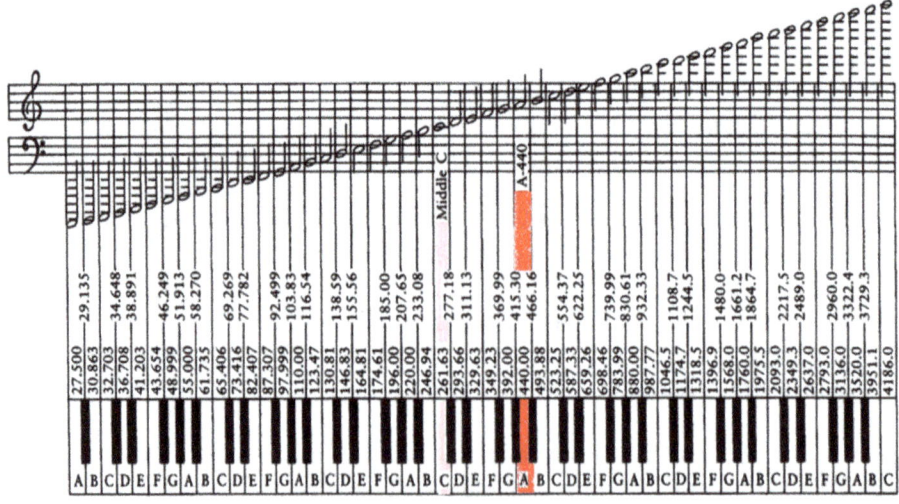

Figure 14.1. A frequency–pitch diagram of a piano keyboard.

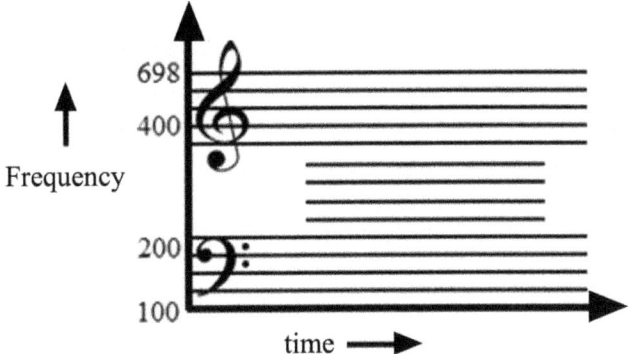

Figure 14.2. An example of a musical staff in terms of frequency.

reference points that help in finding different locations within a piece of music. Measures also make written music easier to follow for composers and players.

A musical staff is always read from left to right. Generally, it consists of five horizontal, evenly spaced lines. The lower part of the staff is designated for lower notes and the higher part is for higher notes. Notes are written on and between staff lines, but when they fall outside the staff, they are placed on ledger lines that lie below or above the staff.

There is one issue that arises when using a music staff: there are a lot more pitches than those that can be fitted into any five lines and four spaces. So, to overcome this limitation and to incorporate a wide variety of frequencies, the musicians' log paper (musical staff) groups pitches together with symbols called *clefs*.[1] Every musical staff has a *clef* symbol, which is placed at the beginning of each line or space to show the exact location of a particular note (A, B, C, D, E, F, or G). The most commonly used clefs are the treble and bass clefs shown at the left side of figure 14.2 at the top and bottom, respectively.

Piano music is written in two staves that have treble and bass clefs, and the two staves are connected by a brace. The two staves span a frequency range of just under three octaves, from 98 Hz to 698 Hz, and the five lines of a musical staff are separated by either three or four semitones.

14.2.2 Important symbols on a musical staff

The most important symbols on the staff are the clef symbol, the key signature, and the time signature, as seen in figure 14.3, and they all appear at the beginning of the staff. Notes are represented either on a line or in the space between lines. If a note lies above or below the staff, ledger lines are added to show how far above or below the staff the note is.

---

[1] Some clefs used in musical notation match the names given to different voices, e.g. soprano, alto, tenor, and bass.

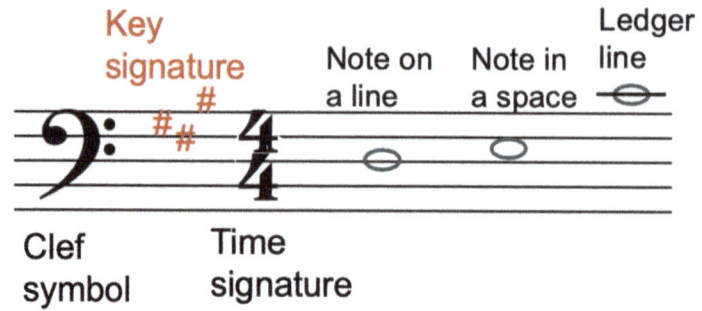

Figure 14.3. Details of a musical staff.

Figure 14.4. (a) The G-clef, (b) notes on the G-clef.

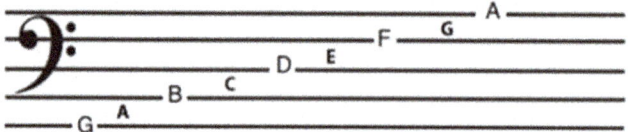

Figure 14.5. Notes on the F-clef.

1. **Clefs**: The most commonly used clefs are the treble and bass clefs:
   (a) **G-clef:** the G-clef, also called the *treble clef*, is the large musical symbol at the beginning of the piano's top staff. A treble clef tells us that the second line from the bottom is **G**. It is known as the **G-clef** because the symbol wraps around the G note located at the second treble staff-line from the bottom, as shown in figure 14.4(a). The notes in the G-clef are: A-C-E-G, as seen in figure 14.4(b).[2]
   (b) **F-clef:** the F-clef, also called the *bass clef*, is the first large symbol on the bottom staff in piano music. A bass clef symbol tells us that the second line from the top is an **F**. On the staff, it is the one bracketed by the dots in the symbol. In other words, the **F-clef** wraps around the highest F note on the bass staff. The bass clef lines are: G-B-D-F-A, as seen in figure 14.5.[3]

---

[2] A popular mnemonic used to remember the notes in the G-clef is: A-C-E-G (All Cows Eat Grass).
[3] A popular mnemonic used to remember the notes in the F-clef is: G-B-D-F-A (Good Boys Deserve Fudge Always).

2. **Key signature**: the key signature includes the symbols called the flat (♭) and the sharp (♯).[4] These symbols are placed immediately after the clef symbol on a musical staff at the beginning of the first line, as seen in figure 14.3. The arrangement of sharp or flat symbols on particular lines and spaces of a musical staff indicates to the reader that the notes in every octave are to be raised (by sharps) or lowered (by flats) from their natural pitches, which means they are to be played a semitone higher (sharper) or a semitone lower (flatter) than would be played otherwise. This applies until the end of the piece or until another key signature is indicated on the staff.
3. **Time signature**: the time signature is indicated on the musical staff after the clef and key signatures and is written in the form of two numbers like a fraction, as seen in figure 14.3. The top number (numerator) of the time signature is the number of beats to count, while the bottom number (denominator) tells you whether to count the beats as quarter notes, eighth notes, or sixteenth notes. The top number can be any integral number, but it is generally between 2 and 12, while the most common bottom numbers are 4 (quarter notes), 8 (eighth notes), and 16 (sixteenth notes).[5]

   In figure 14.3, the time signature is written as 4/4, which is the most used time signature. A 4/4 time signature means that the numerator is four and the denominator is four, which implies we must count four quarter notes to each measure and all the notes in each measure must add up to 4 quarter notes. Any combination of rhythms can be used as long as its notes add up to four quarter notes. For example, a measure could contain one half note, one quarter-note rest and two eighth notes, which is fine because these all add up to four quarter notes. So, for 4/4, the beat is counted as 1, 2, 3, 4, 1, 2, 3, 4, and so on.

   Other examples of commonly used time signatures include the 3/4 time signature and the 6/8 time signature. A time signature of 3/4 means that three quarter notes should be counted in each measure, i.e. 1, 2, 3, 1, 2, 3, 1, 2, 3, and so on. A time signature of 6/8 means means that six eighth notes should be counted in each measure, and the beat is counted as 1, 2, 3, 4, 5, 6, 1, 2, 3, 4, 5, 6, and so on.

14.3 Musical scales

The word *scale* is derived from a Latin word (*scala*), which means a ladder or a staircase. A musical scale is a group of musical notes collected in ascending or descending order used to represent part or all of a musical work. Scales are ordered in pitch, and their order provides a measure of musical distance. In short, a scale is a kind of a *musical ladder* that climbs from a starting note to a note one octave higher. This can be done in a virtually infinite number of ways. The notes of these scales are then used to create melodies and harmonies. Scales are typically listed from low to

---
[4] The key signature also sometimes includes the natural (♮).
[5] The other number you may see as the bottom number (the denominator) is 2 = a half note.

high; in traditional Western music, they generally consist of seven notes and repeat at the octave.

When thinking about the construction of any musical piece, it is very important to keep two key concepts in mind, namely melody and harmony. *Melody* is a succession of different pitches in time that are perceived as a continuing line, whereas *harmony* is the combination of several pitches at one time. Melody gives horizontal structure to music, i.e. progression from one note to another, and harmony gives vertical structure, a cooperative effect due to several notes sounding simultaneously. The pitches used to make a melody include:

- **The chromatic scale:** the chromatic scale consists of the twelve familiar pitches per octave found on a piano keyboard.
- **Semitones:** the step from each note to the adjoining one is called a semitone or half step.
- **Tones:** two adjoining white keys on the piano that have a black key between them are a tone apart (also known as a whole tone or whole step).

14.3.1 Building a musical scale

A musical scale is based on two assumptions about the human hearing process:

1. The ear is sensitive to the ratios of two frequencies in a musical interval rather than the differences in the musical intervals themselves.
2. The intervals which are perceived to be most consonant are composed of small integer ratios of frequencies. The octave, fifth, and fourth are the intervals which have been considered to be consonant throughout history by essentially all cultures, so they form a logical base from which to build up musical scales. A typical strategy for using these universally consonant intervals is the circle of fifths, discussed elsewhere.

14.3.2 The scale step

The distance between two successive notes in a scale is called a *scale step*. The notes of a scale are numbered by their steps from the first degree of the scale. There are two main steps used in Western music, half steps and whole steps.

1. **Half steps:** the smallest interval from one note to the next closest note higher or lower is called a *half step* (figure 14.6). On a musical staff, half steps are usually represented by adding a sharp (♯) or a flat (♭) symbol to the note. Half steps are also sometimes called *semitones*. On a piano, a chromatic scale plays all notes on both the white keys (natural notes) and black keys (sharp (♯) or flat notes (♭)) and goes up or down by half steps. For example, the notes C2 (which has a frequency of 65.4064 Hz) and C2♯ (which has a frequency of 69.2957 Hz) are a half step apart.
2. **Whole steps:** whole steps or whole tones are notes that are two half steps apart. For whole steps, you either have to go up two half steps or go down two half steps, but it must be always two half steps (figure 14.7). For

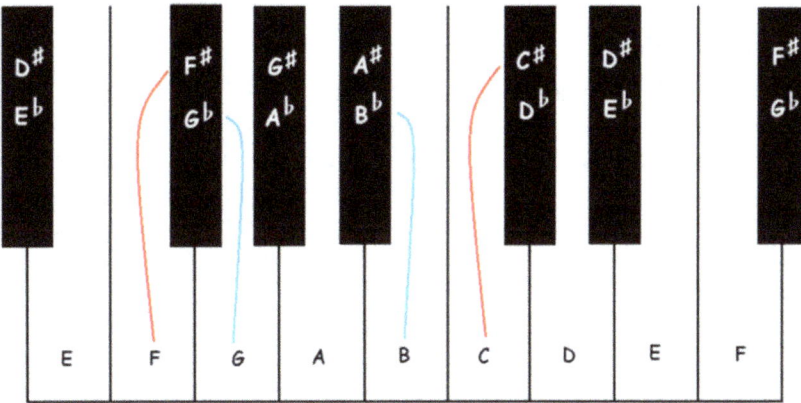

Figure 14.6. Half steps between F and F sharp (or G flat); between G and G flat, between A sharp (or B flat) and B; and between C and C sharp (or D flat).

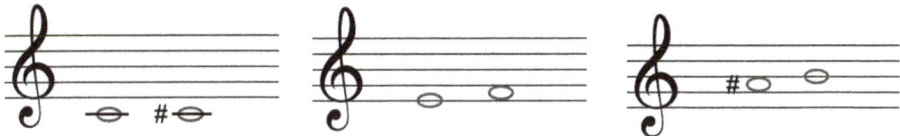

Figure 14.7. Examples of half steps on a staff.

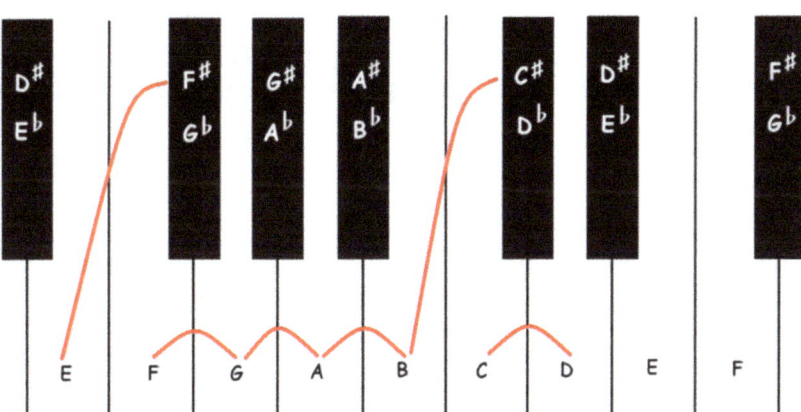

Figure 14.8. Selected examples of whole-step intervals: between E and F sharp (or G flat), between F and G, G and A, and A and B, between B and C sharp or D flat, and between C and D.

example, the notes C2 (which has a frequency of 65.406 Hz) and D2 (which has a frequency of 73.416 Hz) are a whole step apart. A whole-tone scale sounds very different from a chromatic scale since it is a scale made only of whole steps (figure 14.8).

14.3.3 Harmonic series

The notes that musicians play are generally not a single pure frequency, but are a blend of frequencies of a particular harmonic series. Each instrument may play notes of a specific harmonic series; for example, the notes that can be played on a bugle do not have the same range as those that can be played on a piano.

This means that harmonic series play a very important role in the production of musical sounds and hence in the actual construction of musical instruments. A harmonic series primarily depends on its fundamental frequency and is built up from that. Because a harmonic series can have any note as its fundamental, there are many different harmonic series. However, the relationships between the frequencies of a harmonic series are always the same. The second harmonic always has exactly half the wavelength (and twice the frequency) of the fundamental, and the third harmonic always has exactly a third of the wavelength (and thrice the frequency) of the fundamental, and so on.

14.4 Musical intervals

In music theory, intervals are the building blocks of scales, chords (or harmonies), and melodies. An interval may be loosely defined as the difference between two pitches, that is, the perceived spacing between two pitches is called an *interval*. In other words, the distance between pitches is called a *musical interval*. An interval can be a step up or a step down in pitch and the it specifies the ratio of the frequencies. When the ratios are simple and ideal, the series is harmonic and is called a *harmonic series*, and when the frequency ratios are not simple, they are equally tempered; we call this an *equal temperament scale*. In both cases, whether we talk about harmonic series or equal-tempered scales, low notes have low frequencies and long wavelengths, whereas high notes have higher frequencies and shorter wavelengths.[6]

14.4.1 Naming the intervals

To name any interval, we first have to come up with a proper method of characterizing the interval, so that it can be written on the musical staff. One way to do this is to count the spaces on the staff that correspond to the frequencies of our notes. This includes the lines the notes lie on as well as every space between the notes, which gives the number for the interval. For example, the interval between A and C is a third, as shown in figure 14.9 (left), whereas the interval between A and E is a fifth, as shown in figure 14.9 (right).

---

[6] Sometimes it is easier to remember this by an analogy. Suppose a father and his seven-year-old son go out for a run in the afternoon. In order to keep up, the son takes two steps for every one that the father takes. This means that the running steps occur more often for the son, i.e. his steps are more frequent, hence they are at a higher frequency. The father, on the other hand, takes less frequent steps and therefore has a low step frequency and a long wavelength.

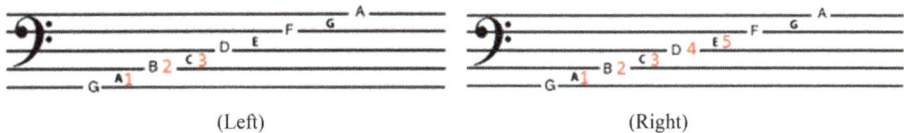

Figure 14.9. Musical intervals: (left) between A and C and (right) between A and E.

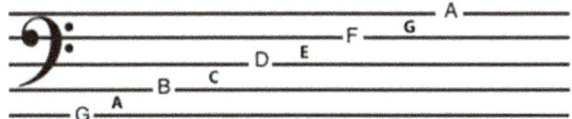

Figure 14.10. Musical octave.

Example 14.1. Compare the frequencies of two notes, one at 512 Hz and the other at 1024 Hz, and reduce the fraction to the lowest term.
Solution:
To compare the frequencies of two notes, 512 Hz and 1024 Hz, we use their ratio: 512:1024 :: 1:2, or:

$$\frac{\text{frequency 1}}{\text{frequency 2}} = \frac{512}{1024} = \frac{1}{2}. \tag{14.1}$$

Now suppose that you switch the frequencies in the expression, the ratio also switches from 1:2 to 2:1, or:

$$\frac{\text{frequency 2}}{\text{frequency 1}} = \frac{1024}{512} = \frac{2}{1}. \tag{14.2}$$

Basically, it does not matter which frequency is put first, both answers are correct as long as you know and keep track of the frequencies.

If you were to compare two other frequencies, one at 630 Hz and the other at 840 Hz, their ratio would be: 630:840 :: 3:4.

14.4.2 The octave

An octave is a musical interval defined by the ratio 2:1 regardless of the starting frequency. This means that the frequencies from 100 Hz to 200 Hz are an octave with a ratio of 1:2, and frequencies in the interval from 2000 Hz to 4000 Hz are also an octave, since their ratios 1:2 are the same. Two notes that are exactly one octave apart sound good together, whereas two notes that are not exactly in the ratio 2:1 do not sound good together. In fact, even the untrained ear is able to tell that two notes that are, say, 2.2 times apart are not in tune with each other and sound quite unpleasant (figure 14.10).

Please note that no one actually has ever scientifically proven why the simple-ratio combinations sound pleasant to the human ear. In his book *Music, the Brain, and Ecstasy: How Music Captures Our Imagination*, Robert Jourdain covers some aspects of how and why some sounds are beautiful and others unpleasant.

Example 14.2. Consider four notes, A2, A3, A4, and A5, with their corresponding frequencies of 110 Hz, 220 Hz, 440 Hz, and 880 Hz.
1. Which note has the longest wavelength? Which note has the shortest wavelength?
   Solution:
   We know that $v = \lambda f$. Hence, the note that has the longest wavelength corresponds to the lowest frequency, therefore A2 has the longest wavelength.
   The note that has the shortest wavelength note corresponds to the highest frequency, therefore A5 has the longest wavelength.
2. Using a ratio, compare the number of waves that make up the note A4 with those of the note A2. (Hint: A4 has a frequency of 440 Hz and A2 has a frequency of 110 Hz.)
   Solution:
   A4 has a frequency of 440 Hz, while A2 has a frequency of 110 Hz. This means the ratio is 4:1, which implies that there are four waves in A4 for each wave in A2.
3. Write the ratio of the frequencies of A3 and A2. What is this ratio when expressed in the lowest terms possible?
   Solution:
   Consider the frequency ratio between A3 and A2:
   frequency A3:frequency A2 :: 220:110.
   In the lowest possible terms, this reduces to:
   frequency A3:frequency A2 :: 2:1.
4. Write the ratio of the frequencies of A4 and A3? What is this ratio when expressed in the lowest terms possible?
   Solution:
   Consider the frequency ratio between A3 and A2:
   frequency A4:frequency A3 :: 440:220.
   In the lowest possible terms, this reduces to:
   frequency A4:frequency A3 :: 2:1.

14.4.3 Octave-repeating scales

A scale is called an *octave-repeating scale* if every pitch in the scale appears in every possible octave. An octave-repeating scale can be represented as a circular arrangement of pitch classes ordered by increasing (or decreasing) pitch class. For instance, the increasing C major scale is C-D-E-F-G-A-B-[C], in which the bracket indicates that the last note is an octave higher than the first note. Alternatively, one could write C-B-A-G-F-E-D-[C], in which the bracket denotes an octave lower than the first note in the scale. This single scale can be manifested at many different pitch levels. For example, a C major scale can be started at C4 (middle C) and ascend an octave to C5, or it could start at C6 and ascend an octave to C7.

14.4.4 Scalar transposition

Composers often transform musical patterns by moving every note in the pattern by a constant number of scale steps; thus, in the C major scale, the pattern C-D-E might be shifted up, or transposed, by a single scale step to become D-E-F♯. This process is called *scalar transposition*. Since the steps of a scale can have various sizes, this process introduces subtle melodic and harmonic variation into the music. This variation is what gives scalar music much of its complexity.

14.4.5 Musical intervals

In music theory, the perceived spacing between two pitches is called an *interval*. We may listen to any interval in two different forms: (i) when we measure horizontally on the musical staff, the two notes occur one after the other and we call this a *vertical interval* or a *melodic interval*; (ii) when we measure vertically, the two notes sound simultaneously and we refer to this as a *vertical interval* or a *harmonic interval*.

The most basic interval in music is the *octave*, which is defined as the distance between one pitch and another that has half or double its fundamental frequency. Starting with this basic interval, one can define notions of other intervals by considering the frequency relations between harmonics (also called the physical approach), geometric relations (also called the mathematical approach), or note relations (called the musical approach). These approaches lead to slightly different notions of intervals, which are, however, referred to by the same interval names.

The size of an interval, also known as its width or height, can be represented using two alternative but equally valid methods, each appropriate to a different context: frequency ratios or cents.

1. **Frequency ratios:** from a physical point of view, the concept of intervals considers frequency relations between harmonic partials. In the physical approach, the intervals are derived from frequency relations between partials that occur within the same harmonic series. 'Pure intervals' are simple frequency ratio intervals found in the harmonic series with very simple frequency ratios. Pure intervals are nice simple ratios such as 2:3, e.g. those found in vibrating strings. Musical tuning based on harmonics is known as 'pure' or 'just intonation.' Some common pure intervals and their frequency ratios are as follows:

    (a) An octave has a frequency ratio of 2:1.
    (b) A pure fifth has a frequency ratio of exactly 3:2.

Table 14.1. Frequency ratios in the equal temperament system.

Interval name	Frequency ratio	Decimal
Unison	$(\sqrt[12]{2})^0$	= 1.0000
Minor second	$(\sqrt[12]{2})^1$	= 1.0595
Major second	$(\sqrt[12]{2})^2$	= 1.1225
Minor third	$(\sqrt[12]{2})^3$	= 1.1892
Major third	$(\sqrt[12]{2})^4$	= 1.2599
Perfect fourth	$(\sqrt[12]{2})^5$	= 1.3348
Tritone	$(\sqrt[12]{2})^6$	= 1.4142
Perfect fifth	$(\sqrt[12]{2})^7$	= 1.4983
Minor sixth	$(\sqrt[12]{2})^8$	= 1.5874
Major sixth	$(\sqrt[12]{2})^9$	= 1.6818
Minor seventh	$(\sqrt[12]{2})^{10}$	= 1.7818
Major seventh	$(\sqrt[12]{2})^{11}$	= 1.8897
Octave	$(\sqrt[12]{2})^{12}$	= 2.0000

(c) A fourth has a frequency ratio of exactly 4:3.

(d) A pure major third has a frequency ratio of exactly 5:4.

Equal Temperament:

Western musicians ordinarily do not use pure intervals; instead, they use a tuning system called *equal temperament*.[7] In the equal temperament system, an 'octave' (say, from C1 to C2) is divided into twelve equally spaced notes. The words 'equally spaced' mean that the ratio of the frequencies of the two notes in any half step is always the same.

This tuning system is very convenient for instruments such as the piano, and it also makes it very easy to change keys without retuning an instruments. In the equal temperament system, the ratios of the notes or frequencies are based on the twelfth root of two, mathematically written as $\sqrt[12]{2}$), as seen in table 14.1. This makes the math a bit more complicated; however, it evens out the intervals between the notes so that the final scales are more uniform.

2. **Cents:** the standard system for comparing interval sizes is the cent, a musical unit discussed in chapter 13, section 13.4.3. Twelve-tone equal temperament (12-TET) is a tuning system in which all semitones have the same size and the size of one semitone is exactly 100 cents.

---

[7] Equal temperament was advocated by theorists as far back as the 16th century and came into its own during the 18th and 19th centuries. This system is the one universally used today for music. Mean-tone temperament has pure (or very nearly pure) major thirds.

Example 14.3.
1. What is the fundamental frequency of a note an octave above 200 Hz?
   Solution:
   We know an octave has a frequency ratio of 2:1; therefore, the fundamental frequency of a note an octave above 200 Hz is: ($\frac{2}{1}$ × 200 Hz) = 400 Hz.
2. What is the fundamental frequency of a note a fifth above 200 Hz?
   Solution:
   We know a pure fifth has a frequency ratio of 3:2; therefore, the fundamental frequency of a note a fifth above 200 Hz is: ($\frac{3}{2}$ × 200 Hz) = 300 Hz.
3. What is the fundamental frequency of a note a fourth above 200 Hz?
   Solution:
   We know a fourth has a frequency ratio of 4:3; therefore, the fundamental frequency of a note a fourth above 200 Hz is: ($\frac{4}{3}$ × 200 Hz) = 267 Hz.

Example 14.4. Suppose that the frequency of one note is 1300 Hz and the frequency of a second note is 1120 Hz. What equal temperament interval do these two notes sound like?
Solution:
To find the equal temperament interval these two notes sound like, we have to calculate the ratio of frequencies and then compare the answer to the frequencies in the equal temperament system:

$$\frac{1300}{1100} = 1.1818$$

Consulting the table 14.1, we see this is very close to 1.1892, hence this interval sounds like a minor third.

Example 14.5. Compare a pure major third from the harmonic series to an equal temperament major third.
Solution:
The pure major third in a harmonic series has the ratio 5: 4 = $\frac{5}{4}$ = 1.25, whereas from table 14.1, we know that a pure major third in the equal temperament system has a ratio of 1.2599.

14.4.6 Types of intervals

Intervals are divided into two types based on how far apart they are in terms of frequency:
1. **Simple intervals:** intervals that are one octave or less are known as simple intervals. Simple intervals are the prime, second, third, and so on up to the octave, as shown in figure 14.11.
2. **Compound intervals:** intervals that are larger than one octave are known as compound intervals. They are the ninth, tenth, and higher intervals, as shown in figure 14.12.

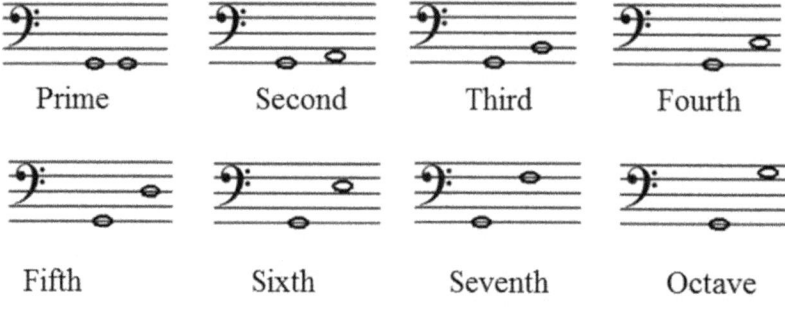

Figure 14.11. Examples of simple intervals.

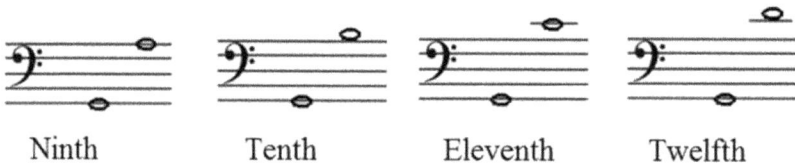

Figure 14.12. Examples of compound intervals.

To find the interval between two notes, find the pitch of the lowest note and start counting until you reach the top note. Keep in mind that when counting intervals, you must start at the bottom note and count both the bottom and top notes. For example, to find the interval between C and G, begin on C and count up the scale until you reach G, in other words, count [CDEFG]. When you do this, you find that the interval between C and G has seven half steps; then, from table 14.2, we discover that it is a perfect fifth.

Example 14.6.
1. What is the size of the interval from C up to E in half steps?
   Solution:
   from C up to E is [CDE] = three notes. These have four half steps; hence, from table 14.2, it is a major third.
   C to E_b has three half steps; hence, table 14.2 indicates it is a minor third.
2. What is the size of the interval from C down to E in half steps?
   Solution:
   from C down to E is [CBAFGE] = six notes. These have eight half steps; hence it is a minor sixth.
3. What is the interval between A and C?
   Solution:
   from A up to C is [ABC] = three notes. These have three half steps (which is a minor third), whereas from A down to C is nine half steps (which makes it a major sixth).

Table 14.2. Interval names and half steps for notes above C.

Note above C	Half step	Interval name
C	0	Unison
C# / D$_b$	1	Minor second
D	2	Major second
D# / D$_b$	3	Minor third
E	4	Major third
F	5	Perfect fourth
F# / G$_b$	6	Tritone
G	7	Perfect Fifth
G# / A$_b$	8	Minor sixth
A	9	Major Sixth
A# / B$_b$	10	Minor seventh
B	11	Major seventh
C	12	Octave

4. What is the interval between E and A?
   Solution:
   The interval from E up to A has five half steps (a perfect fourth), whereas that from E down to A has seven half steps (a perfect fifth).

14.5 Various terms that are important to know

In this section, we briefly discuss some of the notation used in music theory; for more details of these terms, please refer to a thicker textbook.

1. **Intonation:** intonation refers to the degree of accuracy with which pitches are produced. Intonation is defined as a musician's realization of pitch accuracy, or the pitch accuracy of a musical instrument.
2. **Tonic:** a tonic is the first note in a musical scale. Musical pieces are based on the tonic, and it can be thought of as the 'home key.' The melody and harmony of the piece may change to different keys and produce different moods, but the tonic serves a benchmark for comparison. For example, a symphony may start in C minor and switch to E flat major during the exposition or development and conclude in the tonic key of C minor.
3. **Consonance and dissonance:** consonance and dissonance are musical terms that have specific meanings for musicians and physicists. Notes that sound good together when played at the same time are called *consonant*, whereas notes that sound harsh or unpleasant when played at the same time are called *dissonant*. Consonance or dissonance depends on the physics of sound and on the musical preferences of a particular culture or an individual. Please keep in mind that if notes do not sound good together, it might not be due to consonance or dissonance but rather an issue of tuning. The use of proper or improper tuning can significantly affect intervals that sound consonant or dissonant to us.

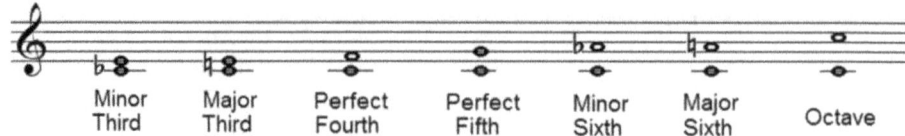

Figure 14.13. Consonant intervals.

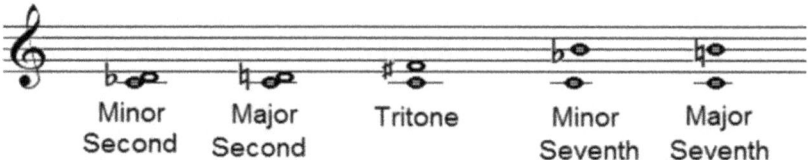

Figure 14.14. Dissonant intervals.

In modern Western music, simple intervals (one octave or smaller) that are considered consonant are the minor third, major third, perfect fourth, perfect fifth, minor sixth, major sixth, and the octave, as shown in figure 14.13. All these intervals are considered pleasing to the ear, and chords that contain only these intervals are considered to be *stable*. The intervals considered dissonant are the minor second, the major second, the minor seventh, the major seventh, and particularly the tritone, which is the interval in between the perfect fourth and the perfect fifth. These intervals are considered to be unpleasant to hear or *tension producing*. Examples of dissonant intervals are shown in figure 14.14.

4. **Chord:** an interval is measured between two notes. More than two notes sounding at the same time form a chord, which contains a combination of several intervals. Fundamentally, a chord is any harmonic set of pitches consisting of multiple notes heard simultaneously.

5. **Just intonation:** in just intonation, the fifth and the third are both based on the pure, harmonic series interval. Because chords are constructed of thirds and fifths, this tuning makes typical Western harmonies particularly resonant and pleasing to the ear, so this tuning is often used by musicians who wish to make small tuning adjustments quickly. This includes vocalists, most wind instruments, and many string instruments.

6. **Temperament:** in musical tuning, a temperament is a system of tuning which slightly compromises, or in other words makes slightly impure, the pure intervals of just intonation in order to meet other requirements of the system. Hence, a temperament is a system of tuning in which the intervals deviate from acoustically pure (Pythagorean) intervals.

    (a) **Pythagorean temperament:** the earliest temperament used in Western music was the Pythagorean temperament. The Pythagorean scale is a scale constructed from only pure perfect fifths (3:2) and octaves (2:1). Pythagorean tuning was used by musicians up to the beginning of the 16th century.

(b) **Mean-tone temperament:** mean-tone temperament is another system of musical tuning. In general, a mean tone is constructed in the same way as Pythagorean tuning, i.e. as a chain of perfect fifths, but in mean-tone temperament, each fifth is narrowed by the same amount (or equivalently, each fourth is widened) in order to make the other intervals such as the major third closer to their ideal just ratios. Basically, mean-tone temperament is obtained by slightly compromising the fifths in order to improve the thirds.[8]

(c) **Irregular or circulating temperaments:** irregular or circulating temperaments were advocated by some Renaissance and Baroque theorists in the late 18th century because of the particular character they gave to each key. As music became increasingly chromatic and ranged more widely in key, irregular temperaments were replaced by equal temperament, previously discussed in section 14.4.6.

14.5.1 False pattern recognition

The human brain tries hard to recognize orderly patterns everywhere. It tries to do this even when patterns do not actually exist. This tendency of the human brain to see patterns that do not actually exist is called *apophenia*. Examples of apophenia include the recognition of faces, animals, and figures in clouds and in patterns with no deliberate design

This happens with incoming sounds as well. For aural phenomena, consider a case in which the incoming sound contains components at 1860 Hz, 2060 Hz, and 2260 Hz. There are two or maybe even more possibilities:

1. The human brain may accept this as the ninth, tenth, and eleventh harmonics of an approximate fundamental at 206 Hz, even though, in reality:
   (a) $9 \times 206 = 1854$ Hz,
   (b) $10 \times 206 = 2060$ Hz, and
   (c) $11 \times 206 = 2266$ Hz.
2. The brain could also interpret it as the tenth, eleventh, and twelfth harmonics of an approximate fundamental at 187 Hz:
   (a) $10 \times 187 = 1870$ Hz,
   (b) $11 \times 187 = 2057$ Hz, and
   (c) $12 \times 187 = 2244$ Hz.

Both possibilities are equally likely.

---

[8] The perfect fifth is the musical interval between a note and the note seven semitones above it on a musical scale. The perfect fourth is a musical interval which spans four scale degrees. It consists of the note and the note five semitones above it on the musical scale.

Review Questions 14.1.
1. What are the 12 keys of Western music?
2. What are the 12 major scales in Western music?
3. What is a JND? Give an example not mentioned in this chapter.
4. What is the difference between harmonic and inharmonic series?
5. What is the difference between consonant and dissonant intervals?

Exercises 14.1.
1. Using properties of logs and the log table in appendix C, find the logarithms of the following:
   (a) log(1024)
   (b) log(3025)
   (c) log(5.36)
   (d) $\log(10^5)$.
2. Given the values of $\log x$, find the number x in each case:
   (a) $\log x = 0.3$
   (b) $\log x = 1.3$
   (c) $\log x = 0.5$
   (d) $\log x = -0.5$.
3. What is the size of the interval from C up to A in half steps? What is the name of this interval?
4. What is the size of the interval from C up to D in half steps? What is the name of this interval?
5. What is the fundamental frequency of a note one octave above 800 Hz?
6. What is the fundamental frequency of a note a fourth above 800 Hz?
7. What is the fundamental frequency of a note a pure fifth above 800 Hz?
8. The frequency of one note is 800 Hz and that of another is 600 Hz. What equal temperament interval do these two notes sound like?

Further reading

- Hall D E 1991 *Musical Acoustics* 2nd edn (Pacific Grove, CA: Brooks/Cole)
- Emanuel D C and Letowski T 2009 *Hearing Science* (Philadelphia, PA: Wolters Kluwer)
- Roy D, Patterson E G and Thomas C W 2010 *The Perception of Family and Register in Musical Tones* in: Riess Jones M, Fay R and Popper A (eds) *Music Perception* (New York: Springer)

- Mari R J, Richard R F and Arthur N P 2010 *Music Perception* (Berlin: Springer)
- Hartmann W M 1997 *Signals, Sound, and Sensation* (Berlin: Springer)
- Wever E G 1949 *Theory of Hearing* (New York: Wiley)

Part VI

Musical instruments

IOP Publishing

The Physics of Sound and Music, Volume 1
A complete course text (Textbook)
Samya Bano Zain

Chapter 15

Stringed instruments

A stringed instrument is any musical instrument that produces sound by means of vibrating strings. When a string in the instrument is set into vibration, it causes sympathetic vibrations to be set up in the air contained within the body of the instrument and causes the air in the body of the instrument to start vibrating, which in turn sets up standing waves in the body of the instrument. All these vibrations together give us the music produced by the instrument. Please note that this is not necessarily true of some electrical instruments, for example, solid-body electric guitars. In these guitars, the sound is encoded into an electrical signal by 'pickups,' which are generally placed below the strings adjacent to the bridge.

Stringed instruments are categorized by the method they use to produce the string vibration. The most common techniques are plucking, bowing, and striking.

1. **Vibrations of a plucked string**: plucking, as the name implies, is the method of plucking the strings by the finger and/or thumb of the player or by some type of plectrum held by the player. The plucking technique is referred by the Italian term *pizzicato*, and it gives a somewhat different quality of sound than you would get by just striking the string. Plucking is used when playing instruments such as the guitar, harp, sitar, etc. In the case of the harp, the string is directly plucked by the musician's finger, whereas in the case of the harpsichord, which has a keyboard like that of the piano, a key mechanism is used to pluck the string. It is worth noting that most of these instruments require one or more separate strings for each note to be played.
2. **Vibrations of a bowed string**: bowing (Italian: *arco*) is a method used with stringed instruments to produce vibrations in the instrument. The violin, viola, cello, and double bass are examples of *bowed instruments*, some of which are shown in figure 15.1. The vibrations in a bowed string are produced by slip-and-stick action (discussed in detail in section 15.6.2) which is the result of running a bow over the string, which makes the string vibrate. The bow consists of a *bowed* stick with many fibers stretched between its two

Figure 15.1. Plucked stringed instruments over the ages. Credit: Hareem Zain at the Met, NYC, 2019.

ends. In higher-grade (and hence more expensive) bows, the fibers are made of horse-tail hair but in lower grade (and hence less expensive) bows, the fibers are made from nylon. The ancestors of the modern bowed string instruments include the rebab of the Islamic empire.

3. **Vibrations in a struck string**: the third common method of sound production in stringed instruments is to strike the string. Violin family stringed instrument players are occasionally instructed to strike the string with the side of the bow, a technique called *col legno*. This produces a percussive sound along with the pitch of the note.

15.1 The history of stringed instruments

As early as several centuries BCE, we find artists' depictions of stringed instruments such as the harp. Greek philosophers such as Pythagoras (around 500 BCE) discussed sounds in a stretched string. By medieval times, the harp and the lyre were associated with the music of the heavens, and we see them depicted with angels in paintings of the time. During this time, hammering and bowing were developed as techniques for creating sound in stringed instruments. In the case of the hammered dulcimer, a small hammer with a head of leather was used to strike the strings to produce sound. This technique was later adapted for the piano. The most important instrument development may have been the lute, shown in figure 15.2 (left). The word 'lute' is related to the Arabic word *ud*, which means wood. These were the earliest instruments with resonant wooden boxes, in direct contrast to those made from animal shells. The lute became by far the most important stringed instrument of the European renaissance period and remained popular for almost 500 years.

Other instruments that developed during the Renaissance period were new varieties of bowed stringed instruments that were fretted just like the modern-day guitar and several other varieties of plucked stringed instruments. Next came the Baroque era, during which some plucked instruments such as the lute decreased in popularity, but others like the banjo, ukulele, and the dulcimer increased in popularity. The only instrument from those eras used in the contemporary orchestra is the harp. It is important to note that no authentic medieval stringed instruments exist today and our primary sources of information about them are paintings and works of art. Some stringed instruments used of the ages in East Asia are shown in figure 15.3.

(Left)   (Right)

Figure 15.2. Stringed instruments: (left) a harp; (right) a lute. Credit: Hareem Zain at the Met, NYC, 2019.

Figure 15.3. East Asian stringed instruments over the ages: (a) the erhu (China); (b) the hegumu (Korea); (c) the sanxian (China); (d) the pipa (China). Credit: Hareem Zain at the Met, NYC, 2019.

15.2 The introduction of energy into a string instrument

The introduction of energy into an instrument (say by striking, bowing, or plucking a string) initially creates energies at many different frequencies that are not related to one another by simple integral multiples. For the brief instant after we strike, bow, pluck, or otherwise cause the instrument to start making sound, the impact itself has a rather noisy quality that is not especially musical. This is called the *attack phase*. The sound during the attack phase is like that of a hammer hitting a piece of wood—basically more noise than music. In one study, recordings of the attack phase of instruments were edited out and played for an audience. It was found that it was nearly impossible for most people to identify the instrument being played. In the attack phase, pianos and bells sound remarkably similar to each other.

Next comes the *steady state*, which occurs after the attack phase is complete. This is a more stable phase, in which the musical tone takes on an orderly pattern of overtone frequencies depending on the material of the instrument. The sound produced in this phase is due to the resonances set up in the instrument itself and it depends on the material of the instrument. For example, a guitar made of wood sounds fundamentally different from a guitar made of plastic.

If we edit the attack of one instrument onto the steady state of another instrument, we get remarkable results. In some cases, it is possible to hear a hybrid instrument that sounds completely different from the original instruments played. In one study, it was found that by adding the attack phase of a violin bow to a steady-state flute tone creates a sound that strongly resembles a street organ!

15.2.1 Allowed frequencies of the standing waves on the string

When we pluck a string of length (L) fixed at both ends in, say, the middle, a special kind of wave is set up in the string. The wave thus set up travels along the string, reflects from the other end, and interferes with the other waves traveling along the string. These waves together create *standing waves*, which were discussed in detail in chapter 5. A standing wave remains in a constant position along the string and from chapter 6, section 6.3, we know that the allowed frequencies of the standing waves on a string are given by the formula

$$f = n\frac{v}{2L}, \qquad (15.1)$$

where v is the speed of sound, L is the length of the string, and n takes values such as 1, 2, 3, ..., depending on the particular mode of vibration.

Example 15.1. Calculate the speed of the standing wave in a 90 cm long guitar string with a fundamental frequency of 400 Hz.

Solution:
Given that the length of the string is 90 cm, we have to convert it to meters. $L = 0.90$ m.

$$f = n\frac{v}{2L}$$

$$\Rightarrow v = \frac{2fL}{n}.$$

Substituting in the values gives

$$v = \frac{2(400 \text{ Hz})(0.90 \text{ m})}{1} = 720 \text{ m s}^{-1}.$$

Hence, the speed of the standing wave in a 90 cm long guitar string is 720 m s^{-1}, which is about twice the speed of sound in air.

15.2.2 Changing the pitch in stringed instruments

There are different ways of changing the pitch of stringed instruments:
1. **Change the string material:** the material of the string changes the quality of the sound produced by the string. The string material influences the overtones that are created in the instrument. So, one way to change the pitch is to change the material of the string, which is why guitar strings are made out of an array of materials ranging from wood to precious metals. Another factor that can affect the sound produced is called the *connectedness* of the string. The connectedness is based on the process used to create the string. Examples include *wound* and *non-wound* strings. Wound strings are made of a wire wrapped around a metal or nylon core (as seen in figure 15.4). They are typically thicker and are held under higher tension, which means they are louder and last longer. Unwound strings, on the other hand, are made of a single thin wire. These are thinner and have a lower tension, which means they are quieter and have a shorter lifespan.
2. **Change the tension in the string:** another way to change the frequency (and hence the pitch) of a note produced by a string is to change the tension the string is held under. The tension can be increased or decreased by tuning pegs that are attached to the string at the *head* of the instrument. The string is tightened or loosened by turning the tuning peg clockwise or counter-clockwise. Higher tension results in an increased frequency of string vibration and hence the production of a sound of higher pitch, while lower tension results in a lower-pitched sound. In general, the playing tension of a violin string is kept in the range between 9 and 20 lb for low range (or 40 to 89 lb at the high range).
3. **Change the length of the string:** the last and the easiest way to change the pitch is to physically change the effective length of the string. The *effective length* of a string is the length of the string that vibrates when it is set into

Figure 15.4. Acoustic guitar string thicknesses vary according to the notes that need to be played. Credit: Hareem Zain at the Met, NYC, 2019.

motion by, say, bowing or plucking the string. A longer effective length results in the production of low-frequency sounds, whereas shortening the effective string length produces higher-frequency sounds. Shortening is usually accomplished by preventing a portion of the string from vibrating. This is done by the player's fingers, which *pin* the strings against the fingerboard. The locations at which a string is pinned should correspond to the harmonics on the string. In the case of the guitar, frets tell the musician where to place their fingers. For bowed instruments such as violins, which do not have frets, there are a certain number of acceptable effective string lengths for each string; the player needs to know what these are to produce musical notes. In fact, for both types of instruments, the set of acceptable effective string lengths are the same for all strings.

15.3 Tuning

Before musicians play together, they have to *tune* their instruments, that is they have to make sure that all the instruments in the band or orchestra are playing at the same frequency for a particular note. This means that they all have to exactly agree on the frequency of an 'a4' versus the frequency of an 'a5' and so on. All tuning systems are based on the physics of sound. Hence, tuning is basically the process of adjusting the pitch of one or many tones produced by musical instruments to establish typical intervals between these tones. In the standard Western musical tradition, musicians usually tune to a4 or $A = 440$ Hz, which is the historically agreed relative frequency, called the *reference pitch,* as defined in section 13.1.6.

Please keep in mind that other cultures not only have different note names and different scales, they may even have different notes based on different tuning systems. These are affected by the history of the various cultures' musical traditions, as well as the instruments used in those traditions.

15.3.1 Some methods of tuning

Before a performance, all instruments must be tuned to each other. Tuning may be accomplished by sounding two pitches aloud and adjusting one of them to match the other or matching it to a frequency of a known tuning fork or an electronic tuning device until the interference beat phenomenon (discussed in section 5.5.6) cannot be detected any more. When beats are no longer heard, the two instruments are said to be *in tune* with each other. The term *out of tune* means that the pitch of the instrument is either too high or too low, i.e. sharp or flat, respectively. While an instrument might be in tune with itself, which is called *relative tuning*, it may not be tuned to the reference frequency ($A = 440$ Hz) when playing with a piano or other fixed-pitch instrument. In this case, musicians will agree that the instrument is out of tune, as it does not match the reference pitch. An instrument may also become permanently out of tune if it is worn down or damaged in some way, and at this point it must either be repaired or replaced.

Different methods of sound production require different methods of tuning or adjustment, some of which are given here:

- A singer can tune to a given pitch with their voice, and this is called *matching pitch*. Most vocalists learn this basic skill during ear training.
- Tuning can be done manually by either turning tuning pegs by hand or by using a wrench to increase or decrease the tension on the strings and thereby control the pitch. The tuning pegs of some instruments, such as the guitar, can be turned by hand, while other instruments such as the harp or piano require a wrench to turn the pegs.
- Tuning can be accomplished by changing the physical length or the width of an instrument. For example, changing the physical length or width of the tube in wind or brass instruments changes the pitch, which can then be used to match the pitch of a given instrument to that of another instrument.

15.4 The guitar

The guitar, as shown in figure 15.5, is a plucked stringed instrument in which the string vibration is induced by a finger or a pick. The oldest known depiction of an instrument that has the essential features of a guitar is included in a 3300 year old stone carving of a Hittite bard. The early period of the ancient Hittite empire was around the 18th century BCE, and it spanned regions that now include parts of the Middle East, Iraq, Arabia, and Turkey. Guitars were traditionally made out of wood and strung with animal gut. The modern era has seen the advent of polycarbonate guitars with nylon or steel strings.

There are two primary families or categories of guitars:

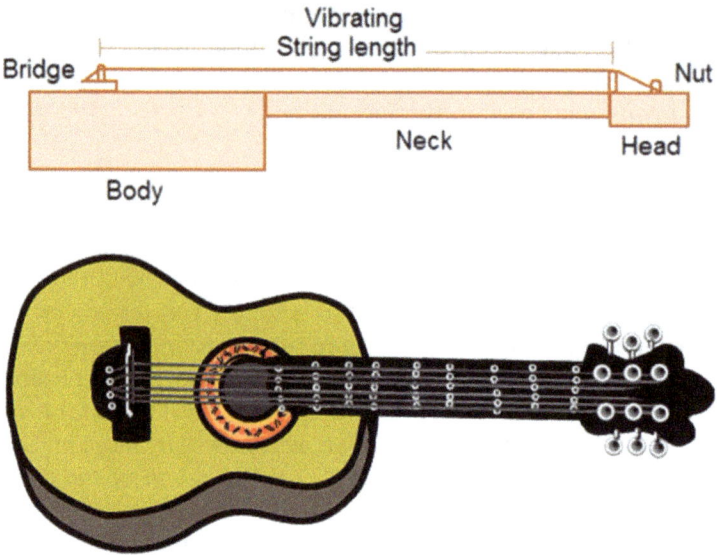

Figure 15.5. The guitar: (bottom) a modern acoustic guitar; (top) a schematic of an acoustic guitar.

1. **Acoustic guitars:** acoustic guitars have hollow bodies and have been played for over a thousand years. Acoustic guitars include many different subcategories, for example, classical or Spanish guitars, Renaissance guitars, and Baroque guitars, which are considered the ancestors of the modern guitar.
2. **Electric guitars:** electric guitars were introduced in the 1930s. These usually have solid, semi-hollow, or hollow bodies and produce little sound without amplification. They rely on an amplifier that can electronically manipulate the tone. Electromagnetic pickups convert the vibration of the steel strings into signals, which are fed to the amplifier.

15.4.1 Guitar strings

Each string in a guitar vibrates at a natural frequency that is directly proportional to the tension the string is held under and inversely proportional to the diameter of the string:

$$f \propto \frac{T}{d}, \tag{15.2}$$

where T is the tension in the guitar string and d is the diameter of the string. The mathematical formula for the tension in a guitar string is:

$$T = (2fL)^2 \mu, \tag{15.3}$$

where f is natural frequency of the string, L is the scale length, and μ is the linear mass length.

From equation (15.2), we see that to produce higher frequencies we have to either increase the tension of the string and/or decrease the diameter of the string. To produce lower frequencies, we have to do the exact opposite, either decrease the tension and/or increase the diameter of the string. The desired string tension on an acoustic guitar, whose typical string length (the distance from the nut to the bridge) is 25.5″, varies from about 70 N (16 pounds[1]) for extra-light strings to around 180 N (40 pounds) for heavy-gauge strings.

- **Lower string tension:** lower tension produces less sound, but the strings are easier to play. At tensions somewhere below 62.27 N (14 pounds), a string may be too loose to have acceptable sound quality.
- **Higher string tension:** higher string tension produces louder sound, but is harder to play and too much tension may damage guitars.

Another thing to remember for all strings is that if we keep the pitch and other factors the same, a heavier string has to be maintained at a greater tension and is harder to bend. It does, however, have more 'sustain' and more volume, and its vibration pattern is lower than the vibration pattern of a lighter string. A lighter string requires less tension; it has a smoother tone and a lower volume but a wider vibrating pattern. For example, on a typical full-size acoustic guitar,

1. A wound string with a diameter of 0.553 mm at a tension of 111.21 N (25 pounds) produces a G3 note, i.e. a frequency of 196 Hz.
2. However, to produce a G3 at a tension of 88.96 N (20 pounds), a plain string with a diameter of 0.475 mm is necessary.
3. Similarly, to produce a G3 using a tension of 66.72 N (15 pounds), we require a plain string with a diameter of 0.411 mm.

Example 15.2. Determine the tension in the E-string of a 25.5 inch guitar; $\mu = 0.0003955$ kg m^{-1}. The string is tuned to E4, i.e. a frequency of 329.6 Hz.

Note played	Frequency	Required tension	Type of string	String diameter
G3	196 Hz			
		133.4 N (30 pounds)	Wound	0.606 mm
		111.2 N (25 pounds)	Wound	0.553 mm
		88.96 N (20 pounds)	Plain	0.475 mm
		66.72 N (15 pounds)	Plain	0.411 mm
E4	329.63 Hz			
		133.4 N (30 pounds)	Plain	0.346 mm
		111.2 N (25 pounds)	Plain	0.316 mm
		88.96 N (20 pounds)	Plain	0.282 mm
		66.72 N (15 pounds)	Plain	0.244 mm

---

[1] The pound (abbreviated lbf) is a unit of force used in English engineering units. The pound-force is equal to the gravitational force exerted on a mass of one avoirdupois pound (≈0.453 kg) on the surface of the Earth. 1 lbf ≈ 4.45 N.

Solution:
Using the formula for the tension in a guitar string, equation (15.3), and putting in our values, we get

$$T = (2fL)^2 \mu$$
$$T = (2 \times 329.6 \text{ Hz} \times 0.6477 \text{ m})^2 (0.0003955 \text{ kg m}^{-1})$$
$$T = 72 \text{ N}.$$

The tension in the E-string of a 25.5 inch guitar tuned to E4 must equal 72 N.

15.4.2 Acoustic guitars

An acoustic guitar produces sound via strings that vibrate above a hollow guitar body. The vibrations are transmitted through the air and do not require electrical amplification. Acoustic guitars have many different subcategories, for example, classical, Spanish, Renaissance, Baroque, etc.

Acoustic guitars are made of hollow wood with a large sound hole. The wood used for the guitars varies, but the most common choices are Sitka spruce for the top panel, spruce and cedar for the bracing, and mahogany for the bottom and side panels. The neck materials also vary, but rosewood and maple wood are popular choices for the neck and fretboard. To achieve the best sound possible, the choice of wood should depend on the sound you are trying to achieve, since each type of wood has its own sound signature.

As for the strings, most acoustic guitars have six strings made of either nylon or steel. The player produces sound either using their fingers or by means of a pick; the choice depends on the the type of guitar, genre of music, or simply the player's personal preference. Nylon string guitars have a mellow tone with strong resonance in the lower-mid frequencies and are typically used for classical styles of music, whereas steel strings are typically used for most other styles.

15.4.2.1 Vibrations of the top, back plate, and air cavity
When a guitar string is plucked, it vibrates. This vibration in turn causes the surrounding air molecules to vibrate at the same frequency. The resulting pressure wave travels outward and is responsible for the sound we hear. However, the sound produced by the plucked string on its own does not produce a very loud sound, since it only disturbs a small volume of air molecules adjacent to the string.

To combat this issue, vibrating strings are attached to a wooden soundboard, such as that of a guitar. The vibration of the strings sets up sympathetic vibrations at the same frequency in the soundboard. This causes the air molecules inside and around the guitar to vibrate at the same frequency. Due to the large surface area of the soundboard, more air molecules are set into vibrational motion, and the sound signal is amplified. These amplified vibrations travel through the air and are what we hear as audible sound coming from the guitar.

15.4.2.2 The resonances of the guitar body
Prior to the 1800s, the ideal volume for the body of a guitar and the size of the sound hole were chosen by trial and error, that is, by simple physical experimentation. It was very difficult to create guitars with the required bass response and sound. However, in the 1800s a German physicist, Hermann Ludwig Ferdinand von Helmholtz did a lot of work in this regard. He used a type of resonator that is now named after him, called the *Helmholtz resonator*. A Helmholtz resonator (previously discussed in chapter 6, section 6.6) is a simple device in the shape of a hollow sphere with a tight opening called the neck. Any empty bottle (glass or metal) can make a Helmholtz resonator in which the sound is created by simply blowing across the top of the bottle. Air passes through the neck and oscillates at the resonant frequency of the Helmholtz resonator, which is called *Helmholtz resonance*. Hence Helmholtz resonance is the sound created by blowing across the top of any empty bottle. The passage of air across the top causes the inside of the bottle to vibrate, which produces audible sound. Helmholtz found that the resonant frequency is given by the following formula:

$$f_n = \frac{v}{2\pi}\sqrt{\frac{A}{VL}}, \tag{15.4}$$

where f is the frequency of resonance (in Hz), v is the speed of sound in air, L is the length of the neck, A represents the area of the tube, and V is the volume of air in the chamber.

Helmholtz used multiple resonators to study the harmonic content of complex sounds during his time. Modern Helmholtz resonators are made of brass. In a guitar, the Helmholtz effect reduces the height of the main resonant peak of the sound produced by the soundboard and extends the bass response of the guitar. Without this, the main resonance of the soundboard could make the instrument sound unpleasant and lacking in bass response.

Example 15.3. Find the frequency and wavelength of a one-liter bottle, given that the area (A) of the tube is three square centimeters and length of the neck (L) is five centimeters. (Hint: use the speed of sound = 344 m s^{-1} at 20 °C.)
Solution:
To solve this question, we first have to make sure that all values are in SI units:
Length:

$$L = 5 \text{ cm} = 5 \text{ cm} \times \frac{1 \text{ m}}{100 \text{ cm}} = 0.05 \text{ m}$$

Area:

$$A = 3 \text{ cm}^2 = 3 \text{ cm}^2 \times \frac{1 \text{ m}^2}{(100)^2 \text{ cm}^2} = 0.0003 \text{ m}^2.$$

Volume:

$$V = 1 \text{ liter} = 0.001 \text{ m}^3.$$

Using equation (15.4), we can then find the frequency as follows:

$$f_n = \frac{v}{2\pi}\sqrt{\frac{A}{VL}} = \frac{344 \text{ m s}^{-1}}{2\pi}\sqrt{\frac{0.0003 \text{ m}^2}{0.001 \text{ m}^3 \times 0.05 \text{ m}}} = 134 \text{ Hz}.$$

We find the wavelength using the sound equation: $v = f_n \lambda$, as follows:

$$\lambda = \frac{v}{f_n} = \frac{344 \text{ m s}^{-1}}{134 \text{ Hz}} = 2.6 \text{ m}.$$

15.4.3 The solid-body electric guitar

The electric guitar takes the weak signal directly from the vibrating string and amplifies it electronically. This amplified signal is then transmitted to a large speaker to be broadcasted. Electric guitars usually have a solid wood body and electronic pickups that are used to amplify the sound, whereas acoustic guitars rely on sound amplification by sympathetic vibrations between the strings, the air inside the guitar body, and the body.

The invention, evolution, and development of the electric guitar involved many inventors and musicians. Some of the earliest experiments from the 1920s and 1930s simply attached a pickup to an acoustic guitar, which was then called an *electric–acoustic guitar* or a *hollow-body electric guitar*. However, the development of electric guitars was not easy, and many of the initial attempts were unsuccessful. The first successful electromagnetic pickup system was developed by George Beauchamp and Paul Barth, and in 1932 it was applied and marketed as the Rickenbacker Frying Pan guitar. These days, pickups are electromagnets mounted under the guitar strings that measure the physical vibrations of the strings magnetically.[2]

In the 1940s, after much experimentation, it was determined that a solid body with relatively high mass was required for the electric guitar. High mass means the guitar has a negligible response to the vibration of the strings and the pickup is able to obtain a cleaner signal of the string's pure tone without interference from the guitar body. Electric guitars are a staple of American musical style and technology and it would not be unrealistic for me to say that they have shaped the direction of modern American music in the latter half of the twentieth century.

15.4.3.1 Pickups

The pickups, shown in figure 15.6, as the name implies, pick up the sound vibrations from a plucked string in an electric guitar and convert them into electrical signals that can be used by a device called the 'amplifier.' Pickups are effectively just electromagnets[3] that usually

---

[2] The pickups magnetize the electric guitar strings by the application of an external field.

[3] The electromagnet, invented in the 1800s by British scientist William Sturgeon and improved in the 1830s by Joseph Henry, is a type of magnet in which a magnetic field is produced by an electric current. A simple electromagnet consists of a coil of wire wrapped around an iron core; the iron increases the strength of the created magnetic field. The strength of the magnetic field generated is proportional to the amount of current that passes through the winding. The main advantage of an electromagnet is that the magnetic field can be quickly changed or reduced to zero by changing the amount of electric current in the wire. Electromagnets are used for picking and moving heavy iron objects, such as iron and steel.

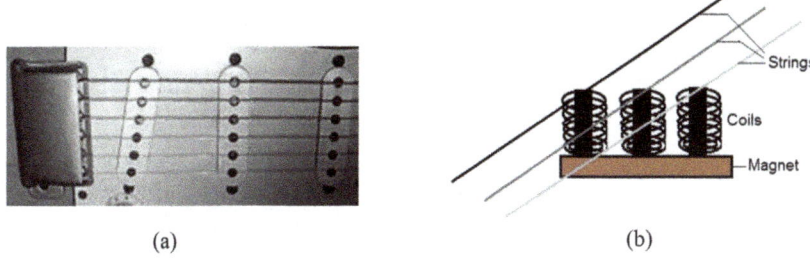

Figure 15.6. (a) The single-coil pickups of the Fender Stratocaster (1963), (b) a schematic of an electric guitar pickup.

consist of a permanent magnet wrapped in a copper wire. The pickup is generally attached to the body of the instrument; however, it may also be attached to other parts of the guitar, such as the bridge or the neck.

The electromagnets convert the vibrational energy of the strings into electrical energy that sets up a vibration in the magnet's magnetic field, which in turn induces an alternating current in the coil. This signal is then transmitted to the amplifier via an attached cable for amplification or to a recorder for recording.

15.5 The piano

The word piano comes from an abbreviation of the Italian word for the instrument, *pianoforte*. In music, the terms piano and forte mean *quiet* and *loud*, respectively.

15.5.1 The development of the piano

The piano is a rather recent instrument. By the 17th century, after centuries of trial and error, the mechanics of keyboard instruments such as the harpsichord, in which the strings are plucked by quills, were sufficiently understood to finally allow the development of the mechanisms that enabled the creation of the piano. Hence, the art of piano-making developed and prospered, and by the end of the 18th century, i.e. during the Industrial Revolution, new technologies were incorporated into piano-making. These resulted in a more powerful and sustained sound for the piano, and the piano came into its own and was modernized (figure 15.7).

These days, a top-of-the-line upright piano can cost upwards of tens of thousands of dollars. The cost of an entry-level brand-new upright piano starts at $3500, whereas an entry-level grand piano, like the one shown in figure 15.8, could cost you anywhere from $700 to $30 000. As a comparison, the cost of a new 5' 1" Steinway[4] S piano is $69,700. Smaller Steinway pianos such as models A, B, L, M, and O cost between $74 300 and $129 000. Larger Steinway pianos, such as the 9' model D, cost $171 000!

---

[4] Steinway pianos made from the best materials using highly skilled labor have a long, rich history of piano manufacture and are considered to be the finest pianos available.

Figure 15.7. The piano: (a) and (b) ancient pianos; (c) grand pianoforte ca 1840 CE. Photo credit: Hareem Zain at the Met, NYC, 2019.

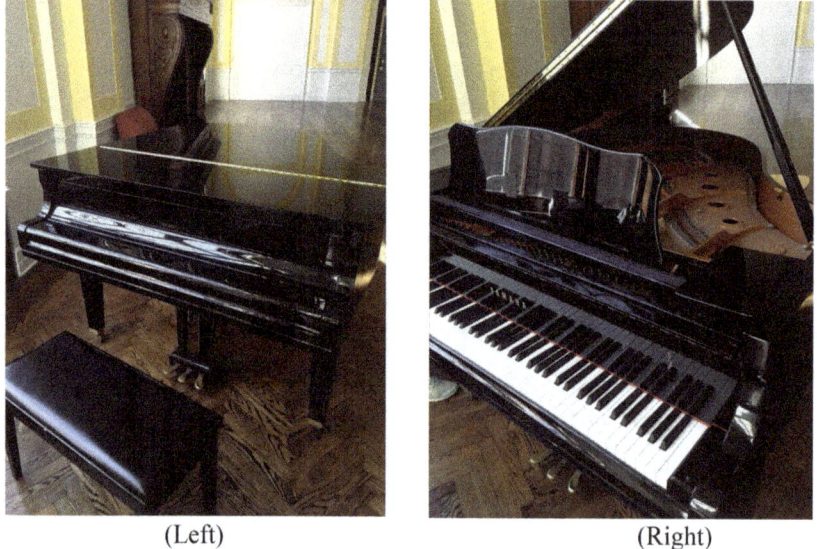

(Left)   (Right)

Figure 15.8. A typical modern piano: (left) closed view; (right) open view.

15.5.2 The piano as a vibrating system

A piano, as seen in figure 15.9, consists of a series of strings which vary in diameter and length. The strings of a piano are struck by hammers rather than being plucked. A piano keyboard has a total of 88 keys, but the total number of strings depends on the model of the piano and is usually around 230 strings. This is because for the tenor and treble (high) notes, three strings are strung for each key, for the bass (lower) notes, there are two strings per note, and for the lowest bass notes, there is usually only one string.

The vibrating piano strings themselves are not very loud, but their vibrations are transmitted to a large soundboard that enhances the sound due to its surface area. Most of the properties of plucked string instruments discussed in section 15.4 also apply to struck piano strings. However, due to the larger size of the piano

Figure 15.9. (Left) Eman playing a piano; (right) hammer response.

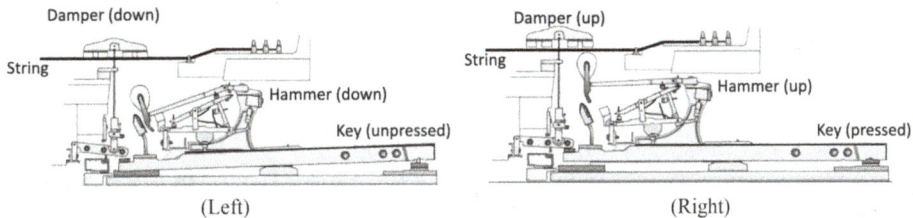

Figure 15.10. The action of a grand piano action mechanism: (left) the key is not pressed and no sound is produced; (right) when a piano key is pressed, the hammer rises, the wippen strikes the string, and sound is introduced into the system.

soundboard, the spectrum of vibration generated in a piano is much different than that generated in a guitar.

15.5.3 The production of sound in a piano

Pressing a piano key causes a hammer to strike one or more strings inside the piano. The mechanical action of a piano, as shown in figure 15.10, involves a very complicated lever system that has many moving parts. All these parts must work together seamlessly to form musical notes. When the key is first pressed, it does two things simultaneously:

1. It raises the wippen, the part of the action that comes into direct contact with the key and is responsible for transmitting the motion of the key to the hammer, and
2. It raises the felt damper from the string and lets it down again when the key is released.

As the wippen is raised, it forces the jack against the hammer roller, which lifts the lever and carries the hammer that strikes the string or strings depending on the key pressed. Vibrations are set up in the strings by this action. Once the piano key is released, the damper falls back onto the strings, which stops the strings from vibrating and hence stops the sound. A couple of things should be kept in mind: (i) striking a piano key with greater force increases the amplitude of the waves and therefore increases the volume of the resulting sound; (ii) if the hammer contact with the string is soft, a duller sound is produced because there is less excitation of the higher modes.

Striking a string displaces it from its equilibrium position and stretches it; however, its inherent elasticity causes it to return toward equilibrium. But it overshoots the equilibrium position because of inertia. On the other side of the equilibrium position, the string slows in its movement, eventually stops, and then makes its way back to the equilibrium position; hence, an oscillation is set up. Due to drag, each oscillation causes less and less displacement, and eventually the string stops moving, a process called damping (chapter 6). This explains why, when you press a piano key, the sound gets softer and softer until it stops. The distance the string covers during each oscillation is interpreted by our brains as loudness, and the rate at which it oscillates is interpreted as the pitch of the note. The further the string travels, the louder the sound seems to us, hence in order to set up louder sounds we need to apply more force.

For modern pianos, striking the piano key with greater force not only increases the volume (amplitude) but also makes the sound brighter. This is because most modern piano hammers are constructed to produce quite soft sounds when they hit the strings slowly and loud sounds when the keys are pressed harder. This effect is achieved by using hard materials on the inside of the hammer but quite deformable material for the exterior of the hammer. So when you press the key harder, a much louder sound is produced because the harder interior materials play a much larger role in this case.

15.5.4 The pitch of a vibrating string in a piano

The rate at which the string vibrates is the frequency of the string and is interpreted by our brains as the pitch of the note. The string frequency depends on factors such as the length of the string and the tension the string is held under but interestingly it does not depend on how hard it is hit.

Provided that all other factors are kept constant, the three factors that influence the pitch of a vibrating sting in a piano include the length of the string (L), its mass per unit length (μ), and the tension (T) the string is held under. Mathematically, these are related as follows:

$$f_1 = \frac{1}{2L}\sqrt{\frac{T}{\mu}}. \tag{15.5}$$

Specifically for the piano, let us consider the different factors individually here:
1. **The length of the string:** if the length of the string is kept too short, it affects the tonal quality of a piano. On the other hand, if the length is made too long, the piano itself becomes bulky and expensive to produce.
2. **The tension on the string:** if the tension is kept small, it decreases the ability of the string to transfer energy to the soundboard properly. On the other hand, the tension must not be higher than the breaking strength of the string, and the breaking strength depends on the material of the string, and its cross-sectional area. To produce the maximum sound, strings must be under the maximum tension they can withstand without breaking.
3. **The diameter of the wire:** strings with large diameters are inflexible. This additional stiffness increases both the tension and the restoring forces of the string. Additional restoring forces raise the frequency of vibration, which negatively impacts the higher modal frequencies. On the other hand, if the diameter of the string is made too small, its mass per unit length is also small. When the mass per unit length of a string is small, it does not couple correctly to the soundboard, which results in low sound output.

In equation (15.5), if we fix any two properties, we can find the ideal value for the third. However, in reality, there are other things to consider, such as the mass production cost of the instrument, the size of the piano, its tonal quality, and the convenience of the piano player. So, to minimize issues and maximize the sound in a piano, we can do the following few things:
1. Tension all strings at close to their maximum tension so that they produce a bright and long-lasting tone.
2. Use triple strings in the treble to transfer adequate energy to the soundboard.
3. Appropriately decrease the lengths and diameters of the strings to avoid excessive stiffness and thus obtain high frequencies for the treble.
4. Appropriately increase the lengths of the strings and make them heavier to obtain low frequencies for the bass.

15.5.5 The decay of piano sounds

The decay of piano tones is very complicated and deserves a lot more space for a complete discussion than I have here. So, in this text I will only briefly discuss it. For more details, please consult a heavier textbook dedicated to piano tones.

When a piano key is struck, the piano tones initially decay very quickly, but then the decay rate becomes slower and the tones last a relatively long time. For example, near C4, the characteristic decay rate at the beginning is approximately 3 to 5 s, but it then changes to approximately 10 to 20 s. The question is: why does this happen? There could be several reasons for this, and some possibilities include the following:
1. Each string is free to vibrate both vertically and horizontally. So, for each natural frequency there exist two natural modes, one vertical, called the 'vertically polarized mode,' and another horizontal mode called the 'horizontally polarized mode.' Now, since the hammer is made such that it should strike the string

vertically, we expect only the vertical mode to occur. However, this is not always true in reality. If the hammer is not completely vertical at the point of impact, both modes occur simultaneously. In a well-constructed piano, the vertical mode is quickly damped, as the string is well coupled to the bridge and soundboard, which allows a quick transfer of energy. On the other hand, the horizontal mode, which is poorly coupled to both the bridge and soundboard, takes some time to dissipate its energy. The result is that the energy of the vertically polarized mode is much larger than the energy of the horizontally polarized mode.

2. A similar effect arises due to multiple stringing. Some piano keys strike a set of three strings simultaneously. Ideally, we would want the hammer to hit all three at the same time, resulting in efficient energy transfer to the bridge and a short decay time. If, however, the hammer is lopsided or does not hit all three in precisely the same way, then the transfer of energy is inefficient, and the modes have a long decay time.

3. Multiple stringing also means that when the hammer hits three strings it produces three times the force, and the soundboard and bridge vibrate at three times the amplitude. This also means that during each cycle, each of the strings must exert its force over three times the distance; hence, it loses its energy three times faster by doing work on the soundboard and we get the rapid decay.

4. In addition, the multiple strings may not be perfectly tuned with each other. Let us say that instead of all being at 440 Hz, they are at 439.8, 440.2, 440.5 Hz. Once struck, they take approximately 1 s to become out of step with each other, after which they come back in step with each other and so on. This produces the interference phenomenon called *beats* (discussed in section 5.5.6), which is an undesirable effect in a piano.

15.6 Bowed stringed instruments

Some examples of bowed stringed instruments include the violin, viola, and cello. One advantage of bowing is that bowing maintains the rich harmonic spectrum longer than simple plucking. The higher-frequency harmonics fade particularly quickly, leaving only the fundamental frequency and some weak lower harmonics. Some bowed stringed instruments are shown in figure 15.11.

15.6.1 The violin

The word violin comes from the Middle Latin word *vitula*, meaning a stringed instrument. The modern violin developed in Italy in the sixteenth century, however it was perfected in the eighteenth century by Antonio Stradivari.

In a violin, the strings are set into vibration by drawing a bow across them. This vibration is transmitted through the bridge and sound post to the body of the violin, mainly the top and back, from which it is radiated into the air. The sound-producing system of the violin body includes the strings, the top, the back, the sides, the bass bar glued to the underside of the top, and the bridge and sound post. The tension, the type of string, the bow, and the construction of the body all contribute to the loudness and tonal quality of the sound produced by a violin.

Figure 15.11. Examples of bowed stringed instruments over the ages. Credit: Hareem Zain at the Met, NYC, 2019.

15.6.1.1 Major parts of a violin
The modern violin consists of about seventy parts. Violins include the top plate (also called the belly), the back, the ribs, inner blocks, inside lining, bass bar, fingerboard, neck, scroll, nut, lower nut, tailpiece, loop, end button, pegs, strings, bridge, and the sound post, among others. The main parts of a violin are shown in figure 15.12 and discussed in this section; more details can be found elsewhere.

1. **The body:** the body of a violin is a hollow box about 35.5 cm in length. It consists of two arched wooden plates, the top and bottom, whose sides are formed by thin curved wooden 'ribs.' The ribs are reinforced at their edges with lining strips to provide extra surface area for the glue-attached plates. The back, neck, ribs, and bridge are usually made of maple wood. The top, linings, and the sound post are usually made of spruce. Meticulous care is taken in selecting the wood and in shaping the belly and back of the violin. They must be fine-tuned to certain frequencies before they are assembled in order for the main wood resonance of the instrument to be of the appropriate strength and pitch.

    One purpose of the shape of a violin is that its 'waist' comes inward to give the bow easier access to the strings. Like many bowed string instruments, the middle of a violin is convex, and its sides have C-shaped bouts so that the bow can easily play each string with a long sustain. To vibrate properly, the top plate must be made very thin (2 to 4 mm thick). Without the sound

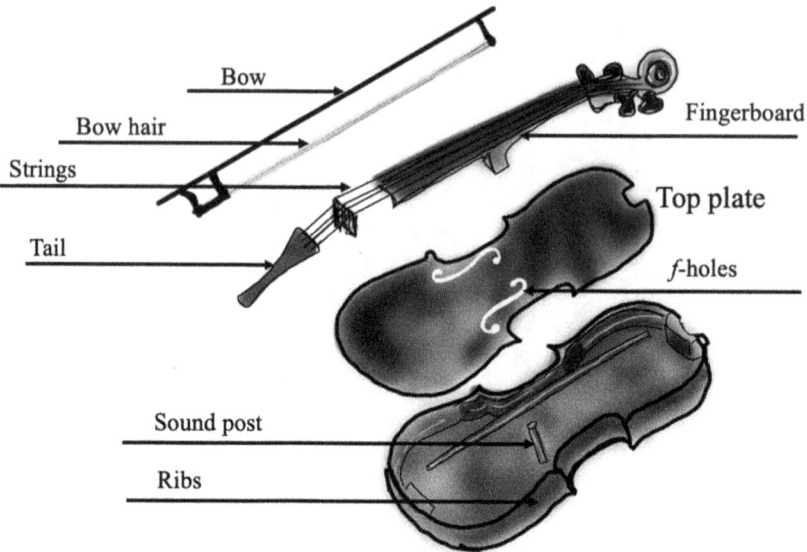

Figure 15.12. Violin construction.

post and the bass bar, this fragile plate would not be able to support the downward force of the bridge and the tensions exerted on it by the strings.

2. **Sound posts:** sound posts are very important in a violin because: (i) they provide structural support and (ii) they help transmit sound produced by the strings to the back plate.
3. **The sound box:** the body of the violin serves as a *sound box*. It resonates with the frequency of the vibrating string and amplifies it to make it audible. Interestingly, the body has resonant modes and the volume of the air enclosed in the body exhibits Helmholtz modes.[5] Resonance must take place in the body and air cavity together for a violin to effectively produce sound.
4. **The *f*-holes:** there are two openings cut into the top plate called *f*-holes or sound holes. They are called *f*-holes because they are physically shaped as the script *f*, as shown in figure 15.13. These openings or *holes* in the body are cut in a symmetrical pair so that they are equidistant from the sides of the strings. Their main purpose is to improve the efficiency of the sound projection.

The majority of the sound emanates from the surface area of the sounding board; however, the *f*-holes allow the sounding board to vibrate more freely and allow some of the vibrations from inside the body of a violin to escape.

---

[5] Helmholtz modes are solutions to the partial differential Helmholtz equation. The Helmholtz equation provides a time-independent description of waves. However, since the study of partial differential equations involves a much stronger math background than that required for this book, I will not use or derive the actual Helmholtz equation here.

Figure 15.13. (Left) violin *f*-holes; (right) a violin bridge.

Without the *f*-holes, the air resonating within the body of a violin would not have a noticeable effect on the sound produced by the violin.

5. **Bridge:** the bridge, shown in figure 15.13(right), is a piece of material generally made of wood that supports the strings on a violin. In addition, it helps to transmit vibrations induced in the strings to the body of the violin, which in turn transmits the vibration to the air cavity inside the body. The bridge is placed perpendicular or nearly perpendicular to the strings and the top plate and is held in place primarily by the tension of the strings pressing down on it. A bass bar is glued to the top plate directly under one foot of the bridge.

6. **Strings:** the strings are attached to the tailpiece of a violin. They pass over the bridge, along the fingerboard, over the nut, and finally end at the peg box. The ends of the strings are kept stationary to produce certain standing waves. Historically, the strings were made from stretched, dried, and twisted sheep gut. At the beginning of the 20th century, strings were made of either gut, silk, aluminum, or steel, whereas modern violin strings are usually made out of solid steel, stranded steel, or other synthetic materials, wound with various metals, and sometimes even plated with silver.

    The frequency of a string and hence the sound it produces are determined by the mass, length, and the elasticity of the string and the tension the string is held under. Heavier strings vibrate slower, whereas increasing tension increases their frequency. In addition, longer vibrating lengths lower the pitch of a string.

15.6.1.2 Vibrations of the violin body
The vibrations of the violin body are described in terms of normal modes of vibration. The normal modes of vibration are largely due to the coupled motion of the top and bottom plates along with the enclosed air. The contributions due to the neck, fingerboard, and ribs are present but they are insignificant when compared with those due to the plates and air.

The presence of the sound post makes it more difficult to model the normal modes; hence, most of our knowledge is based on experimental studies. The normal modes of vibration are experimentally determined by applying an oscillating force to the bridge and then observing (either optically or electronically) the subsequent motion of the various parts of the violin. The most consistent mode is the '*f*-hole resonance' also called the *air resonance*, which occurs at around 260–300 Hz. Other modes include:
- Top modes (T1, T2, T3, ...): motion primarily due to the top plate
- Body modes (C1, C2, C3, ...): motion primarily due to the back plate
- Air modes (A0, A1, A2, ...): motion primarily due to the enclosed air

Example 15.4. Calculate the second harmonic of a vibrating violin string with a length of $(L) = 0.3$ m and a mass of $(m) = 0.2$ grams that is under a tension of $T = 60$ N. Assume that the mass and tension of the string are consistent along the length of the string.
Solution:
We can use equation (15.5). Given that $\mu = m/L$ and $n = 2$,

$$f_n = \frac{n}{2L}\sqrt{\frac{TL}{m}}.$$

Substituting in the values gives:

$$f_2 = \frac{2}{2(0.3 \text{ m})}\sqrt{\frac{(60 \text{ N})(0.3 \text{ m})}{(0.2 \times 10^{-3} \text{ kg})}} = 1085 \text{ Hz}.$$

15.6.2 Slip–stick action

Examples of slip-stick action include things such as squealing chalk or fingernails dragging on a blackboard, door hinges squeaking upon opening, or the action of the bow which drives the vibration of the strings and produces sound in bowed instruments. Another easy way to demonstrate slip-stick motion is by dragging fingers across an inflated balloon. When you drag your fingers, the balloon is set into vibration, producing sound. The frequency of the sound thus produced depends on the speed of the slip-stick action, your finger speed with respect to the balloon surface, and the amount of pressure applied by your fingers. In this section, we discuss why this occurs and the sound produced.

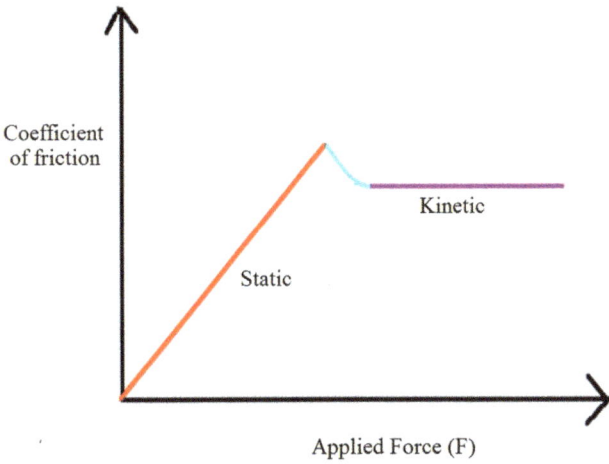

Figure 15.14. The coefficient of friction as a function of applied force.

When two objects slide past one another, the two sliding surfaces alternate between sticking to each other and sliding over each other. During these changing conditions, there is a change in the force of friction from the static friction (f_s) case (in which the objects do not move with respect to each other but rather stick to each other) to the kinetic friction (f_k) case, in which the objects slide over each other. In general, the coefficient of static friction, which applies when the objects are stationary relative to each other, is larger (just before the objects start to slip) than the coefficient of kinetic friction, as shown in figure 15.14.[6]

Between the static (sticking) and kinetic (slipping) cases, there is a transitional period, marked by the blue curve in figure 15.14. This means that when the applied force is large enough to overcome static friction, there is a reduction in friction, which results in a sudden increase in the speed of the movement. The maximum static friction force that can be sustained without movement is given mathematically by

$$f_s = \mu_s N, \qquad (15.6)$$

where μ_s denotes the coefficient of static friction and N is the normal force. The value of μ_s depends on the surfaces in contact and is always numerically less than or equal to one. For situations in which we want things to slide easily over one another, say in between moving parts of car engines, we need to keep μ_s small. On the other hand, to

---

[6] Static friction between two solids is due to the formation of partial chemical bonds between the surfaces of the objects while they are at rest with respect to each other. If the objects have a large differences in their velocities, the molecules of the adjacent surfaces do not interact for long enough to form partial chemical bonds. Kinetic friction is due to the microscopic deformations of the object surfaces. Kinetic friction between objects is present regardless of the relative movements of the objects. There is nearly always some level of surface deformation on an atomic level, as we saw in chapter 4, figure 4.8.

obtain good traction between the road and car tires, we need to keep the value of μ_s large.[7] The mathematical formula for kinetic friction is

$$f_k = \mu_k N, \tag{15.7}$$

where μ_k denotes the coefficient of kinetic friction and N is again the normal force. Remember that in general, the coefficient of static friction is larger than the coefficient of kinetic friction:

$$\mu_s > \mu_k.$$

15.6.3 Regular stick–slip–stick–slip cycles

Violins are a part of the stringed instrument family that uses bows for the production of sound. The advantage of using a bow is that it allows a violin player to input energy continuously and play long sustained notes, which is in direct contrast to plucking and/or striking, in which notes fade away rather quickly. We saw above that sticking (static friction) is greater than sliding (kinetic friction). This means that the coefficient of static friction is higher, while the coefficient of sliding friction is lower. This holds true for most surfaces that are in contact and hence is also true for the action of a bow on a string. The action of the bow that sets the strings into vibrations is a 'regular stick–slip–stick–slip cycle.'

Due to static friction, when the bow is rubbed against the string it tends to *stick* to the string and drags the string in the direction of its motion. This introduces a bend into the string, which travels down the string to the fixed end where it is reflected back. Upon reflection, the initial bend returns to the contact point, where the tensional force of the string pulls it off the bow. Once the strings pulls free of the bow, it is able to *slide* with very little kinetic friction—in other words, it *slips*. The string does not stop here; rather, due to inertia, it continues on until it comes to the other end. The string bend is then reflected back from the other fixed end. The *slip* stage ends once the string catches the bow again, the stick cycle begins, and static friction takes over. This repeating action of slip and stick is called the *stick–slip–stick–slip cycle* and is shown in figure 15.15. Please do not confuse this stick–slip–stick–slip cycle with the forward and backward motion of the bow. While the player moves the bow in one direction, many stick–slip–stick–slip cycles may occur.

---

[7] Keep in mind the laws of thermodynamics. Friction cannot just lessen the amount of force (energy) applied to an object. Friction just transforms motion energy into heat (some energy may also be used on the microscopic level to rearrange the various nanoscale surface deformations). This makes the need for minimal engine friction, as mentioned in the last example, even more apparent, as a conventional combustion engine obtains its energy from the explosions of readily explodable chemicals under conditions designed to promote their volatility.

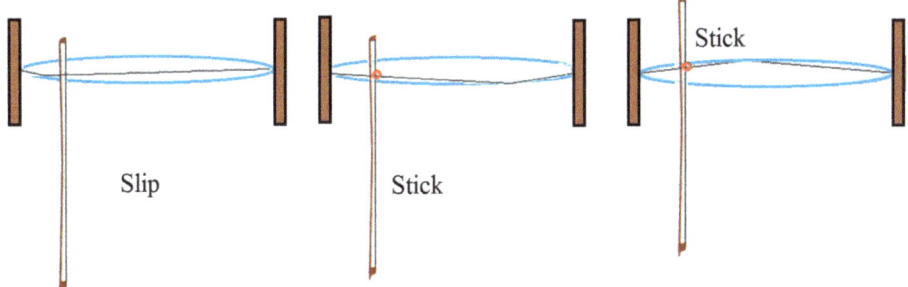

Figure 15.15. A regular stick–slip–stick–slip cycle.

Review Questions 15.1.
1. Why are piano strings wrapped in wire?
2. How is sound produced in a piano?
3. Why do violins have *f*-holes?
4. Why are there no frets on a violin?

Exercises 15.1.
1. A violin string is tuned to concert pitch A (440 Hz) and has a length of 0.33 m. (a) What is the speed of sound in the string?
2. Assuming that the speed of sound is 344 m s^{-1}, find the resonant frequency of a Helmholtz resonator that has a volume of 0.05 liters if its neck has a cross-sectional area of 0.02 m^2 and a length of 0.03 m.
3. Calculate the fourth harmonic of a vibrating steel bass string with a length of (L) = 1.22 m and a mass of (m) = 26.52 grams that is under a tension of T = 60 N. Assume that the mass and tension of the string are consistent along the length of the string.
4. What does inharmonicity mean? What makes piano notes inharmonic? Does this have any effect on tuning and tone quality?

Further reading

- Berg R E and Stork D G 1994 *The Physics of Sound* 2nd edn (Englewood Cliffs, NJ: Prentice-Hall)
- Rossing T D, Moore R F and Wheeler P A 2002 *The Science of Sound* 3rd edn (Upper Saddle River, NJ: Pearson Education)
- Hall D E 1991 *Musical Acoustics* 2nd edn (Pacific Grove, CA: Brooks/Cole)
- Fletcher N H and Rossing T D 1998 *The Physics of Musical Instruments* (Berlin: Springer)

IOP Publishing

The Physics of Sound and Music, Volume 1
A complete course text (Textbook)
Samya Bano Zain

Chapter 16

Percussion instruments

Percussion instruments include tuning forks, chimes, xylophones, marimbas, etc. A percussion instrument is any object which produces a sound when it is hit, shaken, rubbed, scraped, or struck in any way which sets the object into vibration.

The history of percussive musical instruments is as old as the history of human culture. Most early instruments were made during the upper Paleolithic age (about 40 000 years ago). These included percussion instruments made from stones, sticks, and rocks. The next steps in the evolution of percussion music were wooden logs and stones cut into different shapes and designs to change the quality and pitch of sound. The earliest known drum, made from elephant skin, was discovered preserved during the ice age in Antarctica (figure 16.1).

16.1 Rhythms in everyday life

Rhythms can be sounded out using pretty much anything, your body, mouth, hands, fingers, or feet. You can also use ordinary objects like pencils, keys, or even trash cans to make music. STOMP is a famous musical group known for its use of everyday objects to create exciting choreography with rhythms. In the introduction to the video STOMP Out Loud, founder and dancer Luke Creswell states that the group's goal is to invite people to 'listen to the world in a different way and hear music where maybe they did not think there was music before.'

Percussion instruments are made up of both definite and indefinite instruments. Definite instruments are called *pitched* or *tuned* percussion instruments and they are used to produce musical notes with one or more pitches. Examples of pitched percussion instruments include the tympani and the glockenspiel. Indefinite instruments, also called *non-pitched* or *untuned* percussion instruments, are instruments that produce sounds with indefinite pitches. Examples include instruments such as the snare drum.

Sounds with indefinite pitch do not have harmonic spectra or they have harmonic spectra altered to produce a characteristic known as *inharmonicity*. It is worth noting

Figure 16.1. Various drums from around the world. Credit: Hareem Zain at the Met, NYC, 2019.

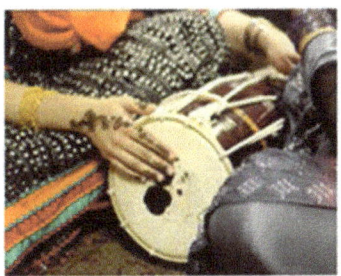  

Figure 16.2. Vibrating membranes. Credit: (left) S.E. Jihad Levine (2014).

that many untuned percussion instruments can and frequently do make sounds that could be used as pitched notes in an appropriate context. Some percussion instruments, for example, many types of bells, are used as both pitched and non-pitched instruments.

16.2 Various percussion instruments

Percussion instruments may be broadly divided into two categories:

1. **Rhythmic percussion:** rhythmic percussion instruments are subdivided into smaller categories according to the manner in which the instrument produces sound.

    (a) *Idiophones:* idiophones produce sound when their bodies are caused to vibrate, e.g. the marimba, cymbals, etc, as seen in figure 16.2.

(b) *Membranophones:* membranophones produce sound when the membrane is put into motion, e.g. bongos, snare drums, and timpani.
(c) *Electrophones:* all electrophones require a loudspeaker; this is sufficient to assign electrophones to the percussion family, e.g. drum machine, radio, etc.
2. **Melodic percussion:** a melodic percussion instrument is a percussion instrument used to produce several different notes of different pitches. Melodic percussion instruments are examples of pitched percussion and include mallet percussion and keyboard percussion. Three main types of instruments that utilize melodic percussion are:

(a) Collections of tuned percussion instruments in different pitches, e.g. hand bells.
(b) Instruments that produce different pitches when struck in different places, such as the steel drum.
(c) Instruments that contain a collection of sounding objects tuned to different notes, such as the xylophone.

16.3 Vibrations in a bar

Rods may vibrate in longitudinal, transverse, or torsional (twisting) modes. Transverse vibrations of rods are mainly responsible for the sound production in tuning forks, chimes, xylophones, etc. However, longitudinal and torsional vibrations are also produced when a hammer or a ball (made from wood or felt) hits a rod or cylinder. This is especially important for glancing blows, which set up all three modes, with the result that the tonal composition of the emitted sound is a composite of all three vibrations. Please note that even in this case, the intensity of the transverse vibration dominates.

The simplest case of transverse vibration in a rod occurs when the rod is clamped at one end and struck perpendicularly to the axis on the other end, as shown in figure 16.3. Instruments such as the marimba and xylophone are supported at two places towards the edges, thus the rod is struck in the middle.

Figure 16.3. Vibrating bars in a mbira, which is a traditional African instrument played with the thumbs dating back thousands of years.

Unlike the vibrations set up in strings (chapter 15), the overtones of a bar are not harmonics of its fundamental. This means that the sound produced by a bar has a metallic property because of the presence of high-frequency overtones. Luckily, these decay quickly, and the sound settles to a nearly pure tone of the fundamental. We may consider the two prongs of a tuning fork as free bars clamped at their lower ends to produce a very close approximation of the sound for which they are tuned.

The longitudinal mode can be set up in the following ways:

1. **Clamp the rod at one end:** one way to set up the longitudinal mode is to clamp the rod at one end and with the help of a rosined chamois leather, rub the end along the axis of rod, as we saw in section 6.6.3. Rubbing produces sound because of the frictional force between two surfaces. The result is a high-pitched sound, which is caused by the longitudinal stationary wave that occurs in the rod. In this case, the rod acts like an organ pipe closed at one end. For a rod length of (L), the fundamental wavelength is $\lambda = 4L$ and the fundamental frequency (f_o) is

$$f_o = \frac{v}{4L}, \tag{16.1}$$

   where v is the speed of sound in the rod.

2. **Clamp the rod in the middle:** the other way to set up the longitudinal mode is to clamp the rod in the middle and rub one of the sides. This also results in a longitudinal wave, but here the rod acts like an organ pipe open at both ends. For a rod length of (L), the fundamental wavelength in this case is $\lambda = 2L$ and the fundamental frequency (f_o) is

$$f_o = \frac{v}{2L}, \tag{16.2}$$

   where v is again the speed of sound in the rod.

In general, the speed of sound in solids is much higher than in air, so we will note that very high pitches result from the higher frequencies.

For the torsional mode, in the simple case of a uniform hollow cylinder, the results obtained are very similar to the longitudinal vibrations, with one difference: the speed of sound in the torsional mode is 1.63 times less than the speed of sound in the longitudinal mode. If we consider the same length for the vibrations in the longitudinal and torsional cases, we find that torsional frequencies are one fifth of those in the longitudinal case.

16.4 Vibrations in plates and membranes

A plate or a membrane has two dimensions (2D): one of length and the other of width. So plates or membranes vibrate in 2D, as opposed to a string or bar that only oscillates in 1D. This implies that the mathematical analysis for plates and

Figure 16.4. Setting up standing waves in a Chaldini plate.

membranes is more complicated than that for strings or bars. To make life a little less complicated, we divide the discussion into two broad categories:

1. **Case 1: Uniformly stretched circular membranes**, such as drum heads. In this case, the main restoring force is due to the tension of the membrane.
2. **Case 2: Thin circular rigid plates**, like those found in the diaphragms of telephones or even in bells. In this case, the main restoring force is due to the stiffness of the material.

Pitches produced in plates that have same shapes vary inversely to their diameters. The best way to visualize nodes in 2D surfaces or membranes was described by a German physicist and musician Ernst Florens Friedrich Chladni in 1787 CE, as previously discussed in section 6.6.2. Chladni is accredited with the invention of a technique, shown in figure 16.4, which illustrates the various modes of vibrations in a membrane or plate. His technique is based on the principle that during vibration, the surface of the plate remains stationary along nodal lines. He experimented with a piece of metal clamped in the middle whose surface was lightly covered with sand, and he observed that when the plate was bowed, sand moved and collected along the nodal lines at resonant frequencies, outlining the nodes on the 2D surface.

Variations of this technique are still commonly used in the design and construction of acoustic instruments. However, since the 20th century, it has become more common to use a mechanically driven controller or a loudspeaker driven by a signal generator placed over or under the plate. This method gives a more accurate frequency and eliminates the human factor by avoiding the necessity for perfect bowing. Patterns formed in this way are, however, different from those produced by bowing at the edges because in this case the center is vibrated, and hence the antinodes are located at the center.

16.5 Membranophones

Sound is produced in membranophones by the oscillation of 2D membranes. The overtone structure in two-dimensional membranes is complex, the modes are complicated, and their frequencies are not integrally related. For a string, we were able to describe the modes by a single integer called the harmonic number. Because a drumhead is two-dimensional, we required two integers to describe each of its modes. Just as there are nodes at the end of a fixed string, there are nodes in a membrane. There is a nodal ring around the edge of the membrane where it is held fixed. Points along this ring do not move up or down as the nodes are set up. In addition to these outer fixed nodes, there is another nodal ring around the center of the membrane.

The Chladni plate is a rigid, thin piece of metal of a regular shape, usually circular or square. The plate is attached to a source of vibration. The frequency of the modes of vibration of a flat circular surface with a fixed center is a function of the number of linear nodes (m) and circular nodes (n). Mathematically, this is given by

$$f_C = C(m + 2n)^P, \qquad (16.3)$$

where C and P are coefficients that depend on the properties of the plate, as shown in figure 16.5. In flat plates, $p = 2$, whereas in non-flat plates, such as cymbals, bells, etc. it is generally less than two.

Figure 16.6(a) shows the lowest frequency mode on a square membrane. Here, the center of the membrane moves up and down in a manner similar to the movement of the human chest as we inhale and exhale during breathing. Figure 16.6(b) shows the next higher mode and its nodal lines. Figure 16.6(c) shows a mode that is not symmetric around the center. In this case, as the left side moves up, the right side moves down and vice versa.

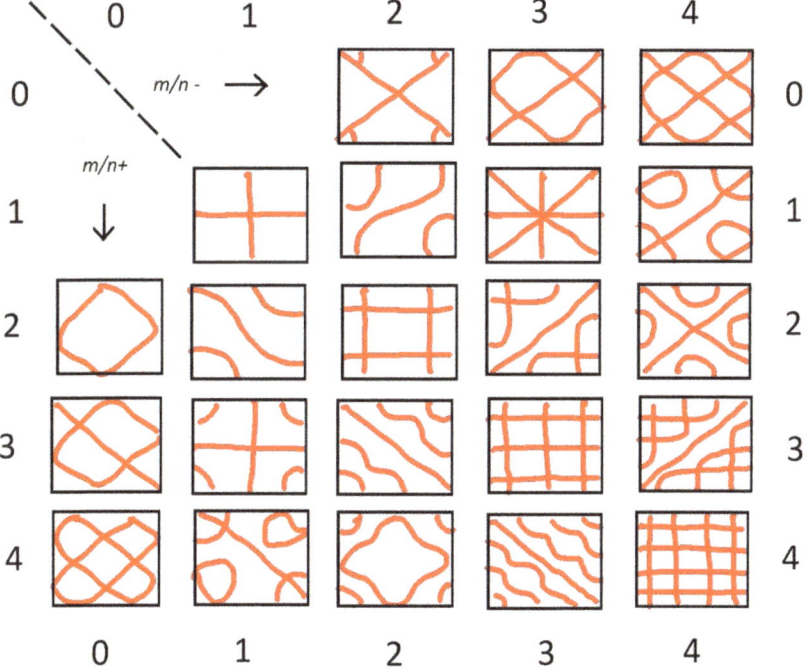

Figure 16.5. Standing waves in a Chaldini plate; m and n refer to the modes in equation (16.3). Chladni figures are arranged according to leading mode, where $m, n \leq 4$.

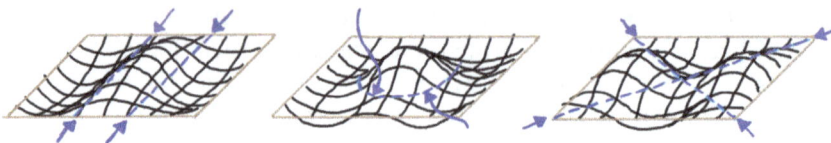

Figure 16.6. Modes on a square membrane: (a) the lowest frequency mode; (b) a higher mode; (c) a mode that is not symmetric around the center. The arrows point to the nodal lines.

Figure 16.7. Examples of drums: (a) kettledrums; (b) a tom-tom; (c) a brass drum.

Just as the tension in a string affects its frequency, the amount of tension in a drumhead affects its frequency. The higher the tension, the higher the frequency, and hence the higher the pitch we hear. This means that a membrane held taut has a higher rate of speed and, for a given wavelength, a higher frequency and hence a higher pitch. The relationship between frequency and tension for the frequencies of various vibrational modes of ideal circular membranes is:

$$f = \frac{\beta}{2r}\sqrt{\frac{T}{\sigma}}, \qquad (16.4)$$

where r is the radius of the membrane in meters, T is the surface tension applied to the membrane (N m^{-1}), σ is the areal density (in kg m^{-2}), and (β) are values for which the Bessel function (a certain mathematical function) becomes zero. This formula shows how the pitch of a vibrating membrane depends on the radius, tension, and the thickness of the membrane.

Some common characteristics of membranophones are:

1. **Kettledrums or timpani**: kettledrums or timpani, as they are called in Italian, shown in figure 16.7(a), are the most important percussion instrument in a standard orchestra. Kettledrums are played in sets of two to six drums, in which each drum is made up of a skin or plastic 'head' stretched over a copper or fiberglass bowl. Tuning pegs located around the bowl can be adjusted so that the head is stretched evenly on all sites. The tension on the head is controlled by a foot pedal that pulls the head uniformly around the rim. A kettledrum is usually struck with a wooden stick whose head is covered with a cloth or felt. Ancient Persians had a bowl-shaped drum that resembled a modern timpani.

2. **Tom-tom**: tom-toms, shown in figure 16.7(b) are drums made of cylindrical shells that are either covered on both ends or are covered on only one end. The motions of the struck head and the bottom head result in a very complicated interaction. This is why tom-toms are generally considered to be indefinite pitch instruments. The length of the cylinder affects the tonal quality of the tom-tom: deeper tom-toms have longer standing waves between their heads and hence produce lower tones.
3. **Brass drum:** a brass drum, as shown in figure 16.7(c), is made of two large heads stretched over a large cylindrical shell. Each head has a loosely defined pitch center but because the heads are tuned to different (but close) pitches, the overall sound of the drum does not have a well-defined pitch.
4. **Snare drum:** snare drums are similar to shallow two headed tom-toms with an additional wire stretched along the outside of the bottom head. These additional wires vibrate against the head when the drum is played and result in a crisper sound.
5. **Tabla:** a south Asian drum called the tabla, shown in figure 16.8, is a special kind of drum. The name tabla comes from the Arabic word *tabl*, meaning drum. The top of a tabla has an additional circular portion of the head called the *seyahi*. The *seyahi* is a paste patch made from rice or wheat starch mixed with a black powder. The seyahi makes part of the membrane heavier than the rest of the membrane and it gives the tabla its distinctive sound by producing harmonic overtones. Tablas are always played as a combination of two drums of slightly different sizes, the right-hand tabla is called the *dayan tabla* and the left-hand drum is called the *bayan tabla*. The tabla is played by striking the membrane with the fingers and sometimes the heel of the hand. The first four *dayan tabla* modes are very nearly harmonic, which creates a clear ringing tone, while the *bayan tabla* has a deeper sound because

Figure 16.8. A tabla, showing (left) the bayan tabla and (right) the dayan tabla. Credit: Hareem Zain at the Met, NYC, 2019.

of its larger size and lower head tension. The tabla is an essential instrument in the religious and devotional traditions of the three major religions of the Indian subcontinent (namely Hinduism, Islam, and Sikhism).

16.6 Bells

Bells, like the ones shown in figure 16.9, are directly struck idiophones that are very similar to plates and can be imagined as a plate suspended from its center and then curved downwards. Historically, bells have been associated with religious rituals and services, from Hindu and Buddhist bells, called *ghanta*, to church bells. These days, bells are used in multiple contexts from announcements to fun activities.

There are many different kinds of bells that are usually made from a type of bronze in varying shapes and sizes, for example, cow bells, electrical bells, bicycle bells, etc. The calculation of a bell's fundamental frequencies is extremely complicated, hence experimental evidence is mostly used to find the fundamental frequency of each bell. What we find is that the pitch of any bell is inversely proportional to both its diameter and the cube root of its weight. For two bells an octave apart, the diameter of the larger bell is twice that of the smaller bell and its weight is eight times that of the smaller bell. Please note that this scale is often used for most bells except ringing bells and the lighter bells of a carillon.

16.6.1 The main parts of a bell

Most bells are made of metal, but some can also be made from glass or ceramic. Bells are mostly constructed in the shape of a hollow cup with either an internal clapper (such as the one seen in figure 16.10) or an external hammer. The shape of a

Figure 16.9. Some types of bells: (left) a cowbell; (right) a handbell.

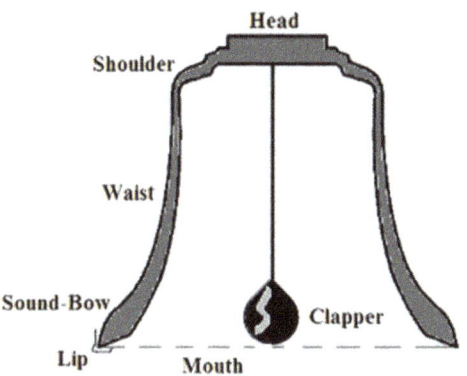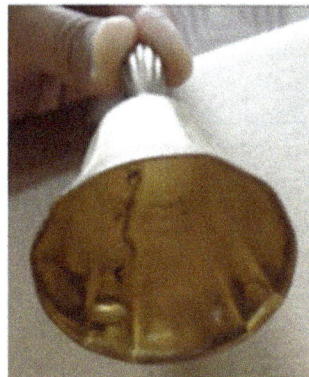

Figure 16.10. Parts of a bell.

bell is responsible for its modes of vibration. The shape determines both the frequencies of the various partials and their intensities. When struck, the hollow cup of a bell behaves like a resonator and the bell vibrates in response.

The top of the bell is called its *head*, which is attached to the curved middle sides of the bell called the *waist* via a *shoulder*, as seen in figure 16.10. The other side of the waist from the head is called the *sound bow*, which ends in the *lip* of a bell. The open part of a bell is called its *mouth*. The *clapper*, also called the *mouth clapper*, is a metal shaft that terminates in a solid sphere. The clapper is free to swing back and forth inside the bell and when it strikes the bell at the sound bow it makes the bell ring. The *sound bow*, shown in figure 16.10, is the most flexible part of the bell and it has the most mass. Thus, it is the most important vibrating part of the bell and it emits the strongest ring.

In earlier centuries, bells also had a crown, often called a *cannon*, which was used to fasten the bell to a wooden beam. However, these days, most bells are constructed without crowns and are simply directly bolted to beams.

16.6.2 Vibrations in bells

When a bell is struck, it rings. This seems simple enough, however, it is very misleading to say the least. Bells are remarkably complex, and the calculations for the frequencies of a bell are extremely complicated, so most studies of bells are experimental.

Based on experimental observations, we know that when a bell is first struck, the basic pitch heard is not the fundamental (also called the *hum note*). Rather, it is a note that has no partial at all, called the *strike note*. Each sound comprises dozens of slightly inharmonic partial tones that die out quickly, giving way to more dominant partials which together form the note immediately perceived by the human ear. The hum note (the lowest of the five partials) is only heard after the initial strike tone dissipates, and it is the slowest overtone to decay. Another thing that we observe is that the nominal (second member of an approximate harmonic series for a bell) is initially strong at the time of impact but decays quickly compared to the other partials. In fact, the intensity relationships of all components change as a function of time. The bell is not unique in this respect; the same phenomenon occurs in vibrating strings.

The rate of a bell's tonal decay is an important factor in the value of the bell itself. The longer it takes the bell's sound to decay, the better the quality of the bell. So in order to determine the quality of the bell, we need to consider the material the bell is made of, the size of the bell, and its shape.

16.6.3 Wavelengths in bells

Bells are unique because when we consider bells, it is essential to know what is vibrating and how the vibration occurs. The clapper, seen in figure 16.10, which is suspended inside the bell, strikes the bell near its rim. This is where resonance and the maximum elastic amplitude occur in a bell.

Many sound frequencies produced by ringing of a bell are associated with characteristic modes of vibrations contained within the bell. These modes are defined by nodal lines at which motion (motion to and fro relative to the bell's axis) is zero for a particular frequency. The nodal lines of a bell are located on the surface of the bell. When they are vertical, they are called *meridians*, and when they are horizontal, basically at different elevations, they are called *parallels*. The pictures in figure 16.11 shows the nodes where the two circles intersect in a bell.

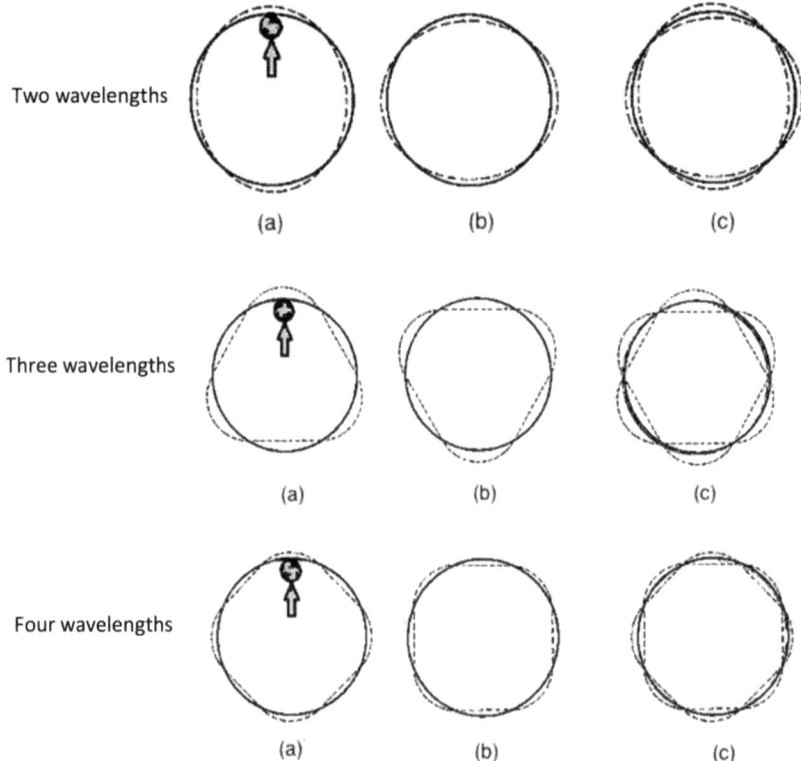

Figure 16.11. Wavelengths in a bell.

For symmetrical bells, the point struck is a loop and the nodal meridians are distributed accordingly. In asymmetrical bells, the nodal positions are governed by their particular asymmetry.

Review Questions 16.1.
1. What is the basic difference between percussion and wind instruments? Are there similarities?
2. What is the ideal placement of the supports to be added to a bar in order to best excite the fundamental mode of vibration?
3. What is the ideal setup for the longitudinal mode in a bar?
4. What are Chladni patterns? How and why are they formed?
5. What is the difference between the vibrations produced in bells and those produced in plates?

Exercises 16.1.
1. Write a formula we can use to find the frequency of the normal mode of vibration of a rod clamped on one side in terms of its length (L) and the speed of sound (v).
2. Find the frequency of the normal mode of vibration of a rectangular brass bar 20 cm long clamped in the middle. (Hint: find the speed of sound in brass, given that the density of brass = 8.4 grams per cubic centimeter and the value of its Young's modulus is 1.10×10^{11} N m^{-2}.)
3. Find the radius of a drum membrane (in meters and cm) held under a tension of 3990 N if the required frequency is 440 Hz. Assume that the areal density σ is 0.01 kg m^{-2} and $\beta = 0.90$. Does the value you obtain for the radius make sense? What diameter does the drum have?

Further reading

- Berg R E and Stork D G 1994 *The Physics of Sound* 2nd edn (Englewood Cliffs, NJ: Prentice-Hall)
- Rossing T D, Moore R F and Wheeler P A 2002 *The Science of Sound* 3rd edn (Upper Saddle River, NJ: Pearson Education)
- Hall D E 1991 *Musical Acoustics* 2nd edn (Pacific Grove, CA: Brooks/Cole)
- Fletcher N H and Rossing T D 1998 *The Physics of Musical Instruments* (Berlin: Springer)

IOP Publishing

The Physics of Sound and Music, Volume 1
A complete course text (Textbook)
Samya Bano Zain

Chapter 17

Wind instruments

The human voice was the first thing that humans used as a musical instrument, for example, for singing. The next set of instruments to be developed and used was the percussion instruments, the most basic element of which may have been body percussion. Wind instruments were undoubtedly the third set in this chronologically ordered saga of instruments.

It is likely that ancient instruments were mostly one-note instruments made out of animal bones or dried wood. Ancient instruments had multiple purposes in ancient times, from commuting with gods to calling for rain and protecting the crops. In this chapter, we will concentrate on wind instruments.

17.1 Wind instruments

In all wind instruments, a coupled system must be formed between the pipe and a column of air, as in figure 17.1. In the case of air reed instruments (such as the flute), the mouth forms one part of the coupled system, while the lips and the mouthpiece form the second part of the system, and the air column forms the third part of the system. For high notes, the cheeks of the player must compress, and the player has to imagine singing the note and adjust the size of their mouth cavity appropriately. Wind instruments may be broadly grouped into three main classes:

1. Brass or lip reed instruments, for example, horns, trumpets, trombones, and tubas. When brass instruments are played, the player's lips vibrate, causing the air within the instrument to vibrate. Air is blown into the instrument by 'buzzing' lips across the cup-shaped mouthpiece, producing vibration and sound through the instrument. Different notes are produced by changing the length of the instrument or the size of the bell-shaped end.
2. Wind instruments with an air reed, for example, the flute or fipple-flute. Sound is created in these instruments by blowing over a fipple, i.e. across an open hole against an edge.

 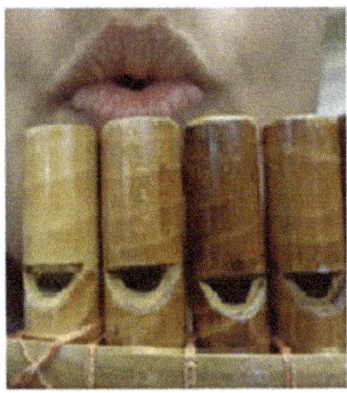

Figure 17.1. Example of a end-blown wind instrument: the pan flute.

3. Wind instruments with a cane reed, for example, the clarinet, oboe, and bassoon. In woodwind instruments with a cane reed, sound is created by causing the reed to vibrate. Air is blown across the reed attached to the mouthpiece of the instrument, which sends vibrating air down the tube and produces sound. Different notes are produced by covering or opening holes in the instrument tube or by changing the reed or the size of the tube.

Woodwind instruments were originally made of wood, just as brass instruments were made of brass, but these days, instruments are categorized based on how the sound is produced, not by the material used to construct them. For example, saxophones are typically made of brass, but are woodwind instruments because they produce sound using a vibrating reed. In this section, we discuss a few properties of sounds in woodwind instruments.

17.1.1 Sound production in woodwind instruments

In woodwind instruments, sound is created either by blowing over a fipple, blowing across an open hole, or causing a reed to vibrate. The tubes used to produce sounds in woodwind instruments may be straight or curved, cylindrical or conical, or a combination of these. The relative strengths of the partials set up in the tube are controlled largely by the shape of the tube, which is responsible for the timbre of the pitch. Therefore, it is important to chose the shape of the tube carefully for the desired effect.

The length of the tube is another factor to be considered: a longer effective length of tube results in a lower frequency. The effective length is changed by opening and closing the keyholes along the side. When all the holes are closed, we get the lowest note. Opening the holes successively from the bottom end gives a chromatic scale. The relation between the tube length and the diameter varies but must also be carefully considered to match the required musical pitches. Please note that throughout the length of the tube, the entire wave moves at the speed of sound.

17.1.2 Sound waves in pipes

Standing sound waves set up in a pipe are different from the standing waves set up in a string. The major differences include the medium, the type of wave, and the types of reflections that the waves undergoes:
1. **Medium:** the medium in which the standing sound waves are set up is air.
2. **Type of wave:** standing sound waves in air are longitudinal waves.
3. **Wave reflections:** sound waves reflected at the two ends of the pipe have two conditions, one for an open end and the other for a closed end.
    (a) *Open end:* for a pipe, the pressure does not oscillate at the open end but remains fixed. Hence, a node exists at the open end which is like the fixed-end string condition.
    (b) *Closed end:* for a pipe, the pressure can vary at the closed end. in fact, the air pressure at this end is maximized. Hence, there is an antinode at this point.

A summary of the results for open and closed pipes is given in table 17.1.

Example 17.1. What are the three lowest resonant frequencies of a tube 20 cm long that is open at one end and closed at the other?
Solution:
Use the equation for the closed pipe condition:

$$f_n = \frac{v}{\lambda} = (2n - 1)\frac{v}{4L}.$$

Solving for three resonant frequencies (f_1, f_2, f_3) gives:

$$f_1 = (2(1) - 1)\frac{v}{4L} = (2(1) - 1)\frac{344 \text{ m s}^{-1}}{4(0.20 \text{ m})} = 860 \text{ Hz}$$

$$f_2 = (2(2) - 1)\frac{v}{4L} = (2(2) - 1)\frac{344 \text{ m s}^{-1}}{4(0.20 \text{ m})} = 3580 \text{ Hz}$$

Table 17.1. Standing waves in closed pipes and open pipes.

Pipes	Closed at one end	Open at both ends
Nodes exist	...at the open end	...at both open ends
Antinodes exist	...at the closed end	...somewhere in the middle
Wavelength	$\lambda = \frac{4L}{2n-1}$	$\lambda = \frac{2L}{n}$
Frequency	Only odd frequencies exist. $f = \frac{v}{\lambda} = (2n - 1)\frac{v}{4L}$ where $n = 1, 3, 5, ...$	All frequencies exist. $f = n\frac{v}{2L}$ where $n = 1, 2, 3, 4....$

$$f_3 = (2(3)-1)\frac{v}{4L} = (2(3)-1)\frac{344 \text{ m s}^{-1}}{4(0.20 \text{ m})} = 4300 \text{ Hz}.$$

Example 17.2. What are the three lowest resonant frequencies of a tube 20 cm long that is open at both ends?
Solution:
Use the following equation for the open pipe condition:

$$f_n = \frac{v}{\lambda} = n\frac{v}{2L}.$$

Solving for three resonance frequencies (f_1, f_2, f_3):

$$f_1 = n\frac{v}{2L} = (1)\frac{344 \text{ m s}^{-1}}{2(0.20 \text{ m})} = 860 \text{ Hz}$$

$$f_2 = n\frac{v}{2L} = (2)\frac{344 \text{ m s}^{-1}}{2(0.20 \text{ m})} = 1720 \text{ Hz}$$

$$f_3 = n\frac{v}{2L} = (3)\frac{344 \text{ m s}^{-1}}{2(0.20 \text{ m})} = 2580 \text{ Hz}.$$

17.1.2.1 Sound waves in an air column
Sound is a longitudinal pressure wave which forms regions of compression and rarefaction in an air column inside a pipe. As the sound wave propagates in the pipe, the air molecules move. At the open end, the air pressure equals atmospheric pressure. Some air molecules are reflected back into the pipe, while others are refracted out into the atmosphere. The reflected and incident sound waves superimpose and produce constructive and destructive interference based on the principle of superposition (discussed in section 5.5). On the other hand, at the closed end, sound waves are fully reflected and they superimpose on the incident waves to form constructive and destructive interference, which also results in the formation of standing waves.

A sound standing wave at a fundamental frequency, i.e. the first harmonic, in an open–open pipe is shown on the left-hand side of figure 17.2, whereas the same is shown on the right-hand side for an open–closed pipe. In the figure, points marked by 'N' are pressure nodes where the amplitude is zero; at these points, there is no pressure difference from atmospheric pressure. Conversely, the points represented by 'A' are points of maximum amplitude, and here the pressure change from atmospheric pressure is a maximum.

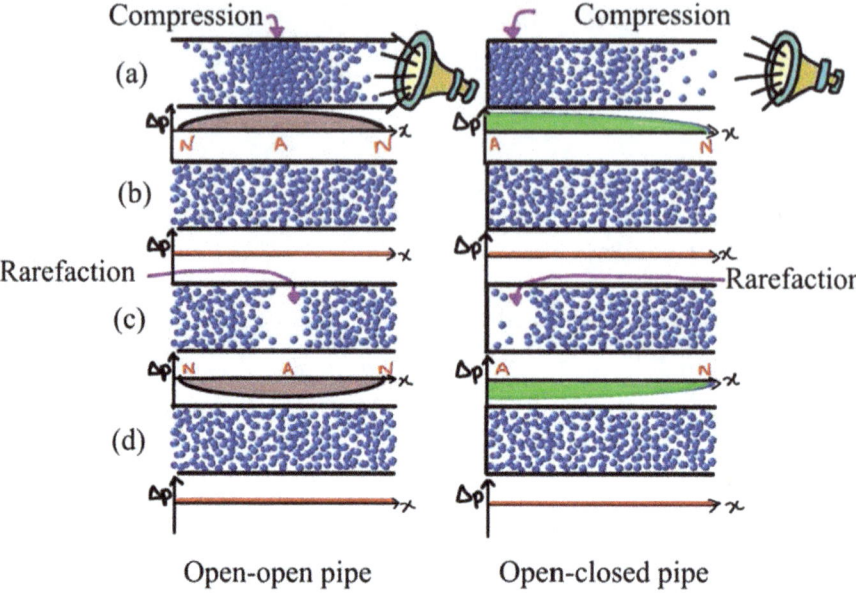

Figure 17.2. A comparison of fundamental frequencies of standing waves in air columns for open–open and open–closed pipe systems. (a) The introduction of the sound source, which sets up pressure compressions in each case. (b) A quarter-cycle later. (c) A half-cycle later, showing the resulting pressure rarefaction. (d) Three-quarters of a cycle later. ΔP gives the absolute value of pressure from atmospheric pressure. 'N' denotes the nodes and 'A' denotes the antinodes.

Let us discuss both cycles in detail here:
1. **The open–open pipe cycle:** suppose that at time ($t = 0$), a sound is introduced into a pipe that is open at both ends, called an open–open pipe, and a compressional sound wave is set up in it as shown in figure 17.2(a). As a result, air molecules in the middle of the pipe are compressed together. A quarter-cycle later, the air molecules move away from each other and reach the normal atmospheric condition, as shown in figure 17.2(b). At this point, all molecules return to their equilibrium positions. However, the molecules do not stop here; rather, because of inertia, they move towards both ends of the pipe. This means that after three-quarters of a cycle, the pipe passes through a minimum as both the air pressure and density in the middle decrease, as can be seen in figure 17.2(c). In the next part of the cycle, restoring forces cause the air molecules to return to the equilibrium position as shown in figure 17.2(d), and the cycle restarts and repeats until the sound is either damped or augmented by additional air to create a musical sound from the instrument.
2. **The open–closed pipe cycle:** an open–closed pipe is a pipe that is open at one end and closed at the other end. In an open–closed pipe, the closed end makes a pressure antinode (A), since the air molecules are compressed at this boundary. At the fundamental frequency of the standing wave, air molecules

at time ($t = 0$) squeeze towards the closed end (in this case the left-hand end, as shown in figure 17.2(a)). In a similar way to the open–open pipe scenario explained above, air molecules try to return to equilibrium condition and pass through steps (b–d) shown in figure 17.2 before returning to step (a). At this point, the cycle restarts and repeats until sound is produced from the instrument.

17.2 The instruments of the woodwind family

Regardless of what the name implies, instruments of the woodwind family are made of either wood, metal, plastic, or some combination thereof. All instruments in the woodwind family have long narrow pipes with side holes, an opening at the bottom end, and a mouthpiece at the top end. They are played by blowing air through the mouthpiece, and the pitch is changed by opening or closing the side holes with fingers or metal caps. In this section, I will briefly discuss a few of the main instruments in the woodwind family; for more details, please refer to a thicker book on acoustics.

17.2.1 The flute

The flute is the oldest of all woodwind instruments and was originally made from wood, stone, clay, or hollow reeds such as bamboo. Modern standard flutes are made of silver, gold, or platinum and are over two feet long. A flute is played by holding it sideways with both hands and blowing across a hole in the mouthpiece. I will define two extreme cases for standing waves in a flute called the *low note condition* and the *high note condition*. The details of both are as follows:
1. **The low note condition**: the low note condition happens when the last hole is kept open and all other holes are covered by the player's fingers. This sets up the longest-wavelength standing wave in the flute, which has the lowest frequency and hence the lowest pitch. The low note condition can be seen in figure 17.3.
2. **The high note condition**: the high note condition occurs when the first hole is kept open and all other holes are covered by the player's fingers. This sets up the shortest-wavelength standing wave in the flute, which has the highest frequency and hence the highest pitch. The high note condition can be seen in figure 17.4.

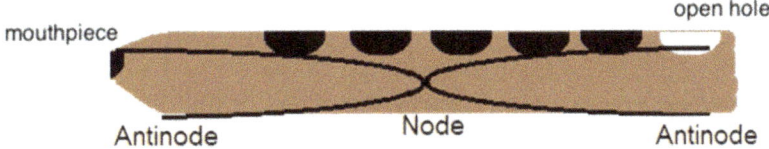

Figure 17.3. The low note condition in a flute.

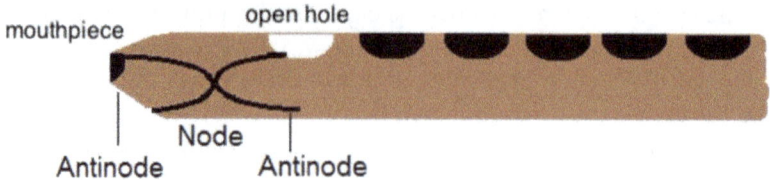

Figure 17.4. The high note condition in a flute.

17.2.2 The piccolo

Sometimes called a 'baby flute,' the modern piccolo is half the size of a flute. Historically, piccolos were made of either wood, glass, or ivory, but most are now made from plastic, brass, silver, or a variety of hardwoods. The piccolo is pitched nearly an octave higher than the flute and is considered harder to play because higher notes are difficult to play and sound unpleasant to the ear. Piccolos are often added to orchestras, marching bands, or wind ensembles to add sparkle and brilliance to the overall sound.

The average piccolo costs anywhere from $500 to $20 000 or more. However, most newer models range between $1500 and $7000. Piccolos are generally more expensive than flutes because being smaller means that the piccolo requires more precision to manufacture. In addition, it is more difficult to set up the piccolo correctly to play in tune and sound good.

17.2.3 The oboe

The oboe is a 2′ long cylindrical instrument which has two reeds that vibrate when you blow air through it. The external air moves the air inside the oboe, which in turn creates sound. The two reeds are the reason why the oboe sounds so beautiful and why it is such a difficult instrument to play. It takes a while until a beginner can even produce sound from an oboe, and even if they do succeed in making a sound, at first it often sounds like a honking noise rather than a musical tone. It takes even longer for a player to develop the ability to control it. The cavity that the air resonates in is quite small in the oboe, and it is rather difficult to direct the air into the cavity correctly.

In addition, pushing small amounts of air very quickly through the small space of an oboe takes a lot of diaphragm control. An instrumentalist has to develop the ability to hold the air in their diaphragm for a long time and let it out slowly. This puts a lot of pressure on the diaphragm, and controlling this process needs time and training.

Oboes of the modern era are usually made from grenadilla wood, but a few are also made from plastic or resin. Oboes generally have 45 keys and come in three levels of complication: student, intermediate, and professional. The cost of a new oboe ranges from $2500 for a student oboe to upwards of $12 000 for a professional-grade oboe. A orchestra usually has from two to four oboes, and the first oboist is generally responsible for tuning the entire orchestra before the concert!

17.2.4 The clarinet

The clarinet, like the one shown in figure 17.5, has a single reed made from the cane of *Arundo donax*, a perennial cane. It is a musical instrument that has a nearly cylindrical bore for most of its tube[1] and a flared bell at the bottom. Clarinets have one of the largest pitch ranges in woodwind instruments and come in a number of different sizes. The standard B-flat clarinet is just over two feet long.

The clarinets is played in a similar way to the oboe, i.e. by holding it upright, blowing air through the reed, and using your hands to change the pitch by opening and closing the keys with the fingers. However, there are some differences between the oboe and the clarinet. The most obvious difference is that the oboe has two reeds and a conical bore, whereas the clarinet has a single reed and a cylindrical bore, as seen in figure 17.6. The two instruments also differ in their bell structure: the bell of the oboe is rounded, whereas that of the clarinet is flared. Most oboes have closed tone holes, while most clarinets have open tone holes.

There are two to four clarinet players in any orchestra, and they play both melodies and harmonies. Clarinets have a dark rich sound in their lower notes, while the upper part of the clarinet's range is bright and resonant. Beginner clarinets start from as low as $200, whereas professional clarinets can be priced at more than $2000.

Figure 17.5. The clarinet: (left) a clarinet in a box, (right) Eman playing an assembled clarinet.

---

[1] In reality, the clarinet has a slight (invisible to the naked eye) hourglass shape. Its thinnest part is below the junction between the upper and lower joint. This hourglass shape helps correct the pitch and responsiveness of the instrument.

Figure 17.6. The reed of a clarinet.

17.3 The pipe organ

The pipe organ (from the Greek word *organon* and the Latin word *organum*, both of which translate as 'instrument' or 'mechanism') is one of the oldest musical instruments. It produces sound by air vibrations created in organ pipes, which are controlled from a keyboard by the player. In reality, an organ is a collection of several organs, each of which serves a different musical function. For example, each pipe organ has some or all of the following: organ pipes, keyboards for the hands (one main keyboard and sometimes two or three keyboards), pedalboards for the feet, plus a number of stops that create the classic 'organ sounds' and imitate the sounds of various instruments in the orchestra.

In summary, we can say that pipe organs are complex with many layers of complication. It takes practice and dedication to learn to play them well. Another downside to pipe organs is that they are very, very expensive. They are expensive to manufacture, expensive to purchase, expensive to transport, house, tune, and maintain![2] A small to medium-sized pipe organ could cost anywhere from $200 000 to $850 000, whereas a full-sized organ could be upwards of $1 000 000!

Organ pipes are generally made out of either metal or wood, and they fall into two main categories: flue pipes, which work like whistles, or reed pipes, which work like clarinets. Let us discuss them in this section:

1. **Flue pipes:** flue pipes, also called *labial pipes*, contain no moving parts and produce sound only through the vibration of air molecules. A flue pipe

---

[2] It is recommended to tune the pipe organ four times per year. Depending on the size of the instrument, a tuning job can take from two hours to two days.

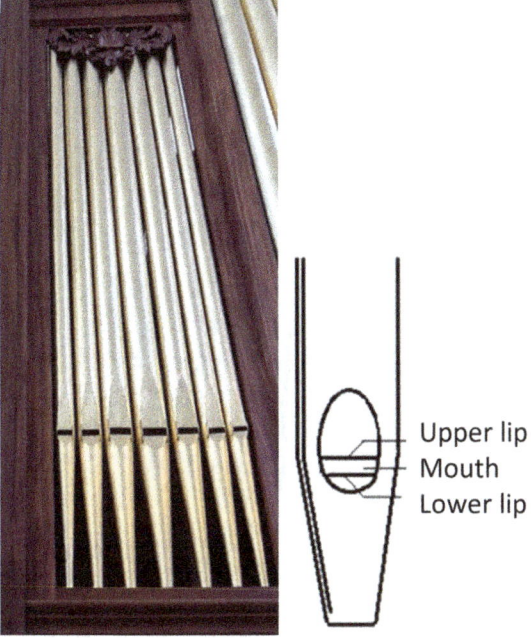

Figure 17.7. (Left) an example of flue pipes; (right) a schematic of a flue pipe.

consists of three main parts: the foot, the mouth, and the resonating length, also called the speaking length. The pipe stands vertically, as seen in figure 17.7(left) and air under pressure (also called blown air or wind) enters at the foot hole. The wind emerges through the flue and strikes the upper lip, producing an audible frequency.

The frequency of a flue pipe depends on many factors, including the length and the diameter of the pipe, the construction material of the pipe, and the pressure of the wind supply. Flue pipes are tuned by increasing or decreasing the resonating length. In modern times, tuning is done by fitting a cylindrical slide over the free end of the resonating column and sliding it up or down.

The diameter of a flue pipe directly affects its tone. Pipes with identical shapes and lengths produce different tones depending on their diameters. As the diameter of the pipe increases, the tone becomes richer and fuller. The material of the pipe is very important to the pipe's final sound. Most musicians agree that a pipe made out of tin or lead alloys sounds very different than one made of zinc or copper metals.

2. **Reed pipes:** reed pipes, also called *lingual pipes*, are a special kind of organ pipe which is primarily driven by a beating reed. Air blown under pressure causes the reed to vibrate. The pitch of a reed pipe is determined mainly by the mechanical properties of the reed and the length of the protruding part. Some of the reed pipes are the loudest pipes in the organ.

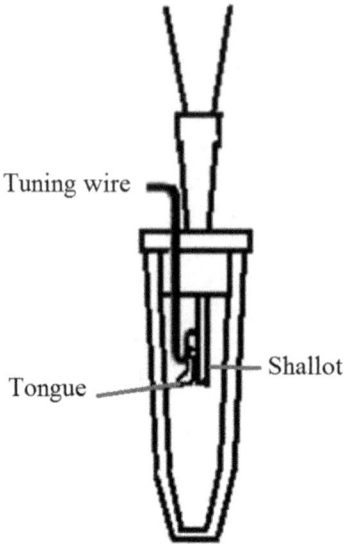

Figure 17.8. A schematic of a pipe–reed system.

Reed pipes work like clarinets, but they have a brass *tongue* instead of a cane reed. A reed pipe, like the one shown in figure 17.8, is made of a reed, a block used to hold the reed and shallot assembly, a tuning wire which can be moved up or down to change the produced pitch, and a resonator.

17.4 The instruments of the brass family

An instrument that produces sound via sympathetic vibrations of air in a tube resonator made from brass is called a 'brass instrument.' The vibrations in the tube are introduced by the vibrations of the player's lips, which is why brass instruments are also called *labrosones*, which literally means *lip-vibrated instruments*. Brass players vibrate their lips by buzzing them against a metal cup-shaped mouthpiece which amplifies the buzzing and produces sound. Most brass instruments have valves (that look like buttons) attached to their long pipes. When the player presses down on the valves, they open and close the different parts of the pipe. The pitch is changed by pressing different valves and the sound volume is changed by buzzing the lips harder or softer.

Brass instruments, like those seen in figure 17.9, are essentially long pipes that widen into a bell-like shape at the far end. Not all pipes are linear; some are curved and twisted into different shapes to make them easier to hold and play. The brass family members that are most commonly used in the orchestra include the trumpet, French horn, trombone, and tuba. They play louder than any other instrument in the orchestra; hence, they can be heard from the farthest distance away (figure 17.9).

Modern brass instruments generally fall into one of two families: valved brass instruments and slide brass instruments. Let us discuss them here:

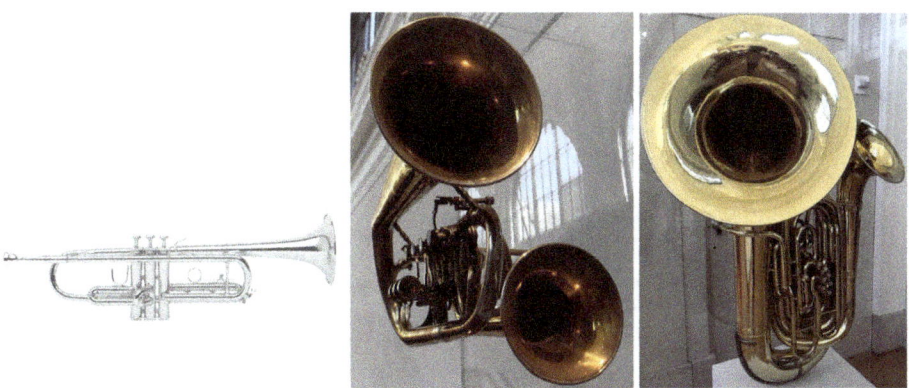

Figure 17.9. Some brass instruments.

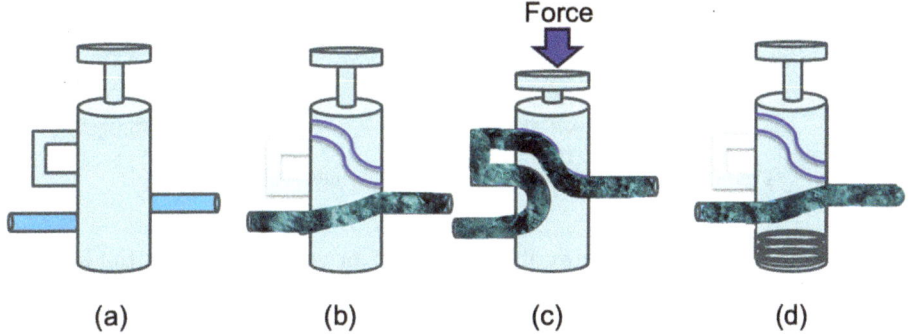

Figure 17.10. The air flow operation of a piston valve.

1. **Valved brass instruments**: valved brass instruments, as the name implies, use a set of valves to produce sound. The valves are pressed by the player's fingers, which changes the overall length of the tubing and hence changes the frequency of sound produced. Valved instruments, including the trumpet, horn, tuba, etc. dominate the brass instruments being played today. Most brass instruments use piston valves, but some also use rotary valves. We will define them very briefly here; for more details, please see other resources.
    (a) **Piston valve:** the operation of a piston valve is shown is figure 17.10. The details of the figure are described here:
     i. Figure 17.10(a). This shows the normal position of the valve. The airflow is on the outside of the valve, as seen here.
     ii. Figure 17.10(b). In the normal position of the valve, the air column passes straight through the valve and the loop is bypassed.
     iii. Figure 17.10(c). When the player presses down the valve, the first (smaller) passage is closed and the other passage aligns with

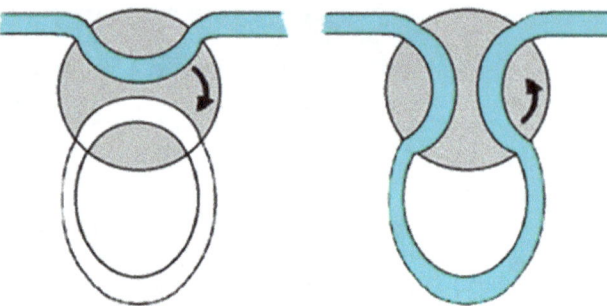

Figure 17.11. Cross-sectional view of airflow through a rotary valve.

Figure 17.12. Slide brass instruments: (left) a trombone, (right) sliders.

        the input air column. The air now flows through the longer air column, which lowers the pitch.
    iv. Figure 17.10(d). When the player releases the valve, a spring at the bottom of the valve returns the valve to the normal position, thereby decreasing the length and raising the frequency again.
  (b) **Rotary valve**: rotary valves, shown in figure 17.11, spin around (by 90 °) instead of moving linearly, but the concept of ports opening and closing to lengthen or shorten the air column is the same as in piston valve instruments. Rotary valves are most common on French horns and rotary-valve trumpets and are regularly used in orchestras from Germany and Austria. Rotary valves have a smoother action than piston valves, which lends itself well to playing more lyrical pieces of music.
2. **Slide brass instruments:** slide brass instruments use a slide instead of a valve to change the length of tubing, as shown in figure 17.12 (right). The main slide brass instruments include instruments from the trombone family, shown in figure 17.12(left). The pitch of a slide brass instrument can be raised or lowered by the use of the tuning slide. Pulling the slide out increases the length of the path that the air travels, thereby lowering the frequency and the pitch. Conversely, pushing the slide in decreases the length of the instrument and raises the frequency (pitch).

17.4.1 The production of sound in brass instruments

The three principal mechanisms that are responsible for the production of sound in brass instruments are:

1. The sound source (the player's lips and the mouthpiece)
2. Wave guides (the air column)
3. Radiation (the bell)

Details of each of the above are given below:

1. **Mouthpiece/the player's lips**: in all brass instruments, the lips of the player form a vibrating double reed that acts as a tone generator. The mouthpiece of the brass instrument acts like the closed end of the resonant air column. This means that only the odd harmonics are present in its spectrum. To generate all (both even and odd) harmonics, a combination of all three components must be present, that is, the mouthpiece, a cylindrical main section leading to a conical section, and a flared end.

    When the player vibrates his or her lips, a small gust of air from the player's lips enters the mouthpiece. The mouthpiece behaves like a pressure-controlled valve that admits the air. This pressure pulse is reflected back from the far end of the horn.

    The lowest resonance of a brass mouthpiece, usually between 50 and 750 Hz, is called the *popping frequency*. It is called the popping frequency because it is usually determined by slapping the rim with the palm of the hand and noting the frequency of the resulting pop.

2. **The air column**: air columns in brass instruments are typically made of a short conical section followed by a long cylindrical section that ends in a flared end. The resonant air tubes are flared at the ends and have no side holes. Each tube is different in the length of the conical tube, the size of the flare, its shape, and the size of the mouthpiece. This makes it very difficult to standardize a mathematical formula for residences in the air column and means we have to make adjustments in order to produce a modal series and can only find the modes experimentally. After multiple years of testing, musicians and scientists have found the harmonic series to be approximately equal to $0.7f_0$, $2f_0$, $3f_0$, $4f_0$, and so on.

3. **The bell**: the bell is flared in shape, which serves several purposes.

    (a) The bell affects the actual resonant frequencies.
    (b) The bell radiates sound by matching the pressure inside the horn to the air pressure outside.
    (c) The bell changes the radiation pattern of the instrument, making it more directional at high frequencies and thus giving the instrument its characteristic bright sound.

17.4.1.1 The trumpet

Trumpets are among the oldest musical instruments. The first known trumpets date back to 1500 BCE. Over the centuries, trumpets have been constructed using different materials ranging from bronze, silver, and ceramic to single sheets of metals. Most of these instruments were employed for signaling, especially in the military, or for important proclamations by kings.

The trumpet is the smallest member of its family and plays the highest pitches with a bright and vibrant sound. The modern trumpet is a long, slender brass tube with three valves. The brass tube is coiled to make the trumpet more manageable for the musician. The trumpet has a mouthpiece at one end and a flared bell at the other end. Each valve increases the length of the tubing when it is pressed. Increasing the length of the resonating column of air lowers the frequency and hence the pitch. Trumpet players are known as 'trumpeters,' and every orchestra generally has two to four trumpeters.

To play a trumpet, the trumpeter holds the trumpet horizontally and blows air through it by closing his or her lips and producing a 'buzzing' sound, which sets up a standing wave in the air column inside the trumpet. By doing this, the trumpeter creates a resonant system. The air can oscillate in and out of the open bell while the air pressure in the mouthpiece oscillates up and down.

When the trumpeter presses the trumpet to his or her lips, it forms a closed tube; therefore, the instrument only produces every alternate overtone of the harmonic series. This means that trumpets naturally only produce odd overtones and most notes in the series are slightly out of tune. Modern trumpets have slide mechanisms built in to compensate for this effect. The shape of the bell makes the missing overtones audible. Standing waves in a trumpet are shown in figure 17.13.

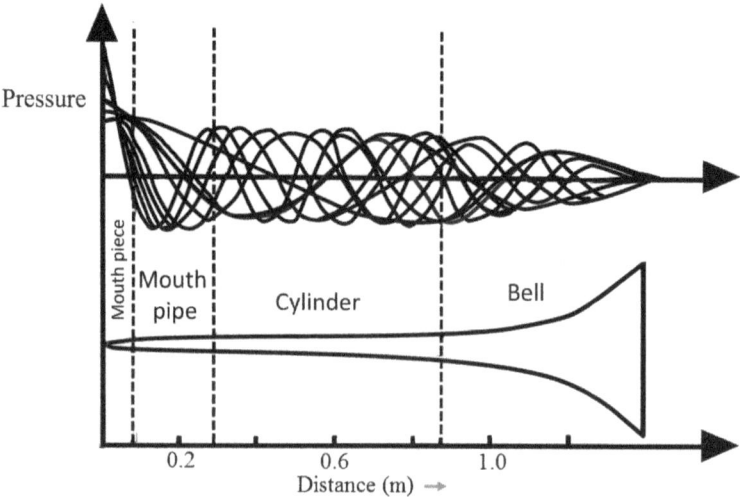

Figure 17.13. Standing waves in a trumpet.

17.4.1.2 The trombone
The trombone, shown in figure 17.12, is the only instrument in the brass family that uses a telescoping slide instead of valves to change pitch. 'Trombone' is derived from Italian and is translated as 'large trumpet.' A standard trombone is made of a long thin brass tube, which, if stretched out, would be about 9' long! There is a mouthpiece at one end of the trombone and a bell at the other end. Trombone bells are usually made of different brass mixtures, but they can also be made from solid sterling silver. Mouthpiece selection is a very personal decision because trombones mouthpieces vary in material composition, length, diameter, rim shape, and other factors.

The two U-shaped pipes are linked at opposite ends to form an S shape. One pipe slides into the other so the total length of the pipe can be extended or shortened to change the pitch. The pitch of a trombone decreases as the length of the pipe is increased. A trombone is played by holding it horizontally, buzzing into the mouthpiece, and using the right hand to change pitch by pushing or pulling the slide to one of seven different positions. There are usually three trombones in an orchestra, and they play pitches in the same range as the cello and bassoon.

Example 17.3. A vibrating column of air inside a tube may be different from a vibrating string, but it still vibrates in a harmonic series just like a vibrating string. It has a fundamental frequency and then higher harmonics. So:

1. Why do different instruments have different timbres?
   Solution:
   The main reason for instrument-dependent timbres is the difference between the relative loudness of all the different harmonics when they are compared to each other. In one instrument, the odd-numbered harmonics may be stronger, in another, the even-numbered harmonics might be stronger, in yet another, perhaps the fifth and eighth harmonics are the strongest. This is what allows us to recognize which instrument is being played and allows you to distinguish between, say, a clarinet and a flute or a French horn.
2. Why do the same instruments have different timbres?
   Solution:
   The relative strengths of the harmonics also change from note to note on the same instrument. This is the difference you hear between the sound of a clarinet playing low notes and the same clarinet playing high notes.

17.5 The bagpipe

The bagpipe was first developed more than 2000 years ago. It is the most unusual of the reed instruments. Bagpipes, as shown in figure 17.14, consist of several single reed pipes without holes called 'drone pipes,' and one single or double reed pipe with a number of finger holes, called the melody or chanter pipe. They are all connected to a large leather bag that is kept inflated by blowing down the mouth pipe.

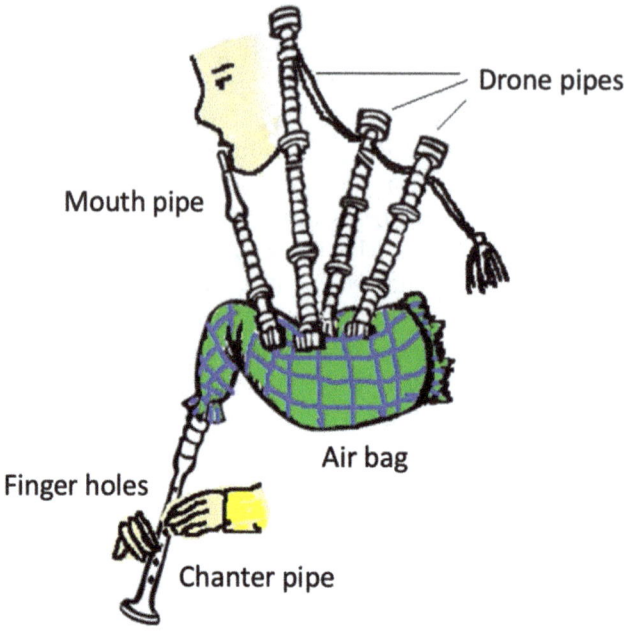

Figure 17.14. A depiction of a bagpipe.

The drone pipes produce a continuous background sound made of several sounds based on the note of 'A' and have a tone that is remarkably similar to that of a clarinet.

Review Questions 17.1.
1. What is the basic difference between men and women that causes a difference in vocal pitch?
2. How does a pipe–reed system work?
3. Why do different instruments have different timbres?
4. Explain why all the harmonics are present in open pipes but only odd harmonics can be present in closed pipes?
5. Does an increase in air temperature affect the pitch of a wind instrument? What is the effect when the temperature decreases?
6. What effect do you think changing the flare and shape of the tube has on the frequencies of a brass instrument?
7. As you increase the water level in a bottle xylophone, does the pitch of the sound increase or decrease? Why?

Exercises 17.1.
1. List some instruments that have conical bores. What instrument has a reverse conical bore? Is there any advantage of disadvantage of the reverse conical bore?
2. What is the lowest-frequency resonance of a tube that is open at both ends and 60 cm long?
3. What is the lowest-frequency resonance of a tube of air that is open at one end, closed at the other, and 2 cm long?
4. Find the length of an air column required to produce a fundamental frequency of 440 Hz if the column is open at both ends.
5. The frequency of the first harmonic above the fundamental of a pipe open at both ends is 880 Hz. Find:
   (a) The length of the pipe
   (b) The frequency of the fundamental
   (c) The wavelength of the fundamental

Further reading

- Baines A 1993 *Brass Instruments: Their History and Development* (New York: Dover)
- Zeng L, Smith C, Poelzer G H, Rodriguez J, Corpuz E and Yanev G 2014 Illustrations and supporting texts for sound standing waves of air columns in pipes in introductory physics textbooks *Phys. Rev. ST Phys. Educ. Res.* **10** 020110
- Berg R E and Stork D G 1994 *The Physics of Sound* 2nd edn (Englewood Cliffs, NJ: Prentice-Hall)
- Rossing T D, Moore R F and Wheeler PA 2002 *The Science of Sound* 3rd edn (Upper Saddle River, NJ: Pearson Education)
- Hall D E 1991 *Musical Acoustics* 2nd edn (Pacific Grove, CA: Brooks/Cole)
- Blaikley D J 1910 How a trumpet is made. I. The natural trumpet and horn *Music. Times* **51** 154

Part VII

Appendix

IOP Publishing

The Physics of Sound and Music, Volume 1
A complete course text (Textbook)
Samya Bano Zain

Appendix A

Review of mathematics

Part 1: Simple mathematics:

1. $12 + 24 =$

2. $-16 + 8 =$

3. $-2 - 6 =$

4. $20/(-4) =$

5. $5 \times 2 \times (-2) =$

Part 2: Least common multiple:
When a number is multiplied by another number we get multiples of the number. The common multiples are those that are found in two different numbers. The least common multiple (LCM) is the smallest of the common multiples.

Example A.1. Multiples of two: 2, 4, 6, 8, 10, 12, …
Multiples of three: 3, 6, 9, 12, …
The common multiples are: 6, 12, …
The LCM of two and three: LCM(2, 3) = 6

1. LCM(3, 6, 18) =

2. LCM(2, 8, 12, 20) =

3. LCM(4, 15, 25) =

Part 3: Fractions:

In a fraction, the top number is called the numerator and the bottom number is called the denominator:

$$\text{fraction} = \frac{\text{numerator}}{\text{denominator}}. \quad (A.1)$$

To add or subtract fractions, we make the denominators of both (or all) fractions the same using the least common number method explained in Part 2. To multiply fractions, we multiply the numerators and denominators of each number and then simplify the resulting fraction as needed. To divide factions, we first write the reciprocal of the second number and then use the rules for the multiplication of fractions to calculate the answer.

1. $\frac{1}{3} + \frac{6}{3} =$

2. $\frac{9}{14} + \frac{4}{7} =$

3. $\frac{3}{8} \times 2\frac{3}{4} =$

4. $\frac{2}{8} \div \frac{3}{4} =$

Part 4: Exponents:

Exponents are used as a shorthand to express how many times a number needs to be multiplied in an expression. Exponents are also called *powers* or *indices*.
Examples: $2^5 = 2 \times 2 \times 2 \times 2 \times 2 = 32$ and $10^4 = 10 \times 10 \times 10 \times 10 = 10\,000$.
Negative exponents tell us how many times to divide a number:
Example: $5^{-3} = 1 \div 5^3 = \frac{1}{5 \times 5 \times 5} = \frac{1}{125} = 0.008$

1. $10^6 =$

2. $10^{-4} =$

3. $(3 \times 10^9) + (2 \times 10^4) =$

4. $(2.1 \times 10^4) - (3.1 \times 10^3) =$

5. $(5.5 \times 10^3) \times (2 \times 10^5) =$

6. $(9 \times 10^4) \div (3 \times 10^8) =$

Part 5: Algebra

Elementary algebra is essential for any study of mathematics, science, or engineering. Algebra is derived from the Arabic word *al-jabar* and is the study of mathematics using letters as stand-ins for numbers. Algebra defines the rules for utilizing these symbols in mathematical operations.

For example:

1. If $a = 2$, $b = 3$, $c = 4$, $d = 5$, then:
    (a) $2a + 4b + 3c =$
    (b) $-8a + 4(3b - c) =$
    (c) $\frac{a + 4b}{c - 3d} =$

2. Solve for the variable:
    (a) $-5x + 7(x - 2) = 0$
    (b) $-8a + 4(3b - c) = 3$
    (c) $\frac{b - 2}{b - 3} = \frac{3}{4}$

Part 6: Problem solving

1. What is 30% of 20?

2. 16% of what number is 12?

3. 17 is what percentage of 51?

4. 25 is what percentage of 300?

5. Find 7/8 of 20.6

6. 80 people took a physics test. 70 of them passed the test. What percentage failed the test?

7. Mary managed to sell 3/5 of her cakes.
    (a) What percentage of cakes was sold?
    (b) What percentage of cakes was left?

8. Pam bought 50 kg of sugar. She used 10 kg of sugar. What percentage of sugar was left?

9. If you are running at 15 miles h^{-1}, how much distance do you cover in two hours?

Appendix B

Unit conversions

Length:
1 mile = 5280 feet = 1.61 km
1 inch = 25.4 mm = 2.54 cm
1 foot = 12 inches = 30.48 cm
1 meter = 1.094 yd =
1 yd = 3 feet = 36 inches

Mass:
1 pound = 16 ounces
1 gallon = 4 quarts
1 quart = 2 pints
1 pound = 454 grams
1 liter (L) = 1.06 quarts
1 ton = 908 kg

Time:
1 year = 12 months = 365 days = 3.15×10^7 s
1 day = 24 hours = 1440 min = 86 400 s

Capacity:
1 fluid ounce (fl oz) = 29.57 ml
1 pint (pt) = 0.473 L
1 quart (qt) = 2 pt = 0.946 L
1 gallon (gal) = 4 qt = 3.785 L

Appendix C

Logarithms

There are five basic important logarithmic properties. For given positive integers A, B, and n:

- **the product property** $\log(AB) = \log A + \log B$,
  for example:
  $\log(15) = \log(5 \times 3) = \log 5 + \log 3 = 0.699 + 0.477 = 1.176$.

- **the quotient property** $\log(\frac{A}{B}) = \log A - \log B$,
  for example: $\log\left(\frac{5}{3}\right) = \log 5 - \log 3 = 0.699 - 0.477 = 0.222$.

- **the power rule** $\log A^n = n \log A$,
  for example: $\log 10^3 = 3 \log 10 = 30$.

- **the change of base rule** $\log_B(A) = \frac{\log A}{\log B}$,
  for example: $\log_2 10 = \frac{\log 10}{\log 2}$.

- **The reciprocal rule** $\log_n A = \frac{1}{\log_A n}$,
  for example: $\log_3 10 = \frac{1}{\log_{10} 3}$.

x	$\log(x)$	x	$\log(x)$
1	0.	6	0.778
2	0.301	7	0.845
3	0.477	8	0.903
4	0.602	9	0.954
5	0.699	10	1.000

Find the logs for the following:

$$\begin{aligned}\log(400) &= \log(4 \times 10 \times 10) \\ &= \log 4 + \log 10 + \log 10 \\ &= 0.602 + 1 + 1 \\ &= 2.602\end{aligned}$$

$$\begin{aligned}\log(2.5) &= \log\left(\frac{25}{10}\right) \\ &= \log(25) - \log 10 \\ &= \log(5 \times 5) - \log 10 \\ &= \log 5 + \log 5 - \log 10 \\ &= 0.699 + 0.699 - 1 \\ &= 0.398.\end{aligned}$$